7th International Conference on Environment and Industrial Innovation (ICEII 2017)

IOP Conference Series: Earth and Environmental Science
Volume 67

Kuala Lumpur, Malaysia
24 – 26 April 2017

ISBN: 978-1-5108-4173-4
ISSN: 1755-1307

Printed from e-media with permission by:

Curran Associates, Inc.
57 Morehouse Lane
Red Hook, NY 12571

Some format issues inherent in the e-media version may also appear in this print version.

Printed by Curran Associates, Inc. (2017)

For permission requests, please contact the Institute of Physics
at the address below.

Institute of Physics
Dirac House, Temple Back
Bristol BS1 6BE UK

Phone: 44 1 17 929 7481
Fax: 44 1 17 920 0979

techtracking@iop.org

Additional copies of this publication are available from:

Curran Associates, Inc.
57 Morehouse Lane
Red Hook, NY 12571 USA
Phone: 845-758-0400
Fax: 845-758-2633
Email: curran@proceedings.com
Web: www.proceedings.com

TABLE OF CONTENTS

ENVIRONMENTAL ENGINEERING AND MANAGEMENT

EVALUATION OF THE ENVIRONMENTAL IMPACT OF PORTION BAG FOR FOOD PACKAGING: A CASE STUDY OF THAILAND 1
Chaniporn Ruangrit, Parnuwat Usapein, Nantamol Limphitakphong, Orathai Chavalparit

EXPERIMENTAL STUDY ON VACUUM MEMBRANE DISTILLATION BASED ON BRINE DESALINATION BY PVDF 6
Xuesheng Yan, Caiyun Wang

ENHANCEMENT OF ENZYMATIC HYDROLYSIS AND LIGNIN REMOVAL OF BAGASSE USING PHOTOCATALYTIC PRETREATMENT 12
P Pattanapibul, S Chuangchote, N Laosiripojana, V Champreda, J Kaewsaenee

EVALUATION OF NANO ZERO VALENT IRON EFFECTS ON FERMENTATION OF MUNICIPAL ANAEROBIC SLUDGE AND INDUCING BIOGAS PRODUCTION 19
Tareq W.M. Amen, Osama Eljamal, Ahmed M.E. Khalil, Nobuhiro Matsunaga

UTILIZATION OF ETHYL CELLULOSE POLYMER AND WASTE MATERIALS FOR ROOFING TILE PRODUCTION 25
Suubitaa Spencer Sam, ChoonAun Ng, Swee Yong Chee, NoorZainab Habib, Humayon Nadeem, Wei Ping Teoh

CRITICAL FACTORS OF IMPLEMENTING INDUSTRIALISED BUILDING SYSTEM IN SARAWAK: A RESEARCH ON SMES 32
N A Hadi, W M N W Muhamad, M K F Othman

GEOEYE IMAGE FUSION VEGETATION INFORMATION EXTRACTION BASED ON BLUE NOISE MEASUREMENT TEXTURE 38
Qiu Luo, Qiming Xiong, Yao Liu, Gui Zhang

ANALYSIS OF BLACK SOIL ENVIRONMENT BASED ON ARDUINO 46
Y Li, Y F Zhang, C H Wu, J F Wang

INTEGRATING ENVIRONMENTAL MANAGEMENT INTO FOOD SAFETY AND FOOD PACKAGING IN MALAYSIA: REVIEW OF THE FOOD REGULATION 1985 52
N H Nordin, H Hara, N Kaida

EXPERIMENTAL STUDY OF THE INFLUENCE OF VARYING CEILING HEIGHT ON THE HEAT RELEASE RATE OF A POOL FIRE 59
Jiahao Liu, Jian Wang, Richard Yuen

PRE-CHLORINATION CONTACT TIME AND THE REMOVAL AND CONTROL OF MICROCYSTIS AEROGINOSA IN COAGULATION 66
T Lapsongpon, S Leungprasert, C Yoshimura

HAZARDOUS WASTE MINIMIZATION CHALLENGE IN AUTOCOMPONENT INDUSTRY, WEST JAVA, INDONESIA 73
L Handayani, S S Moersidik

ENERGY ENGINEERING AND MANAGEMENT

EVALUATION OF MISMATCH LOSSES DUE TO SHUNTS IN INDUSTRIAL SILICON PHOTOVOLTAIC MODULES 79
P Somasundaran, M Shilpi, R Gupta

PERFORMANCE AND ECONOMIC ANALYSIS OF A PLUG AND PLAY REGENERATIVE BRAKE FOR IMPROVING ENERGY EFFICIENCY FOR TRACTION ELEVATORS 87
Chuttchaval Jeraputra, Supun Tiptipakorn

MULTI-OBJECTIVE OPTIMAL POWER FLOW FOR ACTIVE DISTRIBUTION NETWORK CONSIDERING THE STOCHASTIC CHARACTERISTIC OF PHOTOVOLTAIC 94
Bao-Rong Zhou, Si-Liang Liu, Yong-Jun Zhang, Ying-Qi Yi, Xiao-Ming Lin

THE ACTIVE CONTROL STRATEGY ON THE OUTPUT POWER FOR PHOTOVOLTAIC-STORAGE SYSTEMS BASED ON EXTENDED PQ-QV-PV NODE 103
Chen Xu, Bao-Rong Zhou, Jian-Wei Zhai, Yong-Jun Zhang, Ying-Qi Yi

PERFORMANCE EVALUATION OF SOLAR BOX COOKER ASSISTED WITH LATENT HEAT ENERGY STORAGE SYSTEM FOR COOKING APPLICATION 111
V S Vigneswaran, G Kumaresan, P Sudhakar, R Santosh

LOW-COST ENERGY CONTRAPTION DESIGN USING PLAYGROUND SEESAW 119
I A P Banlawe, N J E L Acosta
INVESTIGATION OF INDIRECT BENEFITS OF PV ROOFTOP IN THAILAND 124
T Khumkrong, S Chuangchote, D Chenvidhya, K Kirtikara
SOLAR PV LEASING IN SINGAPORE: ENHANCING RETURN ON INVESTMENTS WITH OPTIONS 130
Shuang Song, K.L. Poh
UTILIZATION OF EGGSHELL WASTE AS LOW-COST SOLID BASE CATALYST FOR BIODIESEL PRODUCTION FROM USED COOKING OIL 136
N P Asri, B Podjojono, R Fujiani, Nuraini
FORMATION OF HYDROCARBON COMPOUNDS DURING THE HYDROCRACKING OF NON-EDIBLE VEGETABLE OILS WITH COBALT-NICKEL SUPPORTED ON HIERARCHICAL HZSM-5 CATALYST 143
L Marlinda, M Al-Muttaqii, A Roesyadi, D H Prajitno
CATALYTIC HYDROCRACKING OF KAPUK SEED OIL (CEIBA PENTANDRA) TO PRODUCE BIOFUEL USING ZN-MO SUPPORTED HZSM-5 CATALYST 151
Y W Mirzayanti, D H Prajitno, A Roesyadi
THE DEVELOPMENT OF CONTROL AND MONITORING SYSTEM ON MARINE CURRENT RENEWABLE ENERGY CASE STUDY: STRAIT OF TOYAPAKEH - NUSA PENIDA, BALI 157
I S Arief, I H Suherman, A Y Wardani, A Baidowi
DAYLIGHT STRATEGIES FOR ARCHITECTURAL STUDIO FACILITIES: THE LITERATURE REVIEW 162
Muhammad Anas Bin Othman, Nur Azfahani Ahmad, Azizah Md Ajis
SOLAR PUBLIC ENGAGEMENT: THE PROSPECTIVE STUDY ON FELDA COMMUNITY IN MALAYSIA 169
Siti Jamiah Tun Jamil, Nur Azfahani Ahmad
THE ENERGY AUDIT PROCESS FOR UNIVERSITIES ACCOMMODATION IN MALAYSIA: A PRELIMINARY STUDY 175
Muhammad Hilmi Dzulkefli

WATER RESOURCES AND ATMOSPHERIC ENVIRONMENT MANAGEMENT

THE STUDY OF RECIRCULATING AQUACULTURE SYSTEM IN POND AND ITS PURIFICATION EFFECT 183
Jiangqi Qu, Qingjing Zhang, Chengxia Jia, Pan Liu, Mu Yang
ASSESSMENT OF AIR POLLUTION OF SETTLEMENT AREAS IN ULAANBAATAR CITY, MONGOLIA 190
Ch Sonomdagva, Ch Byambatseren, B Batdelger
VALIDATION OF TSUNAMI INUNDATION MODEL TUNA-RP USING OAR-PMEL-135 BENCHMARK PROBLEM SET 198
H L Koh, S Y Teh, W K Tan, X Y Kh'ng
INACTIVATION DISINFECTION PROPERTY OF MORINGA OLEIFERA SEED EXTRACT: OPTIMIZATION AND KINETIC STUDIES 206
M A Idris, M S Jami, A M Hammed
ADVANCED WASTEWATER TREATMENT USING MICROALGAE: EFFECT OF TEMPERATURE ON REMOVAL OF NUTRIENTS AND ORGANIC CARBON 215
Mohamad Shurair, Almomani Fares, Simon Judd, Rahul Bhosale, Anand Kumar, Ujjal Gosh, Majeda Khreisheh
RAINFALL-RUNOFF MODELLING OF AJAY RIVER CATCHMENT USING SWAT MODEL 222
Subhadip Kangsabanik, Sneha Murmu
THE EFFECT OF ANTHROPOGENIC ACTIVITIES TO THE DECREASE OF WATER QUALITY 229
N V Sidabutar, I Namara, D M Hartono, T E B Soesilo

THERMAL ENGINEERING

NUMERICAL THERMAL PERFORMANCE STUDY IN A HEAT EXCHANGER TUBE WITH INCLINED ELLIPTICAL RINGS 235
W Sodsri, S Tamna, C Thianpong, P Promvonge
NUMERICAL HEAT TRANSFER STUDY IN A ROUND TUBE WITH 60° V-SHAPED RINGS 242
Y Kaewkohkiat, S Tamna, W Pairoj, P Promvonge

EXPERIMENTAL STUDY ON EVAPORATION CHARACTERISTICS OF A HYDROGEN PEROXIDE DROPLET AT ELEVATED TEMPERATURE 250
Jonghan Won, Hongjae Kang, SeungWook Baek, Sejin Kwon
IMPACT OF TRAFFIC COMPOSITION ON ACCESSIBILITY AS INDICATOR OF TRANSPORT SUSTAINABILITY 256
Nahdalina, S P Hadiwardoyo, Nahry
Author Index

IOP Conf. Series: Earth and Environmental Science **67** (2017) 011001 doi:10.1088/1755-1315/67/1/011001

PREFACE

It is our great pleasure to welcome you to 2017 7th International Conference on Environment and Industrial Innovation (ICEII 2017) which was held in International University of Malaya Wales, Kuala Lumpur, Malaysia during April 24-26, 2017. ICEII 2017 is dedicated to issues related to Environment and Industrial Innovation.

The major goal and feature of the conference is to bring academic scientists, engineers, industry researchers together to exchange and share their experiences and research results, and discuss the practical challenges encountered and the solutions adopted. Professors from Malaysia, Canada, Thailand are invited to deliver keynote speeches regarding latest information in their respective expertise areas. It will be a golden opportunity for the students, researchers and engineers to interact with the experts and specialists to get their advice or consultation on technical matters, sales and marketing strategies.

This proceedings present a selection from papers submitted to the conference from universities, research institutes and industries. All of the papers were subjected to peer-review by conference committee members and international reviewers. The papers selected depended on their quality and their relevancy to the conference. The volume tends to present to the readers the recent advances in the field of Environment and Industrial Innovation and various related areas, such as Advanced Ceramic-porous ceramic, Aerodynamics, Application of Spatial technology, Catalysis and Environment, Enivronmental Biotechnology, Environment management, Environmental Chemistry, Environmental engineering, Environmental Planning of Mines, Environmental Pollution Control, Fermentation Technology, Industrial Environment, etc..

We would like to thank all the authors who have contributed to this volume and also to the organizing committee, reviewers, speakers, chairpersons, sponsors and all the conference participants for their support to ICEII 2017.

Prof. Koh Hock Lye
Sunway University, Malaysia
April. 5, 2017

IOP Conf. Series: Earth and Environmental Science **67** (2017) 011002 doi:10.1088/1755-1315/67/1/011002

Peer review statement

All papers published in this volume of *IOP Conference Series: Earth and Environmental Science* have been peer reviewed through processes administered by the proceedings Editors. Reviews were conducted by expert referees to the professional and scientific standards expected of a proceedings journal published by IOP Publishing.

IOP Conf. Series: Earth and Environmental Science **67** (2017) 012001 doi:10.1088/1755-1315/67/1/012001

Evaluation of the environmental impact of portion bag for food packaging: a case study of Thailand

Chaniporn Ruangrit[1,a], Parnuwat Usapein[2,b], Nantamol Limphitakphong[3,c] and Orathai Chavalparit[1,3,d]

[1]Department of Environmental Engineering, Chulalongkorn University, Bangkok 10330, Thailand
[2]Rattanakosin College for Sustainable Energy and Environment, Rajamangala University of Technology Rattanakosin, Nakorn Pathom 73170, Thailand
[3]Research Unit of Environmental Management and Sustainable Industry, Faculty of Engineering, Chulalongkorn University, Bangkok 10330, Thailand

Email: [a]chaniporn.ruangrit@gmail.com, [b]parnuwat.usa@rmutr.ac.th, [c]nantamoll@gmail.com, [d]orathai.c@chula.ac.th

Abstract. This study applied life cycle assessment methodology in evaluating environmental impact of portion bag. The objective of this study was to identify the hotspot of environmental impact through life cycle of portion bag. The options were proposed for improving environmental performance of the product. The system boundary was defined as cradle-to-grave which included the ethylene production, LDPE and LLDPE resins production, portion bag production, disposal, and transportation. All materials and emissions were calculated based on 1 piece of portion bag which weighed 2.49 g. IMPACT 2002+ was used for assessing environmental impact on SimaPro V8.2 software. The result found that the most of environmental impact was generated from LDPE and LLDPE resins which was used as raw material for producing portion bag. After normalization, non-renewable energy showed the highest potential to concern. This impact related directly to the natural gas drilling, ethane production, ethylene production, resin productions, and energy in all process. In conclusion, it should be suggested that the selection of bio-material for producing portion bag can play an important role to reduce the environmental impact. The research demonstrates the possible way and benefits in improving cleaner raw material and suitable way of product's end-of-life for producing green portion bag in the future.

1. Introduction

Our world has faced many environmental problems since the start of Industrial Revolution in 1975. Many innovations have been created to facilitate the daily life of human, and one of them is plastic. With a wide range of properties, plastics were used in many applications such as packaging, container, film, textile, bag, etc. This is the one reason that the consumption rate of plastic increases every year as well as plastic wastes generation. In Thailand, packaging plastic is an important section in plastic industry. It was produced for supporting other industries, for example, food, chemical, and automotive industries. With the short life cycle of application, however, the volume of packaging plastic waste was increased in each year. Most of the plastic packaging is composite packaging which is difficult to recycle.

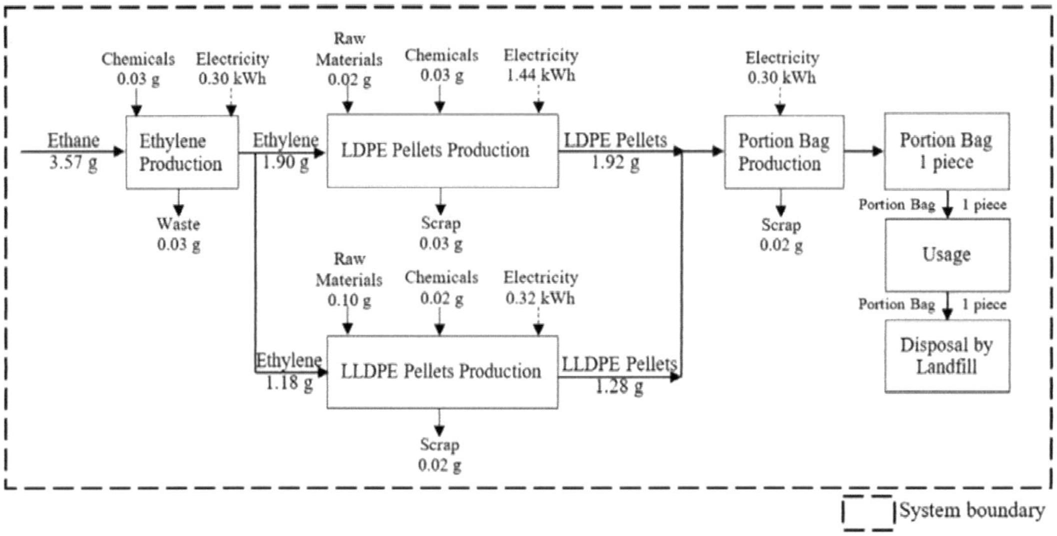

Figure 1. System boundaries of this study.

Although there are some methodologies for recycling packaging plastic, it still has high cost of operation and low cost for resale [1].

Life cycle assessment (LCA) is the methodology for identifying environmental impact associated from cradle-to-grave [2]. The LCA is a technique to evaluate the environmental impacts of product, which included raw material extraction, production, usage, and disposal [3]. Recently, some studies have evaluated the environmental impact of a multilayer polymer bag (made of low density polyethylene (LDPE) and polyamide (PA) for food packaging with LCA methodology [4]. However, the case study of LCA assessment of portion bag product is very rare. Therefore, the objective of this study is to assess the environmental impact of low density polyethylene (LDPE) and linear low density polyethylene (LDPE) by using LCA methodology. The boundary system of this study was cradle-to-the grave. The end of life options for portion bag were proposed and comparison. All environmental impacts were calculated by using SimaPro V8.2 software with background data contained in Ecoinvent V.3 database. Furthermore, feasible measures will be proposed of options for reducing environmental impacts.

2. Materials and methods

This study was investigated the environmental impact of portion bag by using LCA method. The LCA technique used in this study was based on ISO 14040 framework and ISO 14044 guidelines, which consisted of four steps such as goal and scope definition, inventory analysis, impact assessment, and interpretation. The detail in each step can be described as follows:

2.1 Goal and scope

The goal of this study was to assess the environmental impact of the portion bag with zip lock as a case study. The portion bag was made from LDPE and LLDPE resin. The hotspot of environmental impact was identified. The options were proposed for improving environmental performance of the product.

2.2 Functional unit and system boundary

As shown in Figure 1, all environmental impact was calculated based on 1 piece of portion bag which weighs 2.49 g per piece. The scope of determination was started from raw material acquisition (cradle) to the end of life (the grave).

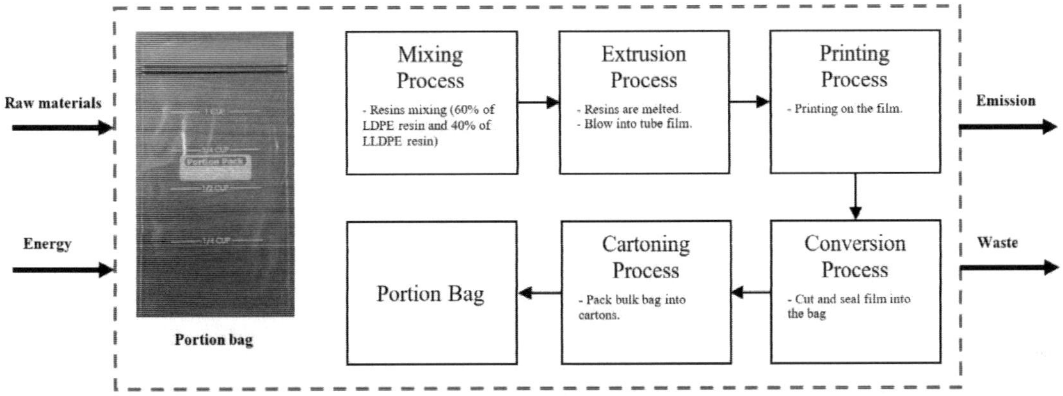

Figure 2. Portion bag production process.

2.3 Data collection and Inventory analysis

Portion bags were produced by blending 60% of LDPE resin and 40% of LLDPE resin. All material inputs and emissions from petrochemical processes (e.g. the ethylene production, LDPE and LLDPE productions) and plastic converter factories (e.g. portion bag production) were collected from the factories in Thailand, as shown in Figure 2. The portion bag production processes can be divided into 6 steps which are, mixing, extrusion, printing, conversion, cartoning and deliver to the customers. All input and output materials from "Portion bag production" were collected in each processes.

2.4 Life cycle impact assessment

All the input and output were allocated to base on 1 piece of portion bag (2.49 g) before assessed the environmental impact with SimaPro software. The processes were extrapolated from Impact 2002+ V.2.12 method and Ecoinvent V.3 database within the SimaPro software in the 8.2 version for identifying all processes and materials [5].

2.5 Interpretation

The results of the impact assessment through life-cycle of portion bag were interpreted to propose the option for improving environmental.

3. Results and discussion

3.1 Midpoint characterization result

In every impact except terrestrial ecotoxicity and aquatic eutrophication, the highest of environmental impact was generated from raw material acquisition (LDPE and LLDPE resins), as shown in Figure 3. Terrestrial ecotoxicity was mainly generated from lorry, while aquatic eutrophocation was come from cargo ship. Long distance to transport product from Thailand to United States was the largest portion of aquatic eutrophication in this study (66.13%). Meanwhile, the disposal of portion bag by landfill was highlighted for land occupation.

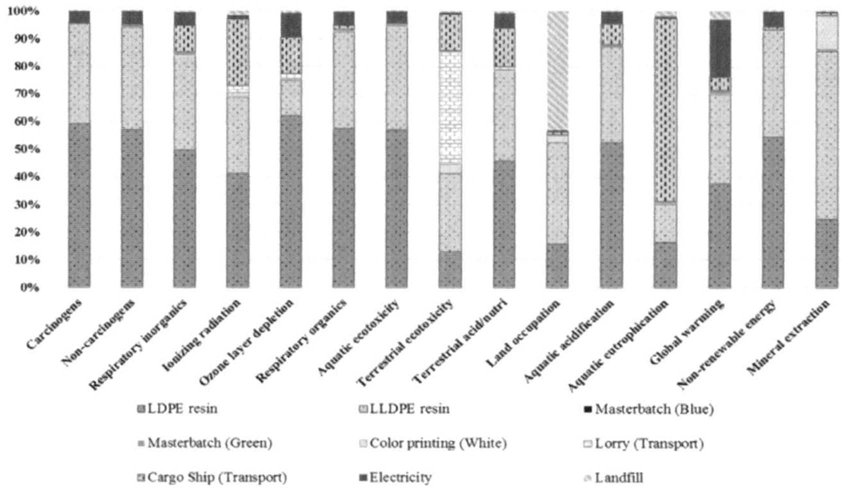

Figure 3. Environment impacts of portion bag.

3.2 Normalization

The result of normalization impact category was shown Figure 4. It can be indicated that there are five most environmental impact categories such as non-renewable energy, carcinogens, respiratory inorganics, global warming, and non-carcinogens, respectively. Environmental impact of raw material (LDPE and LLDPE resins) is highest in every impact like midpoint result. Therefore, the hotspot of environmental impact of portion bag life cycle was LDPE and LLDPE resins.

3.3 Options for reducing environmental impacts

From this study result, the hotspot of environmental impact was LDPE and LLDPE resins. Meanwhile, non-renewable energy impact category is highest value after normalization. Therefore, the goal of proposed option is to decrease the environmental impact of raw material. Options for reducing environmental impacts of life cycle portion bag was shown in Figure 5.

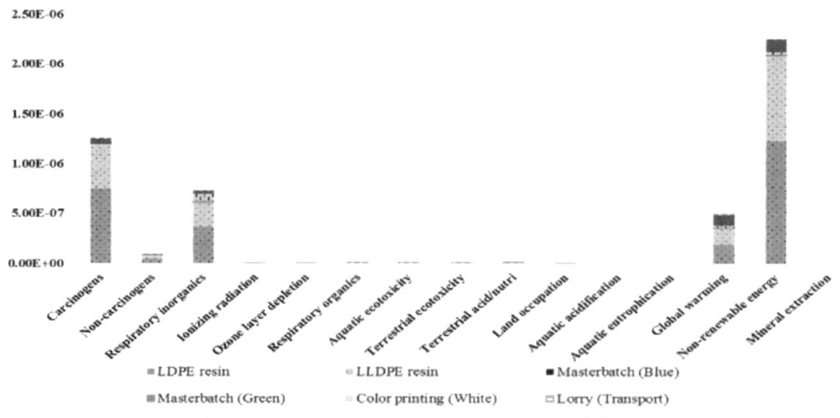

Figure 4. Normalization result of environmental impact from portion bag life cycle.

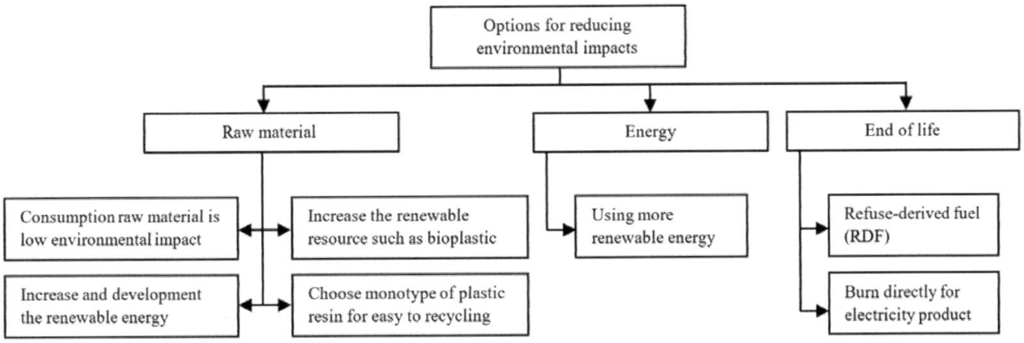

Figure 5. Proposed options for reducing environmental impacts of portion bag.

4. Conclusion

The environmental impact of life cycle portion bag was investigated in this study. The hotspot of environmental impact was LLDPE and LDPE resins. The proposed options are developed based on the result of LCA to minimize the environmental impact through life cycle of portion bag. Therefore, the possible measures to reduce environmental impact are focused on raw material extraction, energy consumption, selection to end-of-life, and replacing conventional raw material by bioplastics or biodegradable plastics. It can be concluded that using cleaner raw material and selecting suitable way of product's end-of-life can lead to produce green portion bag in the future.

5. References

[1] Xie M, Li L, Qiao Q, Qihong S and Tichang S 2011 *J. Cleaner Prod.* **19** 2100
[2] Al-Salem S M, Lettieri P and Baeyens J 2009 *Waste Manage.* **29** 2625
[3] Günkaya Z and Banar M 2016 *Int J Life Cycle Assess* **21** 465
[4] Siracusa V, Ingrao C, Giudice A L, Mbohwa C and Rosa D M 2014 *Food Res. Int.* **62** 151
[5] Usapein P and Chavalparit O 2015 *J. Mater. Cycles Waste Manage.* **17** 303

Acknowledgement

This research was funded by the Ratchadaphiseksomphot Endowment Fund (2014), Chulalongkorn University (CU-57-062-CC). The author also would like to gratefully thank all the factories for providing the collaboration in this research.

Experimental study on vacuum membrane distillation based on brine desalination by PVDF

Yan Xuesheng[1], Wang Caiyun[2]

[1]Institute for Energy Research, Jiangsu University, Zhenjiang 212013, China,
[2]School of Energy and Power Engineering, Jiangsu University, Zhenjiang 212013 China
Email:13951280633@163.com;18094481952@163.com

Abstract Based on polyvinylidene fluoride (PVDF) hollow fiber hydrophobic porous membrane, vacuum membrane distillation desalination test bench was built to study the effects of operating parameters such as temperature of feed solution , flow rate of feed solution, concentration of feed solution and vacuum degree on the performance of vacuum membrane distillation process. Experimental results showed that with an increase in the vacuum degree and temperature of the flux of membrane is increased. While with an increase in the concentration of feed solution of the flux of membrane is decreased and retention rate were essentially the same, with an increase in the flow rate of feed solution of he flux of membrane changed slowly (increase slightly). When temperature of the feed solution was 88℃, the flow rate of feed solution was 270cm/min, the vacuum degree was 0.081MPa, the concentration of feed solution was 5%, the flux of membrane was 14.1kg/ (m^2.h), the retention rate was 99.8%, the electrical conductivity of fresh water was maintained at 12us/cm.

1. Introduction

Membrane distillation (MD), which is a hybrid between thermal desalination and microporous hydrophobic membrane, generally the aqueous solution of non-volatile substances was distilled. When the aqueous solution in different temperature is separated by a hydrophobic microporous polymer membrane, aqueous solution of both sides can't permeate the membrane pores due to the surface tension, while water vapor at high temperature permeates membrane pores into the low temperature side under water vapor pressure difference on both sides of membrane and then is condensed to realize the separation process. Evaporation, mass transfer, condensation process in the process are very similar with conventional distillation , so it is called membrane distillation[1].Comparing with the method of thermal desalination, feed solution is not need to be heated to boiling point in the MD process, it is driven by vapor pressure difference on both sides. Membrane distillation is almost conducted under atmospheric pressure, simple equipment needed, operated easily, without high pressure compared with reverse osmosis desalination device, in the region of weak technical force also has the possibility of implementation.

MD basically has the following four forms: direct contact membrane distillation (DCMD), vacuum membrane distillation (VMD), sweeping gas membrane distillation (SGMD), air gap membrane distillation (AGMD) [2].The study found that one side of the membrane which was called hot side contact with the feed solution directly in vacuum membrane distillation, pressure of the other side is lower than the hot side, water vapor is pumped out from membrane module to condense. The higher vapor pressure difference on both sides of the membrane, the greater flux.

IOP Conf. Series: Earth and Environmental Science **67** (2017) 012002 doi:10.1088/1755-1315/67/1/012002

Despite membrane distillation has many advantages, however its industrialization is effected by low thermal efficiency and low flux[3-6], what' s more the application of membrane distillation on large-scale desalination of sea water is scarce, in order to develop the potential of the desalination process, reduce the cost of desalination, alleviate the water crisis ,it is necessary to carry out research work about low-cost seawater desalination technology. In this paper polyvinylidene fluoride PVDF hollow fiber hydrophobic porous membrane was used, focuses on water desalination via VMD. The effects of vacuum, temperature of feed solution, flow rate of feed solution and the concentration of feed solution on the membrane distillation process were studied.

2. Experiment

2.1. Experiment system

Hollow fiber vacuum membrane distillation assembly used in the present study comed from Tianjin MOTIMO Membrane Technology CO., LTD. The membrane module shell was made of organic glass, the size was , the inside diameter of membrane was 0.5mm,thickness 0.25mm,average pore size was 0.18,the porosity was 86%,the effective area was 0.8m2.

Vacuum membrane distillation experiment system diagram as shown in Figure 1, which was mainly composed of hot side, vacuum side and hollow fiber membrane , thermostatic hot water bath, magnetic pump, rotor flow meter, thermometer, pressure gauge and control valve were included on hot side, the vacuum side includes water ring vacuum pump, freshwater collection tank, electronic balance.

NaCl solution was heated to the predetermined temperature, then was pumped into the hollow fiber membrane module using magnetic circulating pump after the flow was adjusted by rotameter .Due to the vapor pressure difference on both sides, water vapor on the hot side through the hydrophobic porous membrane into the vacuum side, then water vapor was pumped outside from membrane module by vacuum pump to be condensed by the condenser. when the temperature of the feed solution and fresh water on either side of the membrane module is stabilized, the quality of fresh water was weighed by electronic balance.

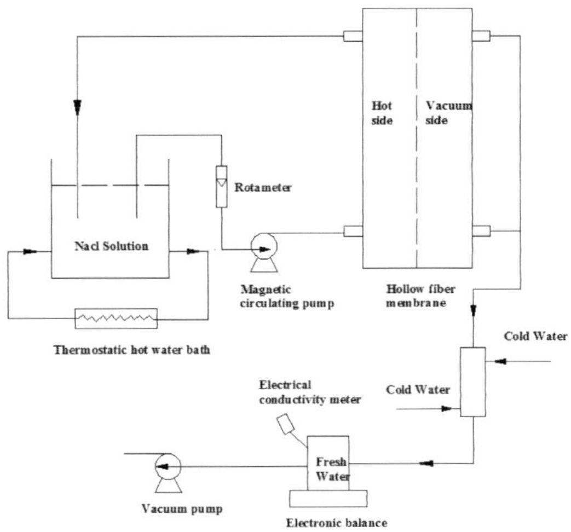

Figure 1. hollow fiber vacuum membrane distillation device schematic

2.2. Calculation parameters

The weight of collected fresh water was measured continuously to calculate the VMD flux,

$$J = \frac{Q}{St} \quad [7] \qquad (1)$$

where J is the VMD flux, kg/(m^2. h), Q is the weight of vapor across the membrane, kg; S is the effective membrane area,m^2,and t is the experimental running time, h.

η describes the salt rejection for the permeation selectivity of aqueous NaCl solution::

$$\eta = \left(1 - \frac{C_P}{C_F}\right) \times 100\% \ [7] \qquad (2)$$

where, η is the salt rention rate %, C_F is the concentration of feed solution,%. C_P is the concentration of permeation solution, %.

3. Results and discussion

3.1. Effect of feed temperature

The concentration of feed solution was set to 0, 6%, 10%, 13%, vacuum degree was 0.080MPa, the flow rate of feed solution was 250cm/min, and the influence of feed temperature on the flux was investigated. The results were shown in Figure 2.The concentration of feed solution was 10%, inlet flow rate was 250cm/min, the vacuum degree was 0.054MPa, 0.063MPa, 0.072 MPa, 0.084MPa, the influence of feed temperature on the flux was investigated, as shown in Figure 3.The concentration of feed solution was 10%, vacuum degree was 0.080MPa, the flow rate of feed solution was 200 cm/min, 250cm/min, cm/min, 270 cm/min, the influence of the temperature on the flux was investigated, as shown in Figure 4.

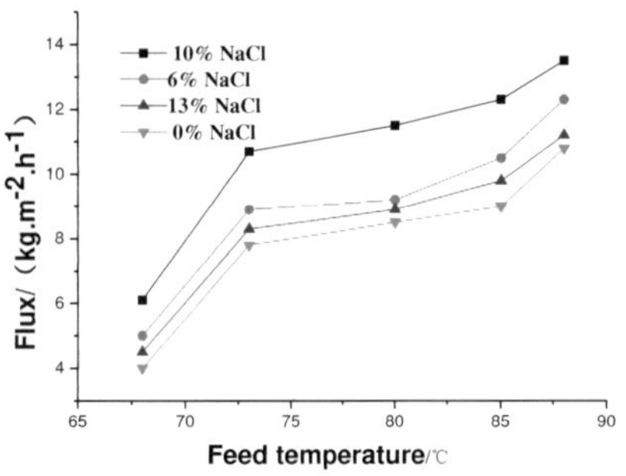

Figure 2. Effect of feed fluid temperature on flux
flux(different feed fluid concentration)

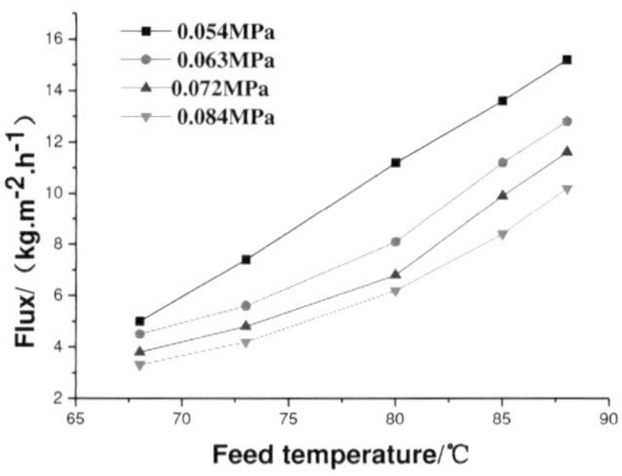

Figure 3. Effect of feed fluid temperature on
(different vacuum)

Figure 2 showed that the flux increased with the increasing of temperature, and decreased with the increasing of concentration of feed solution. Figure 3 showed that the flux increased with the increasing of vacuum, increased with the increasing of temperature of feed solution,

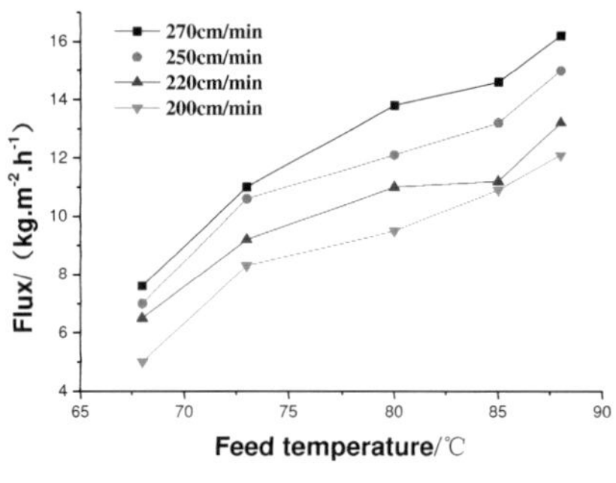

Figure 4. Effect of feed fluid temperature on flux(different flow)

Figure 4 showed that the flux increased with increasing liquid flow rate, and increased with the increasing of temperature of the feed solution.

No matter what kind of operating conditions, the flux increased with the increasing of feed temperature, this is mainly because the saturated pressure of water would increase along with the temperature rises, the driving force was water vapor pressure difference on both sides in the process of vacuum membrane distillation, increasing steam partial pressure on the feed side or increasing the vacuum degree of the permeate side (VMD) both make the vapor pressure difference of both sides increased and increase the driving force, eventually make the flux of membrane also increased.

3.2. Effect of vacuum degree

It can be predicted from Figure 3, the flux would increase with the increasing of vacuum degree. The concentration of feed solution was 10%, inlet velocity was 250cm/min, the feed temperature was 68 °C, 73 °C, 80 °C, 88 °C ,the influence of the vacuum on the flux was investigated, the results were shown in Figure 5. The concentration of feed solution was10%, the temperature was 80 °C, the flow rate of feed solution was200 cm/min, 220cm/min, 250cm/min, 270 cm/min, the influence of the vacuum on the flux was investigated, the results were shown in Figure 6.

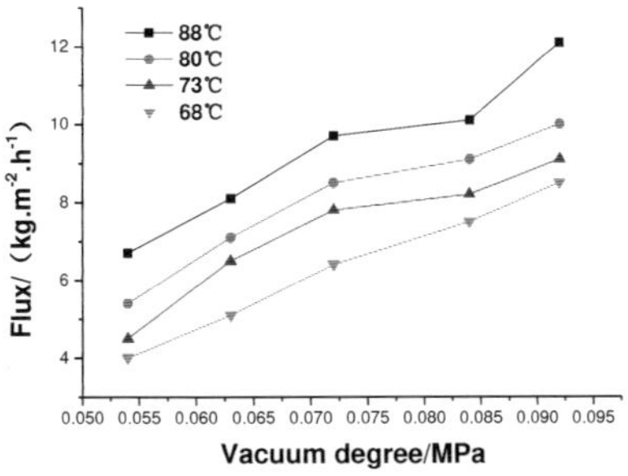

Figure 5. Effect of Vacuum on flux (different feed fluid temperature)

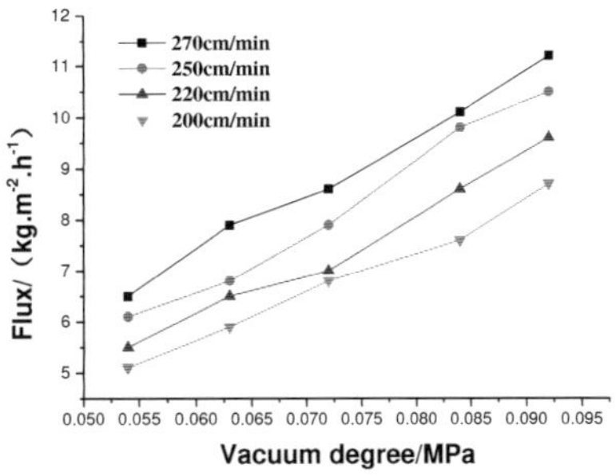

Figure 6. Effect of Vacuum on flux (different feed flow)

The influence of vacuum and feed temperature on flux were the same, the efficiency of the membrane distillation process could be improved by increasing the vapor pressure difference on both sides of the membrane.In the process of industrial applications, increasing the feed temperature maked experimental equipment simpler, the cost lower compared with increasing vacuum,without additional vacuum equipment, feed solution was preheated by condensed steam at the same time. The system could recover the latent heat of vaporization and reduce the heat input.

3.3. Effect of feed solution flow rate

Figure 7 showed that the flux variation with flow rate of feed solution, when the feed temperature was 80,the feed solution concentration was 10%, the vacuum degree was 0.054 MPa,0.063 MPa,0.072 MPa,0.084MPa. Figure8 showed that the flux variation with flow rate of feed solution, when vacuum degree was 0.080Mpa, the temperature was 68 °C, 73 °C, 80 °C, 88 °C .From the two figures above it could be seen the flux increased with the increase of flow rate in any condition, but the curve relatively flat, namely the influence of flow rate of feed solution on flux was very small, increasing feed solution flow rate leaded to boundary layer disturbance enhanced, also enhanced heat and mass transfer Once a stable flow pattern was formed, the influence of velocity was negligible, considering the mechanical properties of the membrane module， the flow rate of the material liquid could't be increased unlimited .

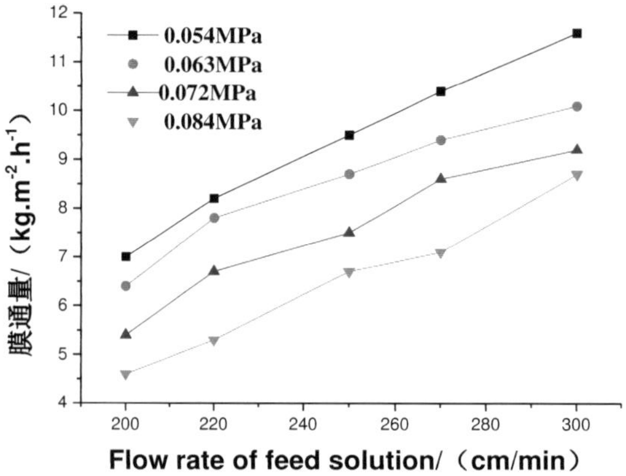

Figure 7. Effect of feed fluid flow on flux(different vacuum)

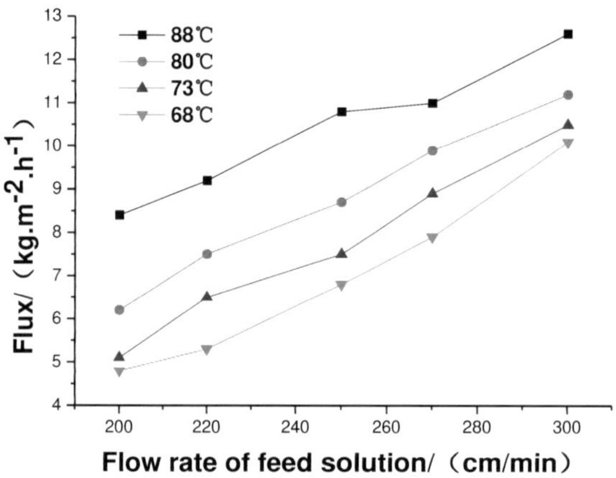

Figure 8. Effect of feed fluid flow on flux(different feed fluid temperature)

3.4. Effect of feed solution concentration

The temperature of the feed solution was 80 $^{\circ}C$, the vacuum degree was 0.080MPa, the feed rate was 250cm/min, the influence of the concentration of feed solution on the rention rate was studied. Results were shown in Figure 9 and Figure 10.With the increasing of concentration of feed solution ,rention rate decreased, because the greater the feed concentration, the smaller the water vapor partial pressure of the hot side, this maked vapor pressure difference decreased and reduced driving force. With the concentration of the feed solution increasing, the curve decreased rapidly, it can be explained that concentration polarization will appear in the membrane separation interface when the concentration of the feed solution up to a certain value. Effect of the concentration of feed solution on rention rate was very small, but if membrane was contaminated large fluctuation of the curve would appear.

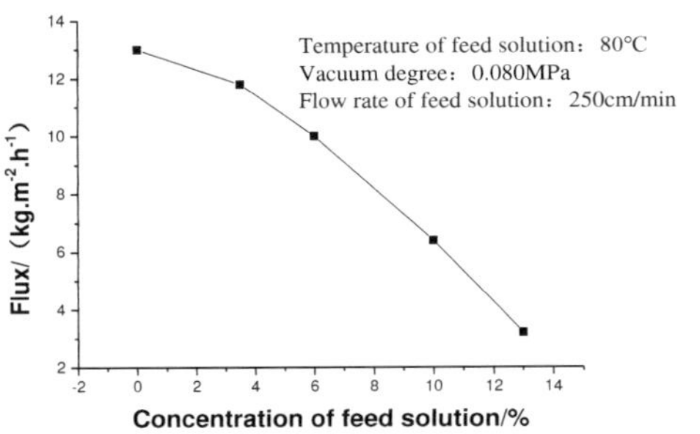

Figure 9. Effect of feed fluid concentration on flux rention rate

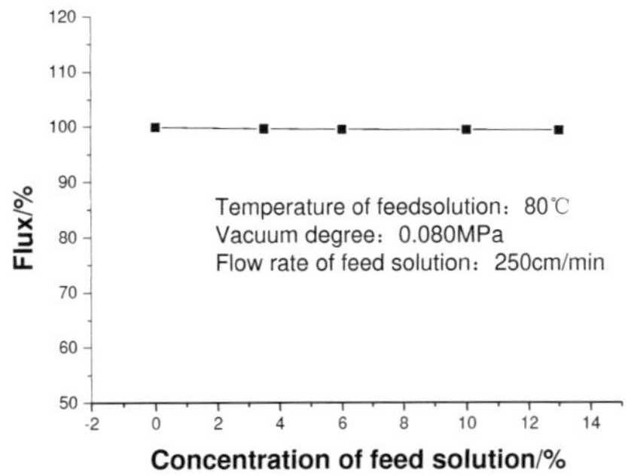

Figure 10. Effect of feed concentration on

4. Conclusion

In this paper PVDF hollow fiber hydrophobic porous membrane was used to carry out experimental research on the process of vacuum membrane distillation, focused on the effect of temperature of feed solution, flow rate of feed solution, vacuum degree and centration of feed solution on the flux and rention rate of VMD process. The results showed that the efficiency of the VMD process can be increased by increasing the temperature of the feed solution, the vacuum degree, flow rate of feed solution appropriately, all of its essence were to enhance the driving force of membrane distillation by increasing vapor pressure difference across the membrane. Which method was used to improve the MD process should be considered comprehensively such as characteristics of feed solution, the material of the membrane module and so on. It was found that the maximum flux reached 14.1kg/(m2.h) in this system, which was larger than reverse osmosis with the same scale, so the membrane distillation has a very broad prospect for the desalination of sea water.

5. References

[1] J. Shi, Q. G. Yuan, C. K. Gao. Handbook of Membrane Technology [M]. Beijing: Chemical Industry Press, 2001.
[2] EL-BOURAWIM S, DINGZ, MAR, etal. A frame work for better understanding membrane distillation separation process [J].Journal of Membrane Science,2006, 285:4 -29
[3] Environmental impact and impact assessment of seawater desalination[J].Desalination.2008 (1)
[4] H. Fang, J. F. Gao, H. T. Wang, C. S. Chen. Hydrophobic porous alumina hollow fiber for water desalination via membrane distillation process[J].Journal f Membrane Science.2012
[5] Y. L. Zhu. Research progress of water desalination technology[J].Energy research and information. 2010 (02)
[6] C. Y. Zhu, Z. P. Zhao, D. Z. Liu, J. Y. Shi , W. F. Liu. Research progress on transfer mechanism of membrane distillation process (III) Vacuum Membrane Distillation [J]. Membrane science and technology. 2009 (06)
[7] C. R. Wu, F. J. Hao, Y. Jia, X. L. Lv, X. G. Jian . Bromine Extraction from seawater with PVDF hollow fiber membrane air-blowing/absorption[J]. Functional materials. 2009 (09)

IOP Conf. Series: Earth and Environmental Science **67** (2017) 012003 doi:10.1088/1755-1315/67/1/012003

Enhancement of enzymatic hydrolysis and lignin removal of bagasse using photocatalytic pretreatment

P Pattanapibul[1], S Chuangchote[1,*], N Laosiripojana[1], V Champreda[2] and J Kaewsaenee[3]

[1] The Joint Graduate School of Energy and Environment, Centre of Excellence on Energy Technology and Environment, King Mongkut's University of Technology Thonburi (KMUTT), 126 Pracha Uthit Rd., Bangmod, Thung Khru, Bangkok 10140, Thailand
[2] National Center for Genetic Engineering and Biotechnology, 113 Phahonyothin Rd., Klong Luang, Pathumthani 12120, Thailand
[3] Department of Physics, King Mongkut's University of Technology Thonburi (KMUTT), 126 Pracha Uthit Rd., Bangmod, Thung Khru, Bangkok 10140, Thailand

*E-mail: surawut.chu@kmutt.ac.th

Abstract. Pretreatment for reduction of biological resistance in a lignocellulosic material, i.e. bagasse, for enzymatic hydrolysis and fermentation was investigated. Photocatalyst (TiO_2) was used as an additive composition to assist this pretreatment process. Reaction time was varied (24, 48, and 72 h) to find the optimum condition for the pretreatment, while concentration of solvent (NaOH, H_2O_2, or NH_4OH), biomass to solvent ratio, and weight ratio of catalyst to bagasse were fixed at 2 M, 1:20 g/ml (typically, solvent = 150 ml), and 1:5, respectively. Batch reaction temperature was at 25°C. After the pretreatment, the enzymatic digestibility of pretreated bagasse was carried out to find the sugar yield. Hydrolysis of pretreated bagasse with photocatalyst show higher sugar yields than the conventional reactions without photocatalyst. The maximum yields of sugars (541.03 mg glucose and 192.79 mg pentose) were obtained at the longest reaction time.

1. Introduction

Nowadays, the increase of petroleum price and the depletion of petroleum resources force many countries to find new resources to replace petroleum. There have been a number of researchers focusing on both adequacy of energy supply long-term and also the environmental implications. A lot of efforts have been carried out to develop clean sustainable renewable energy to be an alternative fuel for the current energy instead of conventional fuels. Ethanol is one of an alternative energy resources [1] since it is cleaner and does not emit any pollutant gas after use.

The utilization of lignocellulosic materials as renewable resources for ethanol production motivates a great deal of interest. The well-known lignocellulosic materials mostly use for bioethanol production are classified into: (1) sugar (e.g. sugar cane, bagasse, and molasses), (2) starch (e.g. cassava, rice, and corn), and (3) cellulose (which is mainly come from the remaining of plants). These raw materials are the most abundant carbon sources and non-food bio-feedstocks which do not compete with the food supply. They also have the benefits on disposal of problematic solid wastes, providing power opportunity to improve energy security, reducing the trade deficit, reducing greenhouse gas emissions, and improving agricultural products [2]. Several countries have been attempting to develop utilization of biomass. Although biomass utilization has been greatly improved by a variety of technologies, there are still challenges that need further investigations [3].

Since Thailand is an agricultural country with many resources of raw biomass, Thailand can highly take an advantage on these materials [1, 4]. Among various local lignocellulosic materials, a variety of agricultural biomass is considered to be an important feedstock for bioethanol production in Thailand as shown in Table 1.

Lignocellulosic biomass is a multi-structure organic compound, which consist mainly of three polymers, i.e. cellulose, hemicelluloses, and lignin, which are associated with each other (with small amounts of acids, salts, and minerals). Lignocellulosic biomass can be converted into bioenergy (biofuel) by chemical processes, i.e. thermolysis, pyrolysis, hydrolysis, fermentation, gasification, and combustion [5, 6].

Table 1. Biomass residues from rice, sugar cane, and palm oil in Thailand, 2004 [7].

Types	Production	Agricultural residues	Residues
Sugar cane	70,101	Bagasse Trash	20,399 21,171
Rice	26,841	Rice husk Rice straw	6,173 11,998
Palm Oil	4,903	Empty fruit bunches Fiber Shells Fronds	1,226 721 240 12,767
Total			74,695

(Unit: 1,000 tons per year)

Theoretically, the main process that is used to break down hemicellulose and cellulose then converted to sugars is hydrolysis. Lignocellulosic material is highly recalcitrant to hydrolyze due to its rigid structure, particular by lignin which acts as physical barrier against hydrolytic enzymes penetration to the biomass microstructure [8]. It is noted that, due to the presence of lignin, the pretreatment of biomass is initially required for degradation and/or removal of lignin from cellulose and hemicelluloses prior to the hydrolysis process. This causes enzymes can penetrate easier to the cell wall and depolymerize celluloses and hemicelluloses leading to increase of fermentable sugars from the enzymatic hydrolysis step [8].

Currently, modifications of pretreatment of lignocellulosic biomass has been investigated by several researchers to maximize the yield of sugar with low unwanted by-product formations [1, 3]. Among the potential pretreatment technologies, alkaline-treatment is known to be an efficient pretreatment process. Nevertheless, the main drawback of this process is very long retention period in the pretreatment step. Alternatively, the use of photocatalysts (i.e. TiO_2 and ZnO) to assist pretreatment can solve this problem [9, 10].

In this work, photocatalysis was used as an alternative pretreatment process of bagasse using a commercial TiO_2 (P25) as the photocatalyst (no alkaline pretreatment). For more practical applications, the use of photocatalyst to assist in the conventional alkaline pretreatment was also investigated and compared.

2. Materials and methods

2.1. Raw material preparation

Bagasse, also called megass, fibers remain after the extraction of the sugar-bearing juices from sugarcane. Bagasse used in this work was obtained from MitrPhol Sugar Refinery in Samut Prakran. It was milled by a cutting miller and sieved through a 0.25-mm-size sieve screen. The physically processed bagasse was dried in a hot air oven at 70°C overnight and stored in a plastic bag at room temperature before use. The composition of the bagasse was analyzed by the National Renewable Energy Laboratory (NREL) method.

2.2. Photocatalyst-Assisted alkaline pretreatment

Bagasse was treated in aqueous solution in the presence of photocatalysts (P25). Each pretreatment experiment was conducted in both with and without the UV irradiation (dark condition). Biomass to solvent ratio is fixed at 1:20 (g/ml). The weight ratio of catalyst to bagasse was 1:5 (20% w/w catalyst to biomass).

Blank experiments were prepared in screw-capped 250-ml bottles (PYREX® No.4980, STOPPER No. 6) and stirred on an Isotemp Basic-stirring hotplate, at room temperature. For the dark trial, surface of the reactor was covered by an aluminium-foil sheet to prevent the irradiation of light.

For the pretreatment experiments, additive solution (2 M) was prepared in distilled water (150 ml). Biomass was mixed with P25 photocatalyst in a bottle. Additive solution was then added in the dark condition and the suspension was continuously stirred for 30 min.

A mercury lamp as a light source to irradiate the sample that located beside the reactor was turned on to start the reaction. The photocatalytic system continuously stirred in the plastic box that was coated with an aluminium-foil sheet inside. The reaction temperature was maintained by a cooling water system. Then, the samples were taken from the photoreactor at specified times for analysis.

The suspension after pretreatment process was filtrated by a vacuum filter. The first liquid product was kept in an amber bottle at room temperature for further analyzed, while the solid product was diluted with distill water until neutral (pH = 7) and dried in an oven at 70 °C overnight and kept in a plastic bag zipper for further product hydrolysis.

2.3. Enzymatic saccharification (Hydrolysis)

Hydrolysis test for each sample was performed to determine the improvement in enzymatic saccharification under the different pretreatment conditions. The pretreated solid product was hydrolyzed in a 1-ml reaction. The reaction contained 5% (w/v) pretreated biomass in 50 mM sodium citrate buffer (pH 4.8) with 20 FPU/g and 5% sodium azide. The sample was incubated at 50°C and shacked at 30 rpm for 72 h. The sample was then taken at specified times. Sugar profiles of the sample were then analyzed by high performance liquid chromatography (HPLC, Shimudzu, LC-20AD pump) and the glucose and xylose contents were calculated.

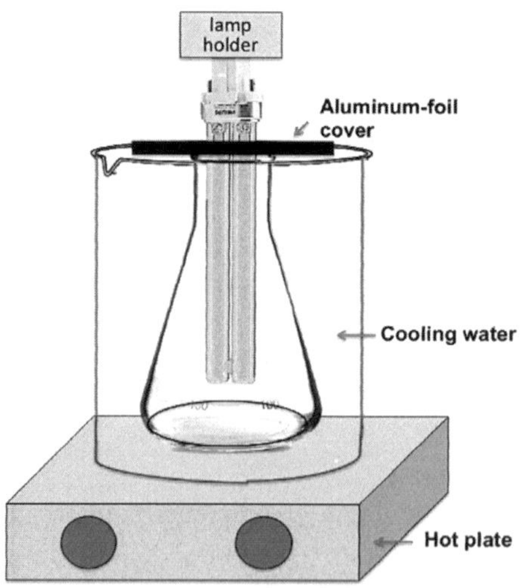

Figure 1. A configuration of photocatalytic pretreatment in the photocatalytic reactor (Mercury lamp, 90 W, λ_{max} = 253.7 nm).

2.4. Products analysis

Liquid product (~5 ml) was filtrated by a 0.22-µm nylon filter before analysis. The generated products were monitored by (HPLC) equipped with a refractive index detector (Shimudzu RID-10A). Separation was performed on an Aminex HPX-87H column (300 x 7.8 mm) (Bio-Rad). The mobile phase was 5-mM sulfuric acid at a flow rate of 0.5 mL/min and an injection volume of 20 µL.

3. Results and discussion

3.1. Photocatalytic treatments

Fig. 2 shows the sugar digestibility plots in cases of: (1) no any treatment (a blank, only water), (2) photolysis (with UV light only, without catalyst), (3) non-active condition (with TiO_2 in dark condition), and (4) photocatalytic pretreatment (with TiO_2 under UV irradiation). It was found that the amount of individual sugar, as well as the total sugar yields, were low in the cases of no catalyst and no UV irradiation. The highest sugar digestibility yields (80.71 mg of glucose and 53.67 mg of pentose based on grams of pretreated biomass) were achieved from the photocatalytic pretreatment. Because, under the UV irradiation, TiO_2 photocatalysts could accelerate the separation of biomass compositions during the pretreatment process. Anyhow, the sugar digestibility yields obtained after the treatment were relatively low compared with the alkaline treatment. Therefore, not only photocatalysis, but use of photocatalyst to assist the conventional alkaline process is also an interesting process.

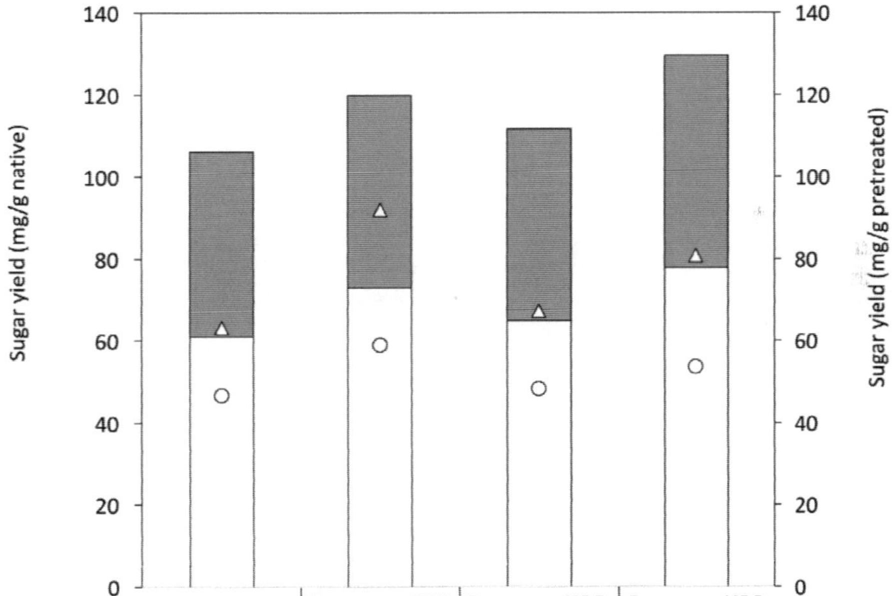

	Bagasse + H2O	Bagasse + H2O + UV	Bagasse + H2O + TiO2	Bagasse + H2O + TiO2 + UV
■ Pentose(mg/g native)	45.09	46.79	46.63	51.71
□ Glucose(mg/g native)	61.02	72.98	64.91	77.76
○ Pentose (mg/g pretreated)	46.71	58.97	48.29	53.67
△ Glucose (mg/g pretreated)	63.22	91.96	67.21	80.71

Figure 2. Sugar digestibility yields in photocatalytic pretreatment (solvent = water) compared with blanks (no catalyst and/or no UV irradiation).

3.2. Effects of solvents on sugar digestibility

Fig. 3 shows sugar digestibility yields in different solvent types. The solvents (NaOH, H_2O_2, and NH_4OH) were used at the same concentration of 2 M. Biomass to solvent ratio and reaction time were fixed at 1:20 g/ml and 24 h, respectively. As illustrated in Fig. 3, it can be observed that the alkaline pretreatment (NaOH and NH_4OH) resulted in the top two ranking in enzymatic digestibility. The highest sugar yields (412.04 mg of glucose and 74.42 mg of pentose based on grams of pretreated biomass) were obtained from NaOH. Compared with the oxidative reagent (H_2O_2), the alkali treatment becomes to be the most powerful method in breaking the crystalline structure of lignocellulosic biomass.

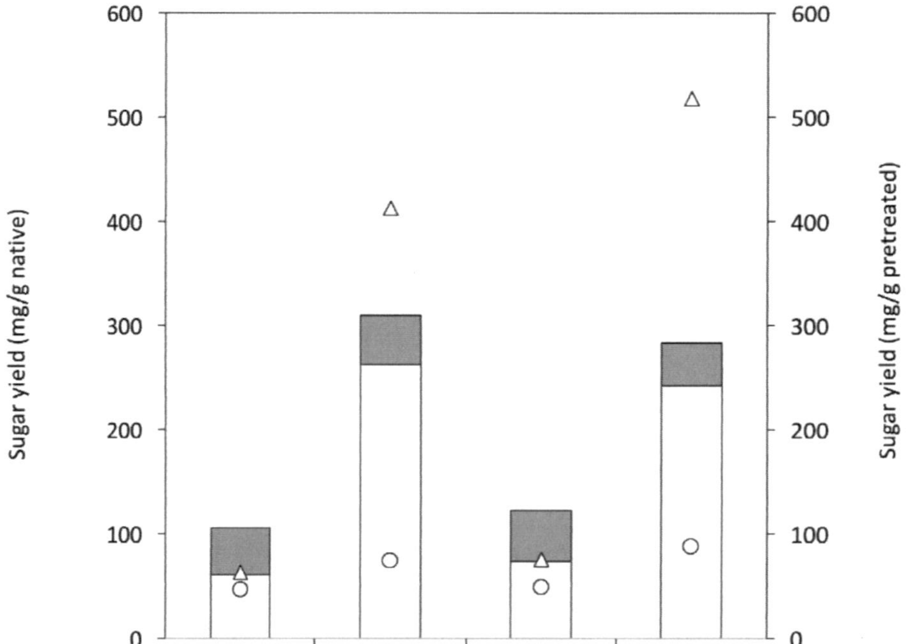

	Bagasse + H2O	Bagasse + NaOH 2M	Bagasse + H2O2 2M	Bagasse + NH4OH 2M
■ Pentose(mg/g native)	45.09	47.36	48.14	41.37
□ Glucose(mg/g native)	61.02	262.21	74.09	242.22
○ Pentose (mg/g pretreated)	46.71	74.42	49.05	88.38
△ Glucose (mg/g pretreated)	63.22	412.05	75.47	517.58

Figure 3. Sugar digestibility yields with different solvent types (NaOH, H_2O_2, or NH_4OH) compared with blank (solvent = H_2O).

3.3. Effects of pretreatment time on sugar digestibility

Fig. 4 shows the sugar digestibility yields from the photocatalyst-assisted NaOH pretreatment of bagasse at different reaction times (24, 48, and 72 h). The photocatalyst-assisted NaOH pretreatment showed 3-times higher sugar yields compared with the photocatalytic pretreatment. It can be observed that increase of pretreatment time led to increasing solubilization period of biomass components, decreasing cellulose crystallinity, and improving enzymatic digestibility. The highest sugar yield (541.03 mg of glucose, 192.79 mg of pentose based on grams of pretreated biomass) at the longest pretreatment time (72 h).

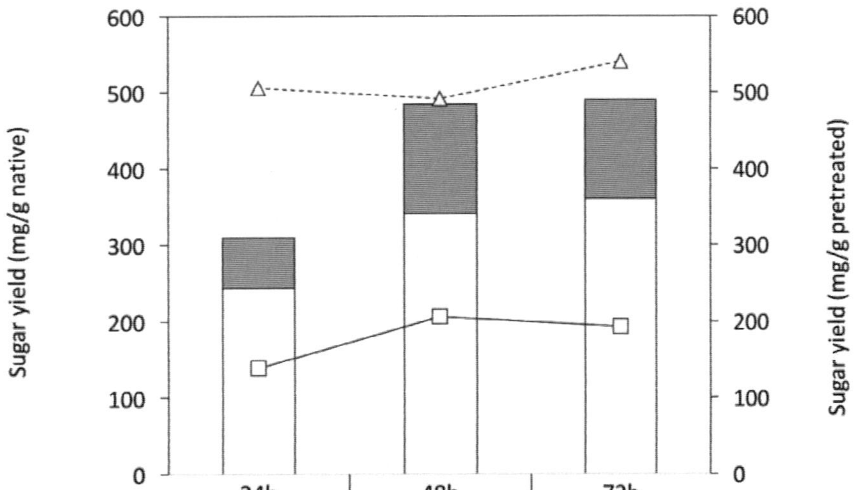

	24h	48h	72h
Pentose(mg/g native)	66.55	142.99	128.75
Glucose(mg/g native)	243.25	341.39	361.30
Pentose (mg/g pretreated)	138.23	206.03	192.79
Glucose (mg/g pretreated)	505.30	491.91	541.03

Figure 4. Sugar digestibility yields in photocatalyst-assisted NaOH pretreatment (solvent = NaOH) at different times.

4. Conclusion

Pretreatment is a pre-requisite step in sugar-platform biorefinery to improve the enzymatic digestibility of the intact native biomass in order to obtain feasible sugar yields for further conversion to biofuels, biochemical, and other value-added products. An experimental study on the enhancement of enzymatic hydrolysis and lignin removal of bagasse was carried out by photocatalytic pretreatment and photocatalyst-assisted alkaline pretreatment. It was found that the use of photocatalyst-assisted alkaline pretreatment could enhance higher sugar production than photocatalytic pretreatment. In addition, the increasing in pretreatment time led the increase in solubilization period, resulting in the higher sugar yields of biomass components. For the future studies, the effects of additive concentration and amount of photocatalyst loading on sugar yields will be investigated.

References

[1] Limayem A and Rick S C 2012 Lignocellulosic biomass for bioethanol production: Current perspectives, potential issues and future prospects *PECS* vol 38 p 449-467
[2] Taherzadeh M J and Karimi K 2008 Pretreatment of lignocellulosic wastes to improve ethanol and biogas production: a review *Int. J. Mol. Sci.* vol 9 p 1621-1651
[3] Chen H and Qiu W 2014 Key technologies for bioethanol production from lignocelluloses *Biotechnol. Adv.* vol 28 p 556-562
[4] Mosiera N, Wymanb C, Dalec B, Elanderd R, Lee Y Y, Holtzapplef M, and Ladisch M 2005 Features promising technologies for pretreatment of lignocellulosic biomass *Bioresource Technol.* vol 96 p 673-686
[5] Chaturvedi V and Verma P 2013 An overview of key pretreatment processes employed for bioconversion of lignocellulosic biomass into biofuels and value added products *J. Biotechnol* vol 3 p 415–431
[6] Golova O P and Nosova N I 2004 Degradation of cellulose by alkaline oxidation *Russ. Chem. Rev.* vol 42 p 327

[7] Papong S, Yuvaniyama C, Lohsomboon P, and Malakul P 2004 Overview of biomass utilization in Thailand *in The Meeting for LCA in ASEAN Biomass Project (Japan)* vol 28
[8] Menon V and Rao M 2012 Trends in bioconversion of lignocellulose: biofuels, platform chemicals & biorefinery concept *Prog. Energ. Combust.* vol 38, p 522-550
[9] Yasuda M, Miura A, Yuki R, Nakamura Y, Shiragami T, Ishii Y, and Yokoi H 2011 The effect of TiO_2-photocatalytic pretreatment on the biological production of ethanol from lignocelluloses *J. Photochem. Photobiol* vol 220 p 195-199
[10] Yoo C G, Wang C, Yu C, and Kim T H 2013 Enhancement of Enzymatic Hydrolysis and Klason Lignin Removal of Corn Stover Using Photocatalyst-Assisted Ammonia Pretreatment *Biotechnol. Appl. Biochem* vol 169 p 1648-1658

Acknowledgement
The authors would like to gratefully acknowledge The Joint Graduate School of Energy and Environment (JGSEE), King Mongkut's University of Technology Thonburi (KMUTT), for the financial support.

Evaluation of nano zero valent iron effects on fermentation of municipal anaerobic sludge and inducing biogas production

Tareq W.M. Amen, Osama Eljamal, Ahmed M.E. Khalil and Nobuhiro Matsunaga

Earth System Science and Technology, Interdisciplinary Graduate School of Engineering Sciences, Kyushu University, 6-1 Kasuga-Koen Kasuga, Fukuoka, Japan.
Email: tareq.amen.040@s.kyushu-u.ac.jp

Abstract. The application of nano size materials on wastewater is going extensive because its high reactivity compared with other materials. As a result, numerous research studies investigated the effectiveness of dosing nano zero valent iron (nZVI) or micro zero valent iron (mZVI) on anaerobic digestion (AD) of sludge and production of biogas as promising renewable energy but inconsistent outcomes have appeared. In this paper, different dosing concentrations of nZVI were applied on anaerobic activated municipal sludge to examine the impact of nZVI on sludge fermentation, biogas generation, and methane (CH_4) content stimulation. The results showed that addition 250 mg/L nZVI nanoparticles could enhance 25.23% biogas production and the methane content reached 94.05% after one week of digestion compared with 62.67% without adding iron nanoparticles.

1. Introduction

Recently, using nano zero valent iron (nZVI) in wastewater remediation is growing worldwide, this potentially might affect the anaerobic digestion (AD) of municipal sludge. AD is a biodegradation process that transforms organic matters into biogas which its main mixture is methane (CH_4) and carbon dioxide (CO_2) under controlled conditions. The anaerobic atmosphere was applied and the temperature degree must be fixed to be appropriate for occurring mesophilic stabilization. Facultative bacteria is responsible for converting part of sludge into biogas [1]- [3]. Generally, AD involves four successive steps hydrolysis, acidogenesis, acetogenesis and methanogenesis [4].

In fact, AD is deemed to be the most energy efficient method for stabilizing sludge and methane by-product as one form of fuel that is promised to reduce the treatment cost [5]. Despite a mass of studies that compared the effect of ZVI in nano-size with that particles in micro-size, different concentrations of nZVI in the literature were not extensively studied. For example, numerous researchers compared different ZVI diameters in same concentration [6] which resulted in unconvincing comparison, further researchers studied the long term of adding same concentration of one nanoparticle on the rate of AD [7], [8].

According to some reports [9], [10] nZVI has a relatively high surface area and superior reactivity, that mean it can work as an electron donor in the AD process which can improve the sludge digestion and increase the biogas yield, this may be will make it an excellent additive to the AD process. Su *et al.* research results showed that adding of 0.1 wt% nZVI to the AD of waste activated sludge for 17 days at the mesophilic temperature increased the methane production by 40.4% [10]. However, Yang *et al.'s* investigation concluded inhabitation properties of nZVI in AD bioreactors compared with mZVI particles that stimulate the methane content. According to their study, 20%, 20% and 70% decrease in methane production were observed when adding 1, 10 and 30 mM nZVI to the municipal sludge, respectively [6].

The conflicting results indicated that further investigations are still needed to assess the impact of nZVI on methane production during AD process. Therefore, the effect of the same size of nZVI with different concentrations during sludge anaerobic digestion is still not yet well documented.

In the research presented here, we study the effect of 50, 100 and 250 mg/L of laboratory-prepared nZVI on the efficiency of anaerobic digestion process by examining the enhancement in the degradation rates and the increases in the methane production.

Three nZVI laboratory scale bioreactors in addition to the control set were tested over a range of nZVI concentrations to determine the optimal nZVI dosing and anaerobic degradation was evaluated by monitoring the chemical oxygen demand (COD) concentration and the biogas production compared to the control bioreactor.

2. Materials and methods

2.1 Iron reagent synthesis

One gram nZVI stock suspension was freshly prepared by reducing five grams of ferric chloride ($FeCl_3$) that were dissolved in 120 mL of deionized water with 5 grams of sodium borohydride ($NaBH_4$) that also were dissolved in another 120 mL of deionized water as presented in Figure 1.

$NaBH_4$ solution was added dropwise to $FeCl_3$ solution under the nitrogen atmosphere while the $FeCl_3$ solution was vigorously stirred at 250 rpm at 27-degree Celsius temperature. The prepared nZVI particles had an average size of 60 nm as they were characterized by transmission electron microscopy (TEM, JEM-ARM 200F, JEOL Co., Japan) and they were analysed by X-Ray diffraction instrument (XRD, TTR, Rigaku, Tokyo, Japan) to check its particles composition.

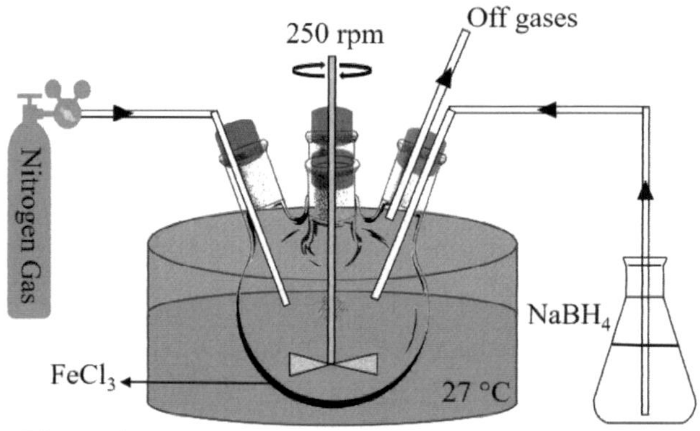

Figure 1. Schematic diagram of nZVI synthesis reactor

2.2 Sludge source

Anaerobic activated sludge was sampled from a two-stage mesophilic digester at the Mikasagawa Wastewater Purification Center, Fukuoka, Japan. Mikasagawa treatment plant treated around of 267,400 m^3/day in 2016 using conventional activated process. The quality of raw sludge is listed in Table 1. The anaerobic sludge samples were kept at 4 °C to maintain its freshness prior to use.

Table 1: Main features of the raw activated anaerobic sludge

Parameter	Reading/Concentration
Total Solids	4.22%
COD	5,453 mg/L
Total Alkalinity	3.53 mg/L
pH	7.1
Organic Fraction	74.86 %

2.3 Experimental setup

The laboratory scale experimental setup of the batch anaerobic system which shown in schematic diagram in Figure 2 was consists of: (1) a bioreactor; 500 mL DURAN neck culture vessel bottle (SANSYO, Japan) plugged tightly with two nozzles perforated cap (SANSYO, Japan), (2) a thermostatic water bath (SANSYO, Japan) as a temperature control, two outlets were connected to the perforated cap, (3) first one was used for examining the sludge characterization by taking liquid sludge samples using (6) syringe while the (4) second outlet was divided into (5) two orifices, one which allows to take gas samples using (7) Hamilton syringe (Hamilton, USA) to inspect the biogas composition and CH_4 fraction using gas chromatography (GC, GL Sciences Inc., Japan), while the second orifice was connected to a water displacement system consisting of 300 mL (8) calibrate glass cylinder to measure the volume of produced biogas. Before the digestion, the sludge samples were flushed with nitrogen gas for 10 min to remove oxygen and to assure the anaerobic condition in the bioreactor at the beginning of experiments.

50, 100 and 250 mg/L concentrations of laboratory prepared nZVI were used to study its effect on biogas production and compared it with control bioreactor. The nZVI particles were added separately to each bioreactor and the operating temperature was maintained at mesophilic conditions 37 ± 3 °C then the bioreactors were reciprocally shaken under 150 ± 5 rpm every 12 hours for 5 minutes long. The bioreactors performance was measured with respect to the daily and cumulative volume of biogas produced and CH_4 yield in the course of time was analyzed. All of the treatments were carried out in duplicate to ensure the reproductively of results.

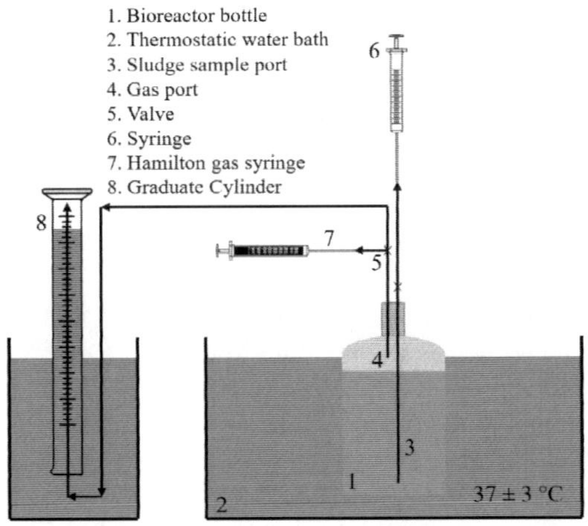

Figure 2. Schematic diagram of experimental biogas production system

2.4 Analytical methods

Raw activated sludge samples were analysed for total solids (TS), total alkalinity and COD. Digested sludge samples were daily withdrawn during the batch experiments using a 10 mL syringe and were then filtered through a 0.45 µm disposable membrane filter (MILLIPORE, USA). The COD concentrations were analysed by spectrophotometer (DR 3900, Hach company, USA) using digestion method according to the standard method for examination of water and wastewater [11].

Gas composition was analyzed using the GC instrument and Argon was used as carrier gas at a flow rate of 30 mL/min. GC was equipped with capillary column with mesh range 60/80 with a molecular sieve (Molesieve 5A, 1/8). Column oven temperature was set at 60 °C whereas the injector temperature was 80 °C. The thermal conductivity detector (TCD) operating temperature was raised to 80 °C and the current was fixed at 60 mA. For analyzing the biogas composition, 1mL of generated biogas was collected using a gastight 10 mL Hamilton syringe and then injected in GC instrument. The CH_4 fractions were defined by analyzing the post-run results obtained from gas chromatographic measurements. All GC post-run measurements were repeated at least two times.

3. Results and discussion

3.1 Iron characterization

TEM image of the laboratory synthesized iron nanoparticles showed a largely spherical form and the single particle size was less than 100 nm, moreover, the particles distribution formed chain-like aggregate as shown in the Figure 3. The XRD spectrum of nZVI revealed apparent peak at 2θ of 44.8° indicating the presence of zero valent of synthesized iron particles and there is no major existence of iron amorphous particles Figure 4.

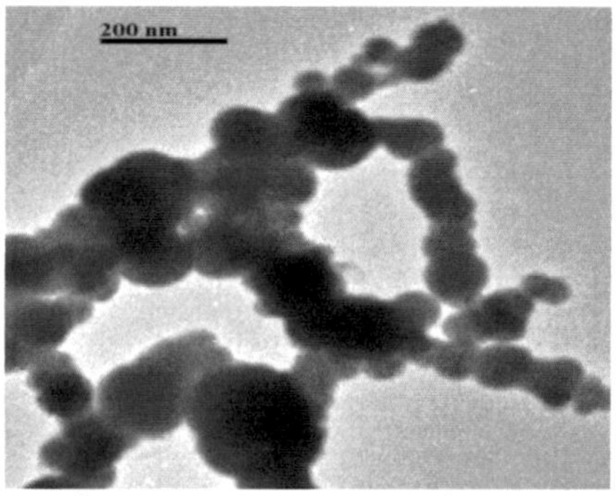

Figure 3. Morphology graphs of aggregates of nZVI particles

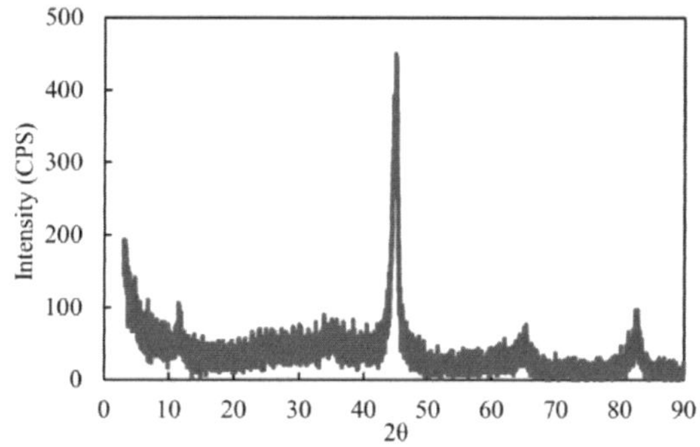

Figure 4. X-ray diffraction results of the synthesized iron nanoparticles

3.2 Biogas production and methane content in the presence of nanoparticles

The cumulatively produced biogas volume for each bioreactor was daily recorded along the study period as shown in Figure 5. It was noted that the biogas started generation at the beginning of digestion duration and at the first four days, the biogas production in all bioreactors had an almost similar tendency. Afterward, the cumulative biogas production increased greatly with nZVI bioreactors.

The daily biogas volume reached the maximum on the day 6th and high concentration of nZVI produced more biogas if compared with low concentrations. By the end of digestion period, 250 mg/L nZVI generated 273 mL of biogas, 100 mg/L nZVI generated 219 mL of biogas and 50 mg/L nZVI generated 192 mL of biogas compared with 218 mL that control generated it. Comparing the produced biogas volume at the day 14th, the biogas volume increased by 25.23% with dosing 250 mg/L of nZVI and this increment did not depict with the concentration of 50 mg/L and 100 mg/L nZVI.

Figure 6 shows the methane fractions in the generated biogas from the municipal sludge with the addition of nZVI. The addition of 250 mg/L nZVI resulted in an increase in the fraction of methane from 22.73% at the second day of digestion to 68.65% in the day 7th and after that, the methane content fractions did not significantly increase and almost remained constant with different nZVI concentrations. Increasing the methane fractions by the addition of nZVI revealed that nZVI have a clear stimulating effect on methanogenic activity during the anaerobic digestion process in comparison with the control.

Based on the consideration above, it was indicated that adding of relatively high concentration of nZVI in anaerobic activated sludge was likely to improve the methanogenic activity and as result increase the biogas production and CH_4 yield which confirmed the positive effects of iron on AD[12].

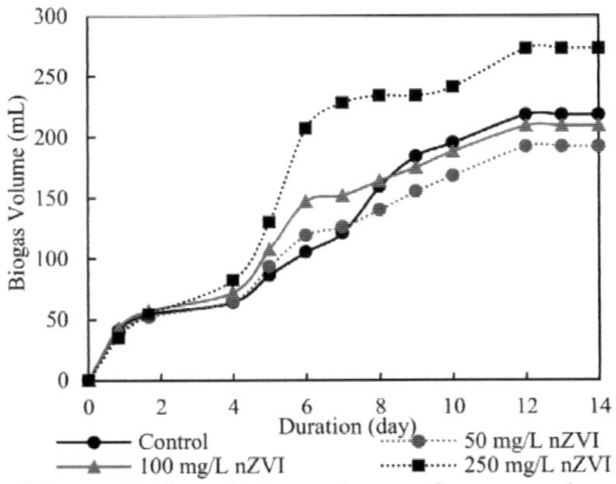

Figure 5. Cumulative volume of generated biogas

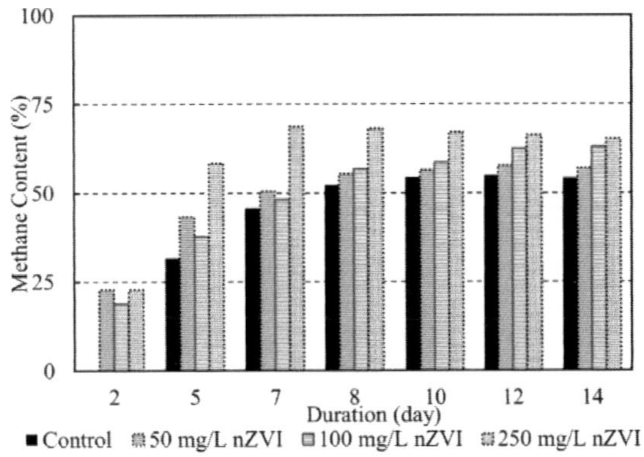

Figure 6. Generated methane content during AD of sludge

3.3 COD changes during anaerobic digestion

The soluble COD was used as an indicator for the change in dissolved/soluble organic matter concentration in the anaerobic sludge [13].

The soluble COD in all reactors was increased through the 14 days of digestion for all treatments indicating that the sludge fermentation process particularly hydrolysis and acidogenesis were took place and the decreasing in COD reading was not depicted due to the long period needed to start decreasing the COD values as reported by Yang *et al.* 2013a [14].

Figure 7 shows the changes of COD during the digestion, with the increase of nZVI dosage the COD concentration increased, which was in accordance with the enhancement of sludge fermentation. These results were in agreement with Feng *et al.*, study that proved the addition of ZVI enhanced the formation of acetate and therefore raised in COD concentrations [5]. The COD concentration was 5453 mg/L before the digestion, and after the 14 days of digestion, its concentration increased to 8722, 8681, 9404 and 11309 at the nZVI of 0, 50, 100 and 250 mg/L, respectively. It pointed out that the sludge fermentation increased significantly with increasing the nZVI concentrations.

According to Figure 7, both of the 50 and 100 mg/L nZVI concentration increased the COD concentrations lower than those obtained by 250 mg/L under same conditions during the digestion duration process means that the reactivity of nZVI on AD could be improved significantly at relatively high dosing concentrations

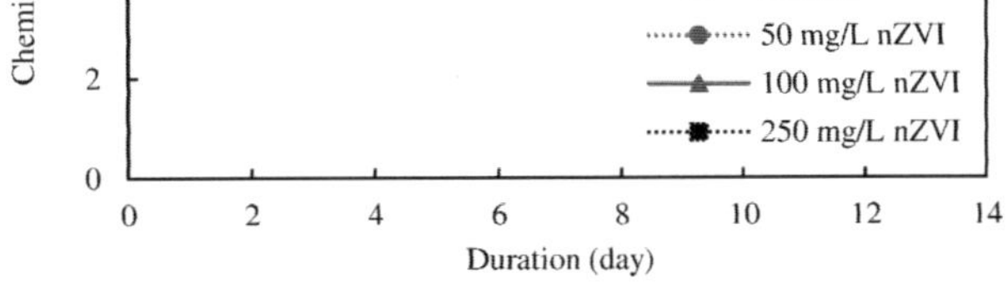

Figure 7. Comparison of COD concentrations during the digestion

4. Conclusion

From the results obtained in this study, it can be concluded that the addition of nZVI to anaerobic activated sludge led to an increase in the fraction of methane in all produced biogas samples through AD. This can be explained by the increase in the production of enzymes by microorganisms that are often detected in the anaerobic digestion environment.

Among the different nZVI concentrations, the relatively high additive was the best one for enhancing the biogas production from municipal activated sludge, which has reached 25.23% higher than the control group.

The addition of nZVI proved to be an interesting alternative for improving energy production from municipal activated sludge which not only may enhance activated sludge fermentation, but also could save a significant portion of wastewater treatment plants operation cost.

5. Acknowledgement

The authors acknowledge the financial support from Kyushu University, Japan.

6. References

[1] F. Suanon, Q. Sun, M. Li, X. Cai, Y. Zhang, Y. Yan, C.-P. Yu, Application of nanoscale zero valent iron and iron powder during sludge anaerobic digestion: Impact on methane yield and pharmaceutical and personal care products degradation, Journal of Hazardous Materials 321 (2017) 47-53.

[2] M.A. Ganzoury, N.K. Allam, Impact of nanotechnology on biogas production: A mini-review, Renewable and Sustainable Energy Reviews 50 (2015) 1392-1404.

[3] Y. Zhang, Y. Feng, X. Quan, Zero-valent iron enhanced methanogenic activity in anaerobic digestion of waste activated sludge after heat and alkali pretreatment, Waste Management 38 (2015) 297-302.

[4] G. Zhen, X. Lu, Y.-Y. Li, Y. Liu, Y. Zhao, Influence of zero valent scrap iron (ZVSI) supply on methane production from waste activated sludge, Chemical Engineering Journal 263 (2015) 461-470.

[5] Y. Feng, Y. Zhang, X. Quan, S. Chen, Enhanced anaerobic digestion of waste activated sludge digestion by the addition of zero valent iron, Water research 52 (2014) 242-250.

[6] Y. Yang, J. Guo, Z. Hu, Impact of nano zero valent iron (NZVI) on methanogenic activity and population dynamics in anaerobic digestion, Water research 47 (2013) 6790-6800.

[7] L. Otero-González, J.A. Field, R. Sierra-Alvarez, Inhibition of anaerobic wastewater treatment after long-term exposure to low levels of CuO nanoparticles, Water research 58 (2014) 160-168.

[8] H. Mu, Y. Chen, Long-term effect of ZnO nanoparticles on waste activated sludge anaerobic digestion, Water research 45 (2011) 5612-5620.

[9] A.M. Khalil, O. Eljamal, T.W. Amen, Y. Sugihara, N. Matsunaga, Optimized nano-scale zero-valent iron supported on treated activated carbon for enhanced nitrate and phosphate removal from water, Chemical Engineering Journal 309 (2017) 349-365.

[10] L. Su, X. Shi, G. Guo, A. Zhao, Y. Zhao, Stabilization of sewage sludge in the presence of nanoscale zero-valent iron (nZVI): abatement of odor and improvement of biogas production, Journal of Material Cycles and Waste Management 15 (2013) 461-468.

[11] W.E. Federation, A.P.H. Association, Standard methods for the examination of water and wastewater, American Public Health Association (APHA): Washington, DC, USA (2012).

[12] B. Yu, Z. Lou, D. Zhang, A. Shan, H. Yuan, N. Zhu, K. Zhang, Variations of organic matters and microbial community in thermophilic anaerobic digestion of waste activated sludge with the addition of ferric salts, Bioresource technology 179 (2015) 291-298.

[13] K. Kennedy, E. Lentz, Treatment of landfill leachate using sequencing batch and continuous flow upflow anaerobic sludge blanket (UASB) reactors, Water Research 34 (2000) 3640-3656.

[14] Y. Yang, S. Gajaraj, J.D. Wall, Z. Hu, A comparison of nanosilver and silver ion effects on bioreactor landfill operations and methanogenic population dynamics, Water research 47 (2013) 3422-3430.

Utilization of ethyl cellulose polymer and waste materials for roofing tile production

Suubitaa Spencer Sam[13], ChoonAun Ng[1], Swee Yong Chee[1], NoorZainab Habib[2], Humayon Nadeem[1] and Wei Ping Teoh[1]

[1]Department of Environmental Engineering, Faculty of Engineering and Green Technology, UniversitiTunku Abdul Rahman, Kampar, Perak, Malaysia
[2]Heriot-Watt University, Dubai International Academic City, P. O. Box 294345, Dubai, U.A.E

Abstract. The aim of this study was to utilize ethyl cellulose, mixture of waste engine oil and waste vegetable oil as a binder in the environmental friendly roofing tile production.The waste engine-vegetable oil wasmix together with ethyl cellulose, fly ash, coarse aggregates, fine aggregatesand a catalyst. The Fourier Transform Infrared (FTIR) analysis showed that the oil mixture added with ethyl cellulose has the relatively high binding effect due to the presence of strong carbonyl group especially after being heat cured at 190^0Cfor 24 hours. The mixed proportion of materials with different amount of ethyl cellulose used was studied in the production of tile specimen. The results showed that the ethyl cellulose composed roofing tile specimens passed the transverse breaking strength, durability, permeabilityand the ultraviolet accelerated test. The shrinkage on the tile can be overcome by adding temperature resistance polymer on the exterior of the tile.

1. Introduction

Knowledge of the use of recycled materials in the production of roofing tiles can facilitate the development of green technologies for building construction.Ethyl cellulose is a derivative of cellulose in which some of the hydroxyl groups on the repeating glucose units were converted into ethyl ether groupsas shown in Figure 1. Ethyl celluloseoffers an attractive range of physical properties and blended with other materials to achieve intermediate characteristics like roofing tile that used asa binder. It provides a green strength to the unfired roofing tile and burnt out clearly leaving no residue after firing. Ethyl cellulose polymer is inert and high purity powders with no caloric value. Ethyl cellulose polymer has been widely used in applications like pharmaceutical, personal care, food, specialty applications and others due to their physical and chemical properties [1]. Ethyl cellulose is used as a binder in some clay products, gives good binding strength, and fires out of the ceramic with an ash residue of only 0.5%. In addition, clays have been used for millennia and can obtain form stability even without being fired. The unfired clays were vulnerable to water and had never been used to form food and beverage containers which have utility in the fast food industry. Nevertheless, if clay is included in large enough amounts, it can impart some degree of binding to the hydrated mixture [2]. Emulsions areobtained at 30^0Cwhereas strong concentrated oil-in-water emulsions are obtained at below 30^0C. Ethyl cellulose functions as a polymeric emulsifier. The high oil-linking tendency of these particles is due to the accelerating temperature of ethyl cellulose in the lipid [3]. A phase separation method is used to mix the rosemary oil with ethyl cellulose microcapsules. A regular spherical shape is observed on the ethyl cellulose microcapsules. The stirring speed of the encapsulation affected the size and degree of sphericity of the microcapsules [4].The polymer ethyl cellulose is mixed with vegetable oil atthe temperature above 140^{0}C to induce gelation. Oil oxidation and thebreakdown of ethyl cellulose occurred at high temperature [5].The vegetable oil is blended with 1 wt % of ethyl cellulose ata temperature between 100 and 150^{0}C at 300 rpm for 5 to 10 h. The mixing is influenced by the concentration and nature of additive. The heat is required to dissolve the ethyl cellulose in the vegetable oil [6].

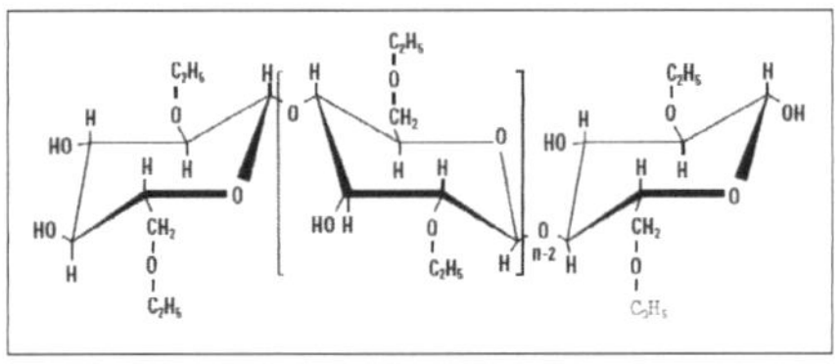

Figure 1. Chemical structure of ethyl cellulose.

Heat sonication treatment is a highly effective way for physical, chemical and enzymatic modification of different starches. Heat sonication process is conducted to promote de-agglomeration and dispersion of the mixture of starch and oil and therefore, producing ahomogenous mixture. In addition, the process modified the molecular structure of the starch without the usage of any chemicals. The oil-polysaccharide mixtures prepared earlier are sonicated using a Cole-Parmer stainless steel ultrasonic tank. The sonicator bath is filled with water until the water level reached one inch from the top of the tank. The oil-polysaccharide mixtures are sonicated with water heated to 60 ^{0}C under a frequency of 40 kHz for 20 min [7].The objectives of this study are to optimize the tile compositions and manufacturing processes. In addition, the mechanical properties, durability and permeability of the tile also have been studied.

2. Materials and method

2.1 Preparation of the materials

Waste engine oil and waste vegetable oil were collected from local car station and restaurant respectively.The waste engine-vegetable oil was blended in the ratio 1:3 and different amount of ethyl cellulose powder were added to the mixture in an orbital shaker at 300 rpm at 100^{0}C for 30 minutes to dissolve it. The ethyl cellulose polymers used werefrom ETHOCEL branded from Sigma-Aldrich.The river sand was used in the mixture to produce tiles. The size river sand used was between 5 to 10 mm and below 5mm which was obtained from sieving. The type of fly ash used was class F fly ash with the size of less than 75μm.Fly ash was obtained from TNB power plant station in Manjung Setiawan, Perak, Malaysia. The chemical substances in fly ash are silica, alumina, magnetite and Fe_2O_3 (Iron oxide). The spherical fly ash has specific gravity and specific surface area between 2.5 and 200 m^2/kg respectively. The percentage of each material in the composition has been chosen due to easy mixing and manufacturing of the tile. The percentage of coarse sand aggregate in the composition should be less than fine aggregate for easy formation of the tile after mixing with appropriate amount of waste oil with ethyl cellulose. The fly ash was used as the filler in the composition. The fly ash has been used in brick making factories in China with low volume between 10% and 30% [8].

2.2 Characterization of the ethyl cellulose mixed with waste engine-vegetable oil in the ratio 1:3 by FTIR analysis

The 0.006%, 0.012% and 0.023% of ethyl cellulose in the mix proportion were blended in a mixture of waste engine-vegetable oil at 300rpm at 100°C for 30 minutes and allowed to cool down. Fourier transforms infrared spectroscopy (FTIR) of model spectrum (Shamadzu, 73093) was used to study the properties of the oil. A drop of the oil was placed on the top of the soaking plate, covered and placed in the holder and analysis performed. FTIR analysis determines the availability of the base oil and other components by spectra. Therefore, the additives, oxidation products, contaminants and breakdown products can be known [9]. The FTIR spectroscopy is important in identifying the molecular structures and the possibility to assign certain absorption bands related to the functional group of the compounds [10].

2.3 Tile production

Different amount of ethyl cellulose had been mixed with 4% of waste engine-vegetable oil in an orbital shaker at 300 rpm at 100^0C for 30 minutes to dissolve it. The amounts of ethyl cellulose used were 0.006%, 0.012% and 0.023% in the compositions. The waste engine-vegetable oil which composed of ethyl cellulose was blended with 81% of river sand and 14.5% of fly ash and with the addition of 0.17% of sulfuric acid. After mixing of the materials, the weight of the materials was measured 400g before compaction by Marshall Compacter to remove the spaces and air in the mixture. The cylindrical shape tile with 100 mm diameter and 20 mm thickness was transferred to the oven for heat curing process at 190 ± 2^0C for 24 h. The composition of different amount of ethyl cellulose in the mixture is shown in Table 1.Triplicate tile specimens were prepared with varying content of ethyl cellulose in the mix proportion and were tested for transverse strength, water absorption and water permeability [11]. A test of shrinkage was carried out according to the CNS 3299 R3071 (Methods of Test for Ceramic Tiles).

Table 1. Compositions of the materials before heat curing (% weight).

Composition	5mm-10mm aggregates	Below 5mm aggregates	Fly ash	Wasteoil	Sulphuricacid	Ethyl cellulose powder	Total
1	5.80	75.46	14.51	4.054	0.17	0.006	100
2	5.80	75.46	14.51	4.049	0.17	0.012	100
3	5.80	75.46	14.51	4.037	0.17	0.023	100

3. Results and discussion

3.1 Water adsorption

The cold water adsorption test was performed according to ASTM C 1167 standard specification for clay roof tiles. The dry tile specimens were weighted and submerged in water for 24 hours and weighted again to calculate the cold water adsorption percentage.

The water absorption was calculated by using the equation:

$$\text{Absorption, } \% = (W_w - W_d) / W_d \times 100\%$$

where, W_w = wet weight (g); W_d = dry weight(g)

Results showed that the tiles had water absorption below than 5%. The water adsorption percentage of 0.006%, 0.012% and 0.023% of ethyl cellulose in the mix proportion were 4.2, 4.3 and 4.3 respectively as shown in Figure 2. The water adsorption values increased as the ethyl cellulose usage increased.The waste oils which functions as a binder prevent the entry of water molecules in the tile specimen and thus resulted in the low water adsorption percentage.

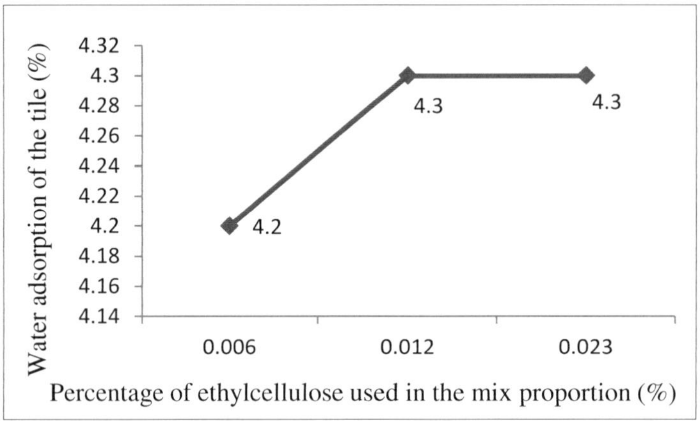

Figure 2. Cold water adsorption of the tile.

3.2 Transverse breaking strength

Transverse breaking strength test was conducted according to ASTM C 1167 standard specification for clay roof tiles. Transverse breaking strength is stated in Newton. Dry transverse breaking strength was performed by exposing the tile to heat at about 110^{0}C for 24 hours before being tested for its strength using materials testing machine (Instron4469). In contrast, wet transverse breaking strength was carried out by immersing the tile intothe water at room temperature for 24 hours before being tested for its strength by using materials testing machine (Instron4469). The effect of ultraviolet on the tiles was carried out by placing the tiles inside the ultraviolet chamber for 30 days. The tiles were sprayed with water five times a week to imitate the real environment of open air. After 30 days, the tiles were tested for their transverse breaking strength.

Results revealed that 0.023% of ethyl cellulose in the mix proportion showed highest dry transverse breaking strength and wet transverse breaking strength of 2510.7 N and 2234.5 N respectively as shown in Figure 3 and Figure 4. The tile specimens composed of 0.023% of ethyl cellulose in the mix proportion showed thehighest transverse breaking strength of 3162 N and 2247 N for with and without the exposure of ultraviolet light respectively as shown in Figure 5. The free radicals created by the ultraviolet radiation react with the oxygen present to form hydro-peroxides that can result in polymer chain breakage [12]. From the FTIR analysis, the carbonyl that found in the ethyl cellulose that mixed with waste engine-vegetable oil in the ratio 1:3 enhance the strong binding with other materials and therefore resulted in the increase in transverse breaking strength of the tile specimen.

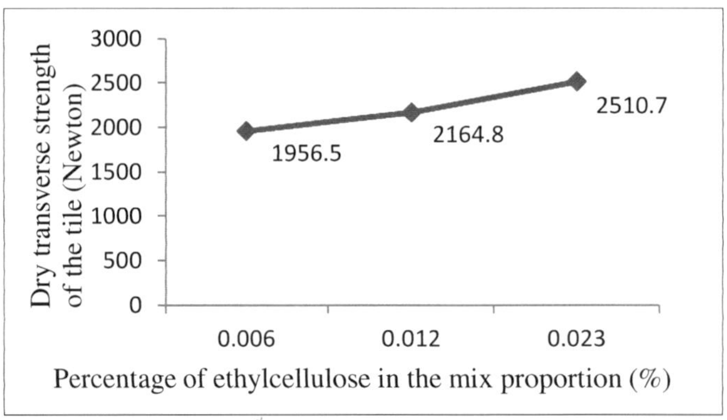

Figure 3. Dry transverse breaking strength of the tile.

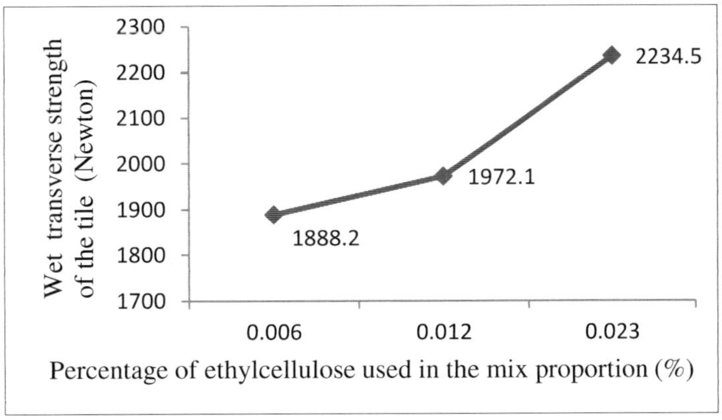

Figure 4. Wet transverse breaking strength of the tile.

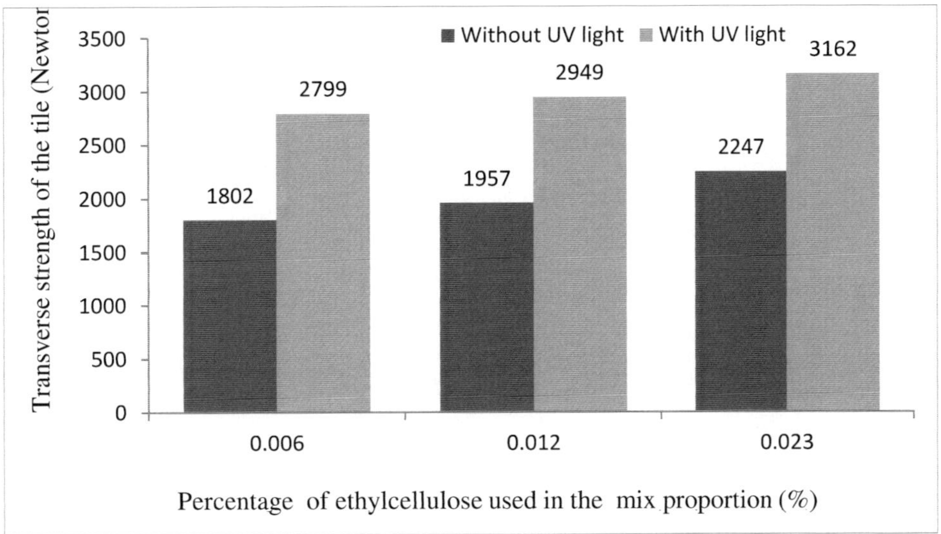

Figure 5. Transverse breaking strength of the tile (with or without ultraviolet light exposure).

3.3 Water Permeability

The water permeability test was carried out according to ASTM C 1167 Standard Specification for Clay Roof Tiles. The tile specimen was placed at the open bottom of a container filled with water at the height of 5 cm. The permeability showed the effectiveness of the concrete to prevent the movement of water into the cement [13]. Results showed that the tile specimens were impermeable to water due to the tile specimens that eliminate water from entering the tile.

3.4 Test of Shrinkage

The production of the tiles composed of ethyl cellulose and other materials requires low temperature of 190^{0}C. However, the shrinkage test of the tiles was measured at temperatures at 1050^{0}C, 1100^{0}C, 1150^{0}C and 1200^{0}C according to CNS Regulation 3299. The shrinkage test usually was performed on ceramic tiles because ceramic tiles require high temperature between 1000^{0}C and 1200^{0}C during production [14]. The ceramic tiles could not produceat low temperature about 200^{0}C. The percentages of firing shrinkage in length of the tiles were between 30% and 40% for the tiles composed of ethyl cellulose and other materials.

3.5 Fourier Transform Infrared Spectroscopy

FTIR analysis showed that the oil mixture added ethyl cellulose has the relatively high binding effect due to the presence of strong carbonyl group especially after being heat cured at the temperature of 190^{0}Cfor 24 hours.The peaks were noticed around 1746 cm^{-1} and 1747 cm^{-1} due to the presence of carbonylssuch as ester, ketones and carboxylic acids as shown in Figure 6. The oxidation of oil occurred with the presence of free radicals during initiation and propagation steps in a chemical reaction [15].

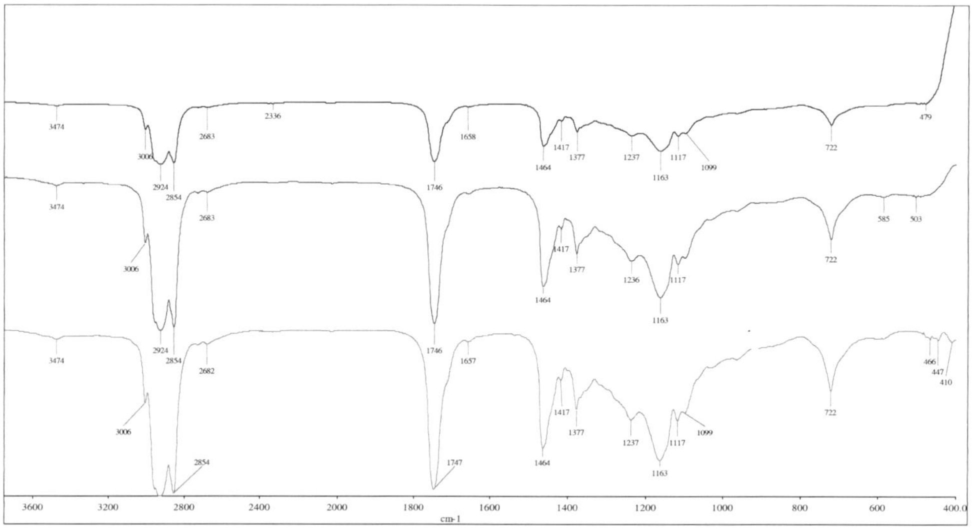

Figure 6. FTIR spectra of 0.023%, 0.012% and 0.006% concentration of ethyl cellulose in ascending order.

4. Conclusion

The tile specimens composed of 0.023% of ethyl cellulose in the mix proportion hadthe highest dry transverse breaking strength and wet transverse breaking strength of 2510.7 N and 2234.5 N respectively. The tile specimens had water adsorptioncapacityless than 5% andalso impermeable to water. The results showed that the ethyl cellulose composed roofing tile specimens passed all the tests as per ASTM requirements and also passed the ultraviolet accelerated test. The ethyl cellulose roofing tile can be commercialized as an environmental friendly roofing tile with proper coating of polymer that has resistance to high temperature.

5. Acknowledgement

The authors would like to extend our gratitude to Ministry of Education for the FRGS fund with project No. FRGS/1/2015/TK06/UTAR/02/1 and UniversitiTunku Abdul Rahman for UTAR Research fund project (IPSR/RMC/UTARRF/2015-C1/N02).

6. References

[1] Dow Cellulosics (2005). ETHOCEL™.Ethylcellulose Polymers Technical Handbook. In T.D.C. Company (Ed.).*Dow Cellulosics*. pp 4-23.

[2] Brady, G. Clauser, H and Vaccari, J. (1997). *MaterialsHandbook*. 1st ed. New York: McGraw-Hill, pp 588-594.

[3] Melzer, E., Daniels, R. and Kreuter. J. (2003).Ethylcellulose : a new type of emulsion stabilizer. *European Journal of Pharmaceutics and Biopharmaceutics,***56**, pp 23–27.

[4] Voncina, B. et al. (2009). Encapsulation of Rosemary Oil in Ethylcellulose Microcapsules. *Textile and Polymer Journal* ,**1**(1), pp 13-19.

[5] Gravelle, A.J., Barbut, S. and Marangoni, A.G. (2012). Ethylcelluloseoleogels: Manufacturing considerations and effects of oil oxidation. FRIN, *Food Research International,***48**(2), pp 578- 583.

[6] Quinchia, L.A. et al. (2012). Low-temperature flow behaviour of vegetable oil-based lubricants.*Industrial Crops & Products,***37**(1), pp 383-388.

[7] Hielscher, (2015). *Ultrasonic modification of starch granules slurries.* Germany: UltrasoundTechnology.

[8] Lingling, X. et al.(2005). Study on fired bricks with replacing clay by fly ash in high volume ratio. *Construction and Building Materials* , 19, pp. 243–247.

[9] Van de Voort, F.R., Pinchuk, D andPinchuk, J. (2001). Quantitative condition monitoring-analyticalwave of the future? In Proceedings of the Society of Tribologistsand LubricationEngineers Condition Monitoring Conference.

[10] Ullah, Z., Bustam, M. A. and Man, Z. (2014). Characterization of waste palm cooking oil for biodiesel production.*InternationalJournal of Chemical Engineering and Applications,* **5**(2), pp 134-137.

[11] ASTM. (2003). Standard Specification for Clay Roof Tiles, *ASTM C 1167-03*. West Conshohocken, PA.

[12] Fan Lee, C. (2015). Thermal breakage of a semi flexible polymer: breakage profile and rate. *Journal of Physics: Condensed Matter,***27** (27), 275101.

[13] Khandaker and Anwar Hossain. (2004). Properties of volcanic pumice based cement and lightweight concrete. *Cement and Concrete Research*, **34**, pp 283-291.

[14] Deng Fong Lin, Huan-Lin Luo, Lisa Y. Chen and Shu-Wen Zhang.(2016). Improvements to ManufacturedTilesusingSewageSludgeAshBased onKilnTemperature and Nano- SiO_2 Additive. *Journal of Residuals Science &Technology,* **13**, 51.

[15] Alistair, G. (1996). *InfraredAnalysis as a Tool for AssessingDegradation in UsedEngine Lubricants*. WearcheckAfrica: WZA-002.

IOP Conf. Series: Earth and Environmental Science **67** (2017) 012006 doi:10.1088/1755-1315/67/1/012006

Critical factors of implementing Industrialised Building System in Sarawak: A research on SMEs

N A Hadi[1], W M N W Muhamad[1] and M K F Othman[1]

[1]Faculty of Architecture, Planning and Surveying, Universiti Teknologi MARA Cawangan Sarawak, Sarawak, Malaysia

Abstract. Industrialised Building System (IBS) have been adopted in Malaysia for over five decades and is expected to grow consistent with Malaysia's economic development. In promoting the adoption of IBS in construction projects, the government has taken several steps such as levy exemption for contractors and developers. Despite its numerous advantages and efforts pulled by the government, the implementation and adoption of IBS are still below the expected figure. Many researchers investigated readiness, setbacks and issues related to the implementation of IBS in Malaysia's construction projects. However, most of the research mainly for projects located in urban areas of West Malaysia. Therefore, this paper aims to close the gap on factors affecting the implementation of IBS for SMEs in Sarawak, where the level urbanisation is low. A semi-structured interview was conducted with 10 SME's contractors to get better insight view. The study found that logistics and infrastructure problems, a limited number of the manufacturer, lack of incentive, conventional payment methods, lack of financial supports are among critical factors affecting the implementation of IBS. Therefore, it was concluded that government plays major role in providing supports, incentives and facilitating the improvement of infrastructure to successfully implementing IBS in Sarawak.

1. Introduction

The terms of Industrialised Building System (IBS) may be limited to Malaysia while in other parts of the world may differently term it as Modern Method of Construction (MMC), Off – site Construction, Modular Construction, Prefabricated Construction, etc. IBS is defined as a construction technique in which components are manufactured in a controlled environment (on or off site), transported, positioned and assembled into a structure with minimal additional site works In Malaysia IBS is divided into five categories which are precast component systems, fabricated steel structures, innovative mould systems, modular block systems and prefabricated timber structures as construction components. [1]

In Malaysia, Industrialised Building System (IBS) was first introduced almost 50 years ago with the completion of the Tunku Abdul Rahman Public Housing Estate; or commonly known as the Pekeliling Flats at Jalan Pekeliling, Kuala Lumpur. Despite many years have passed, the adoption of IBS in the construction sector is still low and represents only 15% of construction projects used IBS in Malaysia [2]. IBS Mid - Term Review in 2007 indicated that approximately only 10% of the complete projects used IBS in the year 2006 as compared to forecasting IBS usage of 50 % in 2006 and 70% in the year 2008 as projected in the roadmap [3].

In contrast, other developed countries such as Finland, Sweden, Japan, Germany, and even Singapore have been successfully implemented IBS and it is very much preferred construction method due to its numerous benefits [4]. The benefits of IBS have been well-documented in the literature. Benefits such as cleaner and safer construction sites [5]; faster construction [5], [6], [7], [8]; high quality and aesthetical value of products [5], [8]; and more organised site management [5]. Due to its many benefits, the government of Malaysia IBS has made a lot of efforts to enhance the current conventional - dirty, dangerous and difficult image of construction industry to a technologically advanced method of construction such as by developing the Industrialised Building Systems (IBS) through the Construction Industry Development Board (CIDB).

CIDB had implemented the IBS Score in construction industry to measure the level of IBS usage in building and become part of the main monitor system of the IBS construction industry. The scoring system is introduced as a structured assessment system that can be used to measure the usage of IBS in a consistent way. In line with IBS Roadmap 2003-2010, a number of strategies and aggressive steps are formulated to promote the adoption of IBS in Malaysia. First, it was announced in Budget 2005 that all new government building projects are required to have at least 50% IBS Score. It was also announced that levy exemption for housing projects that have minimum IBS Score of 50% to attract private developers. The new circular of Surat Pekeliling Perbendaharaan Bil. 7 Tahun 2008 had emphasised on the full utilisation of IBS in government's projects where the use of IBS component in government projects must not be less than 70% IBS Score. However, according to the latest data from the Implementation Coordination Unit (ICU) of the Prime Minister's Office, only 24% of public projects valued at RM 10 million and above have achieved an IBS score of 70 while CIDB reports that only 14% of private projects have achieved an IBS score of 50 [9].

Due to low implementation of IBS in construction projects in Malaysia, many research has been conducted to investigate the issues of IBS. The most significant challenges to the adoption of IBS are higher capital cost [8], [10], [11]; lack of incentive [8], [12], [13]; low number of supplier [14]; high shipping cost [15]; and lack of knowledge and awareness [16].

2. SMEs in IBS projects in Sarawak

Although many previous studies were done in Malaysia related to IBS, only two are found to be conducted in Sarawak. One of the studies conducted a quantitative study where the focus group is large size company. The researcher found that IBS was still considered new to most of the contractors in Sarawak and thus, further indicated that they still lacked exposure [5]. Meanwhile, another researcher revealed that there is no significant difference between setbacks of IBS in rural areas of Sarawak, East Malaysia and urban areas of West Malaysia. These setbacks are resistance to change, negative perceptions of the stakeholders towards IBS designs, expensive overall construction costs, lack of knowledge and exposure to IBS technology and lack of IBS manufacturers in the industry. From the case study done, the researcher found that the lack of local IBS manufacturers and expensive overall construction costs are two significant setbacks [17].

Examining both research done in Sarawak [5], [17], none of them focusing on SMEs. Small - and medium - sized enterprises (SMEs) are at the core of the Malaysian construction industry. Because the SMEs are the majority, their characteristics significantly affect the current practices in the industry. Nonetheless, there is no standard definition of SME in Malaysia. Based on the CIDB's definition of construction SMEs, G1 to G3 contractors fall under the small-size category. G4 and G5 contractors are categorised as medium-sized contractors, and G6 and G7 contractors are categorised as large [18].

In Sarawak, a majority of IBS contractors fall under SMEs. Referring to Table 1, a total of 139 contractors grade G1 to G5 compared to 52 contractors grade G6 to G7 registered under CIDB for various IBS system in Sarawak. Due to majority of SMEs registered in CIDB and lack of research of IBS involving SMEs, this study is intended to close the gap between parties involved in IBS projects.

IOP Conf. Series: Earth and Environmental Science **67** (2017) 012006 doi:10.1088/1755-1315/67/1/012006

Table 1. Number of contractors involved in IBS according to contractor grade of registration in Sarawak.

Number of contractors according to Contractor grade of registration	G1 to G5	G6 to G7
Precast component systems	41	15
Fabricated steel structures	65	30
Innovative mould system	4	1
Modular block systems	25	1
Prefabricated timber structures	4	5
	139	52

Source: Construction Industry Development Board (CIDB)

3. Methodology

The focus of this paper is on getting the insight towards IBS implementation for SMEs. This paper examines critical factors in implementing IBS in Sarawak qualitatively. In order to acquire potential interviewees, a pilot questionnaire survey was conducted in the earlier stage of the study to 250 contractors in Malaysia. From 250, 73 replied to the questionnaire survey. In order to gather more information to increase understanding and expanding knowledge on the issue, semi – structured interview was conducted. From 73 replies, 15 of the contractors were selected to be interviewed. The potential interviewees were selected based on:

1. More than five years experience in managing IBS projects.
2. G1 to G5 contractor
3. Experienced managing projects in Sarawak

From 15 selected respondents, 10 contractors agreed to be interviewed. Data gathered from the semi – structured interviews were analysed qualitatively as the information was in the form of opinions, comments and statements.

4. Findings and Discussions

4.1. Lack of incentive

According to the interview done with contractors, four of the respondents agree on the lack of incentive implementing IBS for small contractor. The incentive has no direct effect meaning to other than the main contractor and developer due to the fact levy exemption only affecting the main party of the IBS project. For small contractor who mostly undertakes installation of IBS components, the respondents further emphasised that the contractor did not receive the said levy exemption due to the amount of contract below RM 500,000. Two of the respondents received the levy exemption but further argued that the value of exemption is too little (based on the 0.125% rates of levy to be paid).
In Budget 2006 announcement, tax incentive was offered through Accelerated Capital Allowance (ACA). IBS manufacturers would be given ACA for expenses incurred in the purchase of moulds used for production of precast concrete components [2]. The tax benefits are expected to motivate industry players to invest more in IBS. However, the respondents argued that they did not aware of the announcement and they further added only one party is benefitting from it, i.e. precast concrete manufacturer. From the result of the interview session, incentives given by the government is unfair in the context to facilitate the stakeholders. The current policy is developed in favour of larger stakeholders i.e. manufacturer and developer who are viewed to be the main driving force of IBS implementation. Hence, the policy maker has to impartially formulate a new strategy which will

benefit each stakeholder in IBS implementation. The respondents further suggested the government should impose a reduction of company's tax for a company that involved in IBS project where all parties will the received the same benefits.

4.2. Limited number of manufacturer of prefabricated elements
Based on the result from interview session, six of the respondents suggested that limited supplier of IBS components in Sarawak, among others affecting the construction cost of IBS. The lack of IBS manufacturer caused no competition on the price, resulting in higher price of IBS components. This is supported by the database developed by CIDB, the number of IBS manufacturers in Malaysia is 169 in 2013 where almost 90% located in West Malaysia [19]. However, there is no current list of manufacturers available. The respondents further added that most of the factory located far from the site. Thus, it results to ineffective distance to a site, causing more on transportation cost.
Three of the respondents suggested database of IBS manufacturer should be compiled and made online for easy reference of industry players. Information such as the location of the IBS factory is important to consider initially before requesting further information i.e. price quotation. Three of the respondents stated that they are interested in investing in IBS factory. However, the chances of securing a continuous and sufficient demand for the product worries the contractor especially to reach breakeven point of the investment.

4.3. Logistics problems
According to the interview done with contractors, five of the respondents stated that logistics problems, especially in the rural area of Sarawak, hindered the usage of IBS. The respondents further added that logistics and shipping costs would result in expensive overall construction costs. Three of the respondents who involved in precast concrete components stated that the components are big, bulky, heavy, and require the use of expensive cranes for hoisting. In the situation where the site was situated in a rural area and can only be accessed through a river, the additional cost incurred to transport the components.
In addition, one of the respondents argued the infrastructure of Sarawak is still 30 years behind the other states. Infrastructure such as road outside the cities which is long, narrow and unpaved needs to be upgraded and not suitable to move large components of IBS and heavy machineries. Inaccessible interior over remote areas is inaccessible with ground transportation and dependent on river access where only shallow-draft vessels and small boats like sampans can manoeuvre. Shallow river of Sarawak depended heavily on tidal movement and the silting of the river. This situation will give impacts on the construction schedule and the construction cost.
All respondents further suggested that the government should consider providing additional incentives for parties in IBS projects located in the rural area. Improvement of the infrastructure i.e. roads will also significantly contribute to the growth of IBS in Sarawak.

4.4. Conventional payment method
According to the interview done with contractors, three of the respondents emphasised that conventional payment method where progress payment is made when materials arrived at the site is not suitable for IBS. In the event where the offsite casting is employed, contractors will not be getting any payments from the employer until IBS components delivered to the site, though payments have to be made to the manufacturer once the components are cast. This issue will cause financial difficulties and complicate the cash flow of the contractors.
The respondents further suggested that a new set of standard payment where off-site casting is employed should be developed and promoted to ensure all parties are made aware of it.

5. Conclusions and recommendations
This paper manages to draw significant factors affecting the implementation of IBS in Sarawak as the barriers and issues of the implementation differs from West Malaysia. Prior to IBS implementation, factors such as site location, location of the manufacturing factory, infrastructure conditions and services available should be assessed beforehand. Sarawak is less developed state compared to other

states in Malaysia and most of the rural areas are underdeveloped. In a rural area, the facilities and infrastructure i.e. road network and river network need to be improved to accommodate the transportation of IBS components and machinery.
Incentive programme developed by the government through CIDB plays important to enhance the adoption of IBS if it is implemented properly. The current incentives programme falls short as it is found to be inadequate and impartial towards all IBS parties. This paper proposes that the incentives should be provided to all IBS parties, without many conditions that have to be met.
Although this paper managed to provide qualitative study, further research is recommended to be conducted in terms of increasing number of respondents. It is also recommended spatial study for future research to be done to locate IBS factory in Sarawak. It will help in manufacturing new location and knowing the flow of transporting IBS component from IBS manufacturer to the site construction.

6. References

[1] CIDB 2003 *Industrialised building systems (IBS) - roadmap 2003-2010*
[2] Shaari S N 2003 Survey on the Usage of Industrialised Building Systems (IBS) in Malaysian Construction Industry *CIDB Malaysia Publ.* 39
[3] Hamid Z A, Kamar K A M, Zain M Z M, Ghani M K and Rahim A H A 2008 Industrialized Building System (IBS) in Malaysia: the current state and R&D initiatives *Malaysia Constr. Res. J.* **2** 1–13
[4] CREAM 2011 Experienced and lesson learnt on Industrialised Building System (IBS) construction in Malaysia *IBS Roundtable Workshop* ed Construction Industry Development Board (CIDB) (Kuala Lumpur)
[5] Bohari A A M, Mahat N and Kipli K 2012 Industrialised Building System (IBS) in Sarawak construction industry *ICIMTR 2012 - 2012 Int. Conf. Innov. Manag. Technol. Res.* 433–7
[6] Idrus A, Hui N and Foo K 2008 Perception of Industrialized Building System (IBS) Within the Malaysian Market *International Conference on Construction and Building Technology (ICCBT)* pp 75–92
[7] Pan W, Gibb A G F and Dainty A R J 2007 Perspectives of UK housebuilders on the use of offsite modern methods of construction *Constr. Manag. Econ.* **25** 183–94
[8] Anuar K, Kamar M, Hamid Z A, Zura M, Zain M, Hazim A, Rahim A, Ghani M K, Nor M, Azman A, Majid T A, Sanusi M, Ahamad S and Azman M N 2012 Drivers and Barriers of Industrialised Building System (IBS) Roadmaps in Malaysia *Malaysian Constr. Res. J.* **9** 1985–3807
[9] MIDA MIDA targets 100 additiona IBS Manufacturing Companies by 2020 *Malaysian Invest. Dev. Auth.*
[10] Abd Rahman A B and Omar W 2006 Issues and challenges in the implementation of industrialised building systems in Malaysia *Proc. 6th Asia-Pacific Struct. Eng. Constr. Conf.* C-45-C-53
[11] Kamar K A M, Azman M N A and Nawi M N M 2014 IBS survey 2010: Drivers, barriers and critical success factors in adopting industrialised building system (IBS) construction by G7 contractors in Malaysia *J. Eng. Sci. Technol.* **9** 490–501
[12] Idrus A, Hui N and Foo K 2008 Perception of Industrialized Building System (IBS) Within the Malaysian Market *International Conference on Construction and Building Technology (ICCBT)* pp 75–92
[13] Lou E C W 2012 Industrialized Building Systems : Strategic Outlook for Manufactured Construction in Malaysia 69–74
[14] Nasrun M, Nawi M, Nor M, Azman A, Baluch N, Anuar K, Kamar M and Hamid A 2015 Study on the Use of Industrialised Building System in Malaysia Private Construction Projects *ARPN J. Eng. Appl. Sci.* **10** 7368–74
[15] Nawi M N M, Elias E M, Hamid M S A and Yusoff M N 2005 A study of IBS formwork usage in the malaysian construction industry *National Seminar on Engineering Support Course* (University Malaysia Perlis)
[16] Baharuddin M N, Bahardin N F, Zaidi M A, Lokman I and Nawi M N M 2016 An exploratory

review on critical factors of IBS formwork implementation for Malaysian construction stakeholders *AIP Conf. Proc.* **1761**

[17] Wong S S and Lau L K 2015 Advantages and Setbacks of Industrialized Building System (IBS) Implementation : A Case Study in Sarawak *Int. J. Sustain. Constr. Eng. Technol.* **Vol 6, No** 52–61

[18] Kamal E M and Flanagan R 2014 Key Characteristics of Rural Construction SMEs *J. Constr. Dev. Ctries.* **19** 1–13

[19] CIDB 2013 IBS Manufacturers/Distributors/Suppliers and on Site Manufacturers Directory Unpublished Directory

7. Acknowledgments

Authors would like to express gratitude for the financial support from Kementerian Pendidikan Malaysia (KPM) under purview of Research Acculturation Grant Scheme (RAGS) 600-RMI/RAGS 5/3 (237/2014) and Research Management Institute (RMI) Universiti Teknologi Mara

GeoEye Image Fusion Vegetation Information Extraction Based on Blue Noise Measurement Texture

Luo Qiu [1,*], Qiming Xiong [2], Yao Liu [3] and Gui Zhang [2,*]

[1] School of Geosciences and Info-Physics, Central South University, China
[2] Central South University of Forestry and Technology, China
[3] Computational Geosciences Research Center, Central South University, China
Corresponding authors' e-mail: 34489492@qq.com (LQ); 595772457@qq.com (GZ)

Abstract. Vegetation high-resolution image segmentation is an important kind of target, while how to fuse image by using spatial and spectral information of GeoEye image, and make effective determination of the scale of vegetation texture segmentation is an important topic. This paper presents an algorithm of GeoEye image fusion vegetation information extraction based on blue noise measurement texture, which the vegetation information spectral response is calculated based on fast Fourier transform, in order to obtain the texture gray scale distribution of vegetation information, thus realizing accurate extraction of vegetation information. The simulation result got by the algorithm shows the deviation between NDVI index and original value index is below 0.05, which is obviously better than those obtained with the two comparison fusion algorithms of PCA transform (smaller than 0.4) and Brovey (smaller than 0.25), and provides the best parameter combination form [120,0.34,0.48]. Hence, it is an effective algorithm for high accuracy GeoEye image vegetation information acquisition.

1. Introduction

Vegetation is a key factor in biogeochemical cycle related to human survival, and is treated as a significant characteristic variable in ecohydrological research [1] – [3]. There are two methods available for vegetation information survey, i.e. conventional site survey and remote sensing survey based on remote sensing images and technologies [4] – [7]. Conventional site survey allows accurate and comprehensive learning of vegetation information in the region. However, such method requires a lot of manpower, material resources and financial resources, thus it is rarely used now. The remote sensing survey can realize vegetation information extraction in a simple manner, while the extraction accuracy and spatial resolution of this method are sometimes relatively low and fail to meet the accuracy required for research. Along with the development of remote sensing technology, remote sensing is developed toward the trend of high spatial resolution, high time resolution, high spectral resolution, multiple platforms, multiple angles and multiple sensors, thus laying a good foundation for accurate vegetation information extraction [8] – [9]. Multispectral remote sensing images (e.g. WorldView, GeoEye) can realize meter resolution, and panchromatic band can realize decimeter resolution. Therefore, the hardware resources are sufficient to meet the demand of accurate vegetation information extraction. How to realize accurate vegetation information extraction with the software becomes a hot and important point of research at present [10] – [11].

The object oriented information extraction method has already become the development direction of high resolution remote sensing image information analysis [12]. There are two critical issues existing in the process of information extraction [13] – [15]: (1) in selection of image segmentation algorithm. The shape factor, scale factor, compactness and other indexes will influence feature segmentation of

IOP Conf. Series: Earth and Environmental Science **67** (2017) 012007 doi:10.1088/1755-1315/67/1/012007

the polygon, and even the accuracy of vegetation information extraction; (2) establishment of classification rules and feature extraction. For the issues above, the statistics, frequency domain and texture structure analysis method is the main research mode at present, of which, accurate extraction of vegetation texture features is a critical step. Indexes of texture scale, shape, gray scale distribution and others are crucial for image segmentation.

Many scholars of combined texture feature extraction research have raised different algorithms, for example, Mashimbye et al [16] put forward the extraction model of high spectrum remote sensing data soil information; Bie et al [17] put forward the object and image fusion oriented remote sensing image vegetation information feature extraction technology; Cui et al [18] put forward the spectral feature and object oriented vegetation information feature extraction technology; and Li et al [19] put forward the remote sensing image edge mathematical morphology detection method. Although certain effects have been obtained with these algorithms, they have a common problem of low vegetation texture feature extraction accuracy. Hence, in this study, according to the specific scale signal frequency feature presented by different vegetation information, an image fusion vegetation information extraction method based on blue noise theory measurement texture is put forward on the basis of GeoEye data, for realizing fast and accurate extraction of vegetation feature information for the detected region [20] – [22].

2. Site description

GeoEye is a well-known spatial geographic information supplier at present, which can provide better regional mapping and monitoring services for the users [23] – [24]. The image data selected in this article comes from GeoEye-1 satellite of the USA, which is equipped with a CCD high resolution camera, and allows resolution above 0.41m for the panchromatic band and 1.65m for multispectral band. These images were taken at 10:30 am on August 5th, 2013. The region under research is Tianlaochi Catchment in the upstream of the Heihe River. It is located in the middle section of Qilian Mountain, with geographic coordinates of 38°23'56"~38°26'47", 99°53'57"~99°57'10"E surveyed. The total area of this region is about 12.8 Km^2.

Tianlaochi Catchment is at the elevation of 2,700~4,440m, with the landform rising from the west to the east. In this region, the annual mean temperature is 5℃, the peak monthly mean temperature 20℃, the minimum monthly mean temperature -18℃, the annual mean rainfall 433mm and the annual mean water evaporation 1,080mm.The main terrestrial objects in this region include: sabina przewalskii, picea crassifolia, bushwoods, steppes, alpine meadow, bare rocks, rivers and so on. Sabina przewalskii and Picea crassifolia cover the largest area in this region. But they are distributed at obviously different locations, i.e. Sabina przewalskii mainly on the sunny slope of Qilian Mountain, and Picea crassifolia mainly on the shady slope of Qilian Mountain. GeoEye image of this region is as shown in Figure 1.

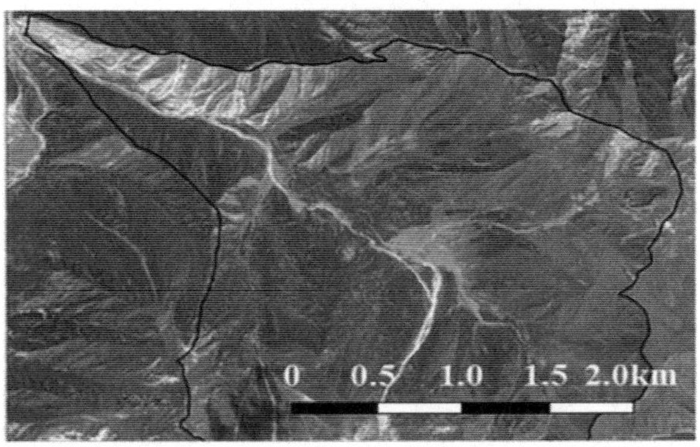

Figure 1. GeoEye image of study area.

3. Blue noise characteristics of GeoEye image vegetation texture

3.1. Vegetation texture analysis

Texture analysis can be divided into microcosmic and macroscopical aspects generally, of which, the former one is mainly related to texture comparison and set shape, and the later one mainly involves texture period, spatial scale, direction and so on. In the process of high resolution GeoEye image vegetation information extraction, the spatial scale of vegetation shall be distinguished macroscopically, so as to improve the recognition efficiency. There are sabina przewalskii, picea crassifolia, grasslands, meadow and others distributed in the research region. They differ from each other in color tones, crown characteristics, plant locations, contrast and so on. In the microcosmic aspect, the vegetation texture presents obvious cluster shape structure, with the contrast at the center obviously different from that at the surrounding part. Additionally, the different contrast shapes presented are related to the plant types.

When GeoEye image is shrunk to a certain scale, the crown shadows of tall trees and highlight regions will be shrunk into single pixel points and present high frequency characteristics. These high frequency, aperiodicity and random features of vegetation texture coincide with the features of blue noise. The blue noise is a type of aperiodic, low frequency random signal without obvious peak in the frequency spectrum. With a limited bandwidth, the power spectrum density is in proportion to the frequency index. In computer graphics, the blue noise is usually used for composition of halftone dithering images, which agree with the visual features of human beings. Therefore, GeoEye image processing mode based on blue noise theory is a new method for vegetation information extraction.

3.2. Vegetation texture region selection of GeoEye images

In order to further increase the detection efficiency, it is necessary to select the blue noise feature regions in a reasonable manner. GeoEye image is equally divided into *N* regions firstly, and each region is scanned for fast selection of the suitable regions. The gray scale and color tone distribution features of the region are utilized for realization of forest region judgment in this research.

In the aspect of color tone distribution features, the vegetation is in green color generally. The deciduous vegetation shows other colors in winter, with the crown shapes varying as well. The algorithm shall be determined based on different conditions. The brightness interference can be eliminated effectively since HSV tonal space feature is unrelated to the brightness. The texture region of GeoEye image is mapped from RGB space to HSV tonal space, and H value in the region is counted and averaged. For HSV tone model, the optimal green color H value approaches 120.

As for the gray scale distribution, the typical vegetation in GeoEye image usually presents regional distribution. The crowns of sabina przewalskii, picea crassifolia and other tall plants present areatus distribution microcosmicly with obvious contrast. Such gray scale distribution is repeated and shows low gray scale feature, and the gray scale probability density presents normal distribution feature. Assuming that the data vector in a region of GeoEye image is X, the data capacity is n, the third moment of X is S and the fourth moment is K, then $\mathrm{Jarque-Bera}$ the detection formula can be expressed as:

$$JB = \frac{n}{6}[s^2 + \frac{(K-3)^2}{4}] \tag{1}$$

Where, in case of normal distribution of gray scale, the third moment of vector X is 0 and the fourth moment is 3, $\mathrm{JB} = 0$. The results of detection formula above $\mathrm{Jarque-Bera}$ are normalized, and the region is distinguished using the average color tone value and normal distribution, in order to select the blue noise feature image region.

4. Image fusion algorithm based on feature classification

4.1. Algorithm description

Equally divide the GeoEye image into *N* ones, and screen the gray scale and color tone of each divided region, so as to obtain the reasonable region for enhanced contrast adaptation processing. At the same time, work out the blue noise feature of image using the iterative algorithm for the region obtained. The iterative algorithm process mainly includes the following steps: 1. geometric transformation for size reduction of the region, 2. fast Fourier transform (FFT) spectral response of the region, 3. Blue noise feature extraction. Perform vegetation information feature calculation for the feature region in GeoEye image, and work out the vegetation texture size using the blue noise feature region size. The algorithm process is as shown in Figure 2.

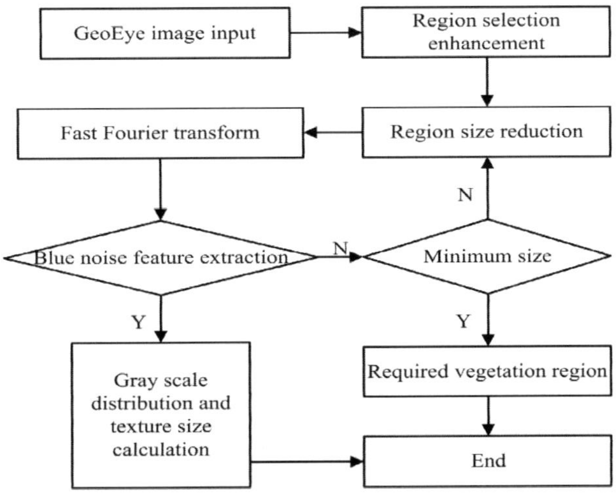

Figure 2. Texture feature extraction process.

When the region selected in GeoEye image is oversized, the vegetation texture of GeoEye image presents obvious microcosmic structure feature, and the vegetation texture feature reflected presents large texture size and smooth gray scale transition. On the aspect of frequency feature, it presents strong low frequency signal and pink noise or red noise feature. When the vegetation texture size of GeoEye image is reduced to the pixel level, the microcosmic low frequency signal will disappear and the high frequency random oscillation feature becomes more obvious. Similar to the blue noise feature, there is no low frequency signal, and the high frequency signal is in proportion to the frequency.

4.2. GeoEye blue noise feature extraction

During pre-processing, smoothing and frequency shifting is required for the region spectrum obtained via fast Fourier transform. For effective judgment of direction, it is necessary to obtain the energy spectrum of *N* directions at equal angle interval with the origin as the center. The calculation formula is $E_{i1}, E_{i2}, \cdots, E_{ik}$. Then, the total energy P of direction i can be worked out, i.e.:

$$P_i = \sum_{j=1}^{K'} E_{ij} \quad (2)$$

Where, K' is the total energy spectrum in a direction, and j is the temporary counting variable. The energy variance of different directions is expressed as:

$$\sigma = \sqrt{\frac{1}{N}\sum_{i=1}^{N}(p_i - p_\mu)^2} \quad (3)$$

Where, p_i is the random energy value in direction of i, and p_μ is the expected value of p_i. If $\sigma = 0$, it indicates that GeoEye blue noise feature has no directionality .

Assuming that the directionality of GeoEye blue noise feature is weak, the energy spectrum in N directions is combined to obtain the overall one-dimensional energy spectrum $E_{i1}, E_{i2}, \cdots, E_{ik}$, and p_r is the sum of energies above, the calculation formula of the third moment is:

$$S = 1/P_T \sum_{x=1}^{K'}(f_x - f_m)E_x \quad (4)$$

Where, f_x is the frequency, f_m is the center frequency, and E_x is the sum of frequency spectrum in *N* direction. Under ideal conditions, when $k(x) > 0$ in the function, the calculation formula of $k(x)$ is:

$$k(x) = E(x)/f(x) \quad (5)$$

Work out the variance of $k(x)$ defined in the formula above. The closer is the value to 0, the stronger is the feature of energy spectrum density proportion and frequency. In conclusion, the condition for judging the blue noise feature of GeoEye region is: $\sigma = 0$, $S > 0$, and the variance of $k(x)$ close to 0.

4.3. Calculation of texture size and gray scale distribution

With the blue noise feature acquired with the algorithm above, calculate the scale of vegetation crown texture based on the sizes of this region and the original region, in order to obtain the pixel quantity of vegetation crown. Since the shadow and highlight part of texture occupy a pixel point respectively, the texture scale calculation formula is as below:

$$wc = 2 \times \frac{C_{Original}}{C_{current}} \tag{6}$$

Where, wc is the texture size, $C_{Original}$ is the original region size and $C_{current}$ the current region size. Gray scale distribution calculation is the foundation of morphological segmentation design for GeoEye image. Repeated appearance of vegetation crown texture can form obvious normal distribution form. The gray scale distribution form of GeoEye image can be calculated based on the histogram distribution form of original gray scale, with the contrast, average value and peak value of gray scale involved. The calculation formula of gray scale contrast is:

$$DB = \frac{1}{M}\sum_{i,j-1}^{M} \delta(i,j)^2 P_\delta(i,j) \tag{7}$$

Where, $P_\delta(i,j)$ is the pixel distribution probability, $\delta(i,j) = |i-j|$ is the adjacent gray scale difference between pixels, and M is the total number of adjacent pixel relation.

5. Experimental parameter selection and comparison analysis

5.1. Parameter selection of fusion algorithm

Define the algorithm parameter index: extract the quality rate (QR), under segmentation rate (UR), over segmentation rate (OR) and distance (ED). The calculation formula is:

$$\begin{cases} QR = 1 - \frac{\sum|r_i \cap s_k|}{\sum|r_i \cup s_k|} \\ OR = \frac{\sum|s_k - r_i|}{\sum r_i} \\ UR = \frac{\sum|r_i - s_k|}{\sum r_i} \\ ED = \sqrt{(OR^2 + UR^2)/\,2} \end{cases} \tag{8}$$

Where, r_i is a polygon of blue noise region, s_k is the polygon of vegetation segmentation region, $|r_i \cap s_k|$, $|s_k - r_i|$ and $|r_i - s_k|$ represent the overlapped, under segmentation and over segmentation part respectively.

GeoEye remote sensing image 1 is taken as an example for explanation, with the reference polygon of sabina przewalskii, picea crassifolia, alpine meadow, steppes, bushwoods, rivers, bare rocks and other landforms formed. Six corresponding reference polygons are formed for each type of landform. With {78, 103, 122, 137, 158} as the segmentation scale, and compactness and shape factor of {0.11, 0.23, 0.34, 0.48, 0.52, 0.65, 0.78, 0.86, 0.97}, there are $5 \times 9 \times 9$ types of parameter combination. The parameter combination and evaluation index value are as shown in Table 1 and Table 2.

According to Table 1 and Table 2, among the segmentation parameter combinations of the vegetations, better effects can be obtained with large scale segmentation for area distribution feature of abina przewalskii, picea crassifolia, alpine meadow, bushwoods and steppes. At the same time, for various vegetation types, the optimal segmentation scale is 120, the optimal shape factor is 0.34, and the optimal compactness 0.48. Hence, the parameter combination adopted in this article is [120, 0.34, 0.48].

Table 1. QR sort parameter combination.

Terrestrial object type	QR sort	Segmentation scale	Shape factor	Compactness
Picea crassifolia	0.153	100	0.34	0.65
Abina przewalskii	0.132	102	0.23	0.52
Bushwoods	0.124	140	0.48	0.86
Alpine meadow	0.157	140	0.34	0.48
Steppes	0.146	120	0.34	0.34
Rivers	0.079	80	0.52	0.34

Table 2. ED sort parameter combination.

Terrestrial object type	ED sort	Segmentation scale	Shape factor	Compactness
Picea crassifolia	0.563	120	0.23	0.48
Abina przewalskii	0.479	140	0.23	0.65
Bushwoods	0.316	120	0.48	0.52
Alpine meadow	0.608	100	0.11	0.86
Steppes	0.491	120	0.34	0.48
Rivers	0.238	80	0.48	0.23

5.2. Comparative analysis of fusion result

The remote sensing fusion algorithm influences the fusion effect largely, and is liable to cause data distortion. The commonly used centralized image fusion algorithm is used herein for comparison experiment with GeoEye remote sensing image in Figure 1 and a standard universal remote sensing image. Experiment hardware configuration: i7-990X 3.73GHz, 4G ddr3-1600. The comparison algorithms are: main component analysis algorithm (PCA transform), Brovey fusion algorithm and FFBN algorithm discussed herein. The experiment comparison result is as shown in Figure 3. Fusion comparison of two GeoEye remote sensing images, i.e. RGB combined image and gray scale image, is provided. The comparison result shows that, the fusion effect of FFBN algorithm discussed herein is better than those of PCA transform and Brovey fusion algorithms visually.

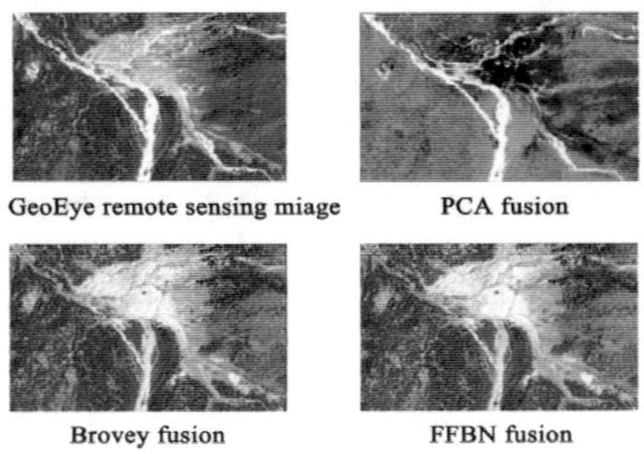

(a) GeoEye remote sensing image 1.

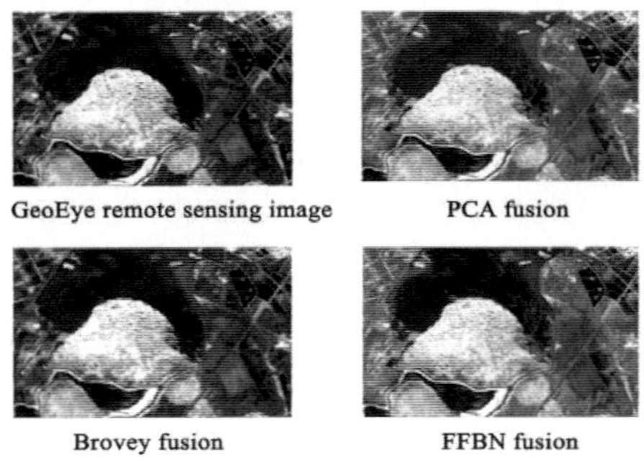

(b) GeoEye remote sensing image 2.

Figure 3. Comparison between two images of GeoEye remote sensing image.

The terrestrial object information can be reflected by defining the vegetation index normalization index after vegetation information feature fusion extraction. The calculation formula is:

$$\text{NDVI} = (\rho_{NIR} - \rho_R)/(\rho_{NIR} + \rho_R) \tag{9}$$

Where, ρ_{NIR} is infrared radiation parameter, and ρ_R is red light radiation.
Figure 4 shows comparison of NDVI indexes of many algorithms. According to NDVI index comparison in Figure 4, the difference between the NDVI indexes of FFBN algorithm and the original value index is smaller than 0.05, better than that obtained with PCA transform (smaller than 0.4) or Brovey (0.25) fusion algorithms, i.e. the NDVI index of FFBN algorithm is the closest to the original value.

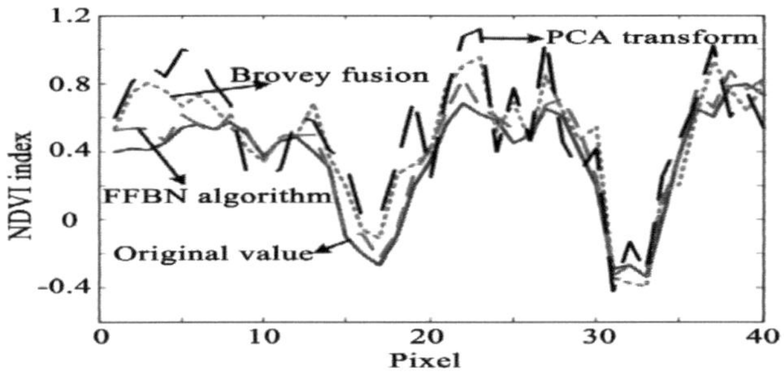

Figure 4. NDVI index comparison.

6. Conclusion

What makes high-resolution remote sensing image different from general natural color images is that the former contains large amount of information, complex relationships and multiple features. Therefore, it is difficult for high-resolution image segmentation to achieve the desired effect by directly using the methods of gray scale segmentation or color image segmentation. The current common technologies of image segmentation are mostly based on texture analysis, and the texture features is the vital step. The consistency of color and gray scale is so poor when extracting the remote sensing image vegetation information, while the texture information is very stable. A new texture describing method based on blue noise measurement texture is discussed herein, which is used to measure texture dimension and gray scale of typical regional vegetation in remote sensing image, thus realized extraction of high resolution GeoEye remote sensing image vegetation information. Effectiveness of the algorithm put forward is verified via experiment and comparison analysis, and experiments for value determination of relevant parameters for the algorithm are given herein, thus providing basis for application expansion of the algorithm.
The measuring result gives support to further texture segmentation of forest vegetation by measuring remote sensing image vegetation information according to blue noise theory. For example, this may involve structure element scale related to texture dimension, and gray scale distribution of structure element related to distribution of texture gray scale when the segmentation makes use of gray scale morphology texture filter method. By measuring remote sensing image texture based on blue noise theory, this paper proves the vegetation texture of remote sensing image has blue noise features in specific spatial scales, which is a reference for processing similar texture of ground feature in remote sensing image. The research will be mainly focused on utility program development, and calculation accuracy and efficiency of the algorithm need to be further optimized, so as to meet the requirements for accuracy and real-time performance in practical applications.

7. References

[1] VF Rodriguez-Galiano, M Chica-Olmo, F Abarca-Hernandez 2012 *Remote Sensing of Environment* **121** 93-107.
[2] K Mansour, O Multanga, T Everson 2012 *Isprs Journal of Photogrammetry & Remote Sensing* **70** 56-65.
[3] XL Gao, XQ Wang 2008 *Resources Science* **30** 153-8.
[4] Liu C, Qiumin L 2010 Research on plant information identification, extraction and change monitoring based on multi-source remote sensing data *Geoscience and Remote Sensing*

(IITA-GRS), 2010 Second IITA International Conference on IEEE **2** 194-7.
[5] Li Chunyan 2012 *Science Technology and Engineering* **12** 1941-44.
[6] Ramdani F 2013 *Journal of Geographic Information System* **5** 404-8.
[7] Zheng Li 2014 Application of active learning in content-based remote sensing image retrieval *Computer Applications and Software* **10** 110-3.
[8] Huiyuan Yang 2014 *Remote sensing image characteristics description method and retrieval application based on content.* Beijing: Universiy of Chinese Academy of Sciences.
[9] Zengguang Zhang, Xiaoli Zhang, Wei Liu 2010 *Research on extraction of vegetation information from remote sensing images based on knowledge rules* The International Conference on Computational Intelligence and Industrial Application.
[10] Wenxin Tian 2012 *Silicon Valley* **15** 22-3.
[11] Liu Y, Ling B, Meng Y H 2012 *ISPRS Journal of Photogrammetry and Remote Sensing* **67** 144-156.
[12] Benz U C, Hofmann P, Willhauck G 2004 *International Journal of Photogrammetry & Remote Sensing* **58** 239-258.
[13] Yunhao Chen, Tong Feng, Peijun Shi 2006 *Geomatics and Information Science of Wuhan University* **31** 316-320.
[14] Deshpande S, Inamdar A, Vin H 2013 Overview of hyperspectral remote sensing of impervious surfaces in urban environment *Communications in Computer & Information Science* **398** 128-140.
[15] Polak M, Zhang H, Pi M 2009 *Image and Vision Computing* **27** 1223-27.
[16] Mashimbye Z E, Cho M A, Nell J P 2012 *Pedosphere* **22** 640-9.
[17] Qiang Bie, Lei He, Chuanyan Zhao 2014 Study on vegetation information extraction based on object-oriented image analysis *Remote Sensing Technology and Application* **29** 164-171.
[18] Yijiao Cui, Lin Zhu, Lijuan Zhao 2013 Abstraction and analysis vegetation information based on object-oriented and spectra features *Acta Ecologica Sinica* **33** 0867-75.
[19] Jie Li, Changyun Miao, Zhigang Wu 2012 *Computer Science* **39** 546-8.
[20] Yanfen Guo, Zhihong Liu, Mingyuan Xie, et al 2009 *Remote Sensing Technology & Application* **18** 581–5.
[21] Jing, L. I., Hanqiu, X. U., Xia, L. I., & Guo, Y 2015 Vegetation information extraction of pinus massoniana forest in soil erosion areas using soil-adjusted vegetation index. *Journal of Geo-Information Science* **7** 17-26.
[22] Yao, F., Luo, J., Shen, Z., Dong, D., & Yang, K 2016 Automatic urban vegetation extraction method using high resolution imagery *Journal of Geo-Information Science* **18** 248-254.
[23] Shengguang Dong 2010 *Spatial information extraction and 3d mapping technology research based on GeoEye-1 image* Changsha: Changsha University of Science and Technology.
[24] M.A. Aguilar, M.M. Saldaña, & F.J. Aguilar 2013 Geoeye-1 and worldview-2 pan-sharpened imagery for object-based classification in urban environments *International Journal of Remote Sensing* **34** 2583-2606.

8. Acknowledgement
This paper is based on the key research project in Hunan, which is "Natural disasters data based on satellite remote sensing monitoring interpretation technology research"(Item number: 2016SK2088). Thanks to Professor Gui Zhang for providing remote sensing data and guidance, and my junior fellow apprentice for inspiration in the written process. Here, I wish to express my heartfelt thanks.

Analysis of black soil environment based on Arduino

Y Li[1], Y F Zhang[1], C H Wu[1], J F Wang[1,*]
[1]Anhui Science and Technology University, No. 9 Donghua Road, FengYang, Anhui, China
[*]Correspondence:jfwang1@aliyun.com;Tel:+86-550-673-2024;Fax:+86-550-673-2094

Abstract. As everyone knows, the black soil of Heilongjiang bred rice is famous in the world. How to use networking technology to detection the growth environment of Heilongjiang rice, and expands it to the local planting environment to our country is the most important topic. However, the growth environment of rice is complex. In current research, some importnat factors such as carbon dioxide, oxygen, temperature and humidity, pH value and microbial content in black soil that affect the growth of plants are selected, and a kind of black land based on data acquisition and transmission system based on the Arduino development environment and the mechanism construction of Kingview has been realized. The collected data was employed to establish the simulation environment for the growth of rice in Heilongjiang. It can be applied to stimulate the rice growing environment of Heilongjiang province, and gives a improvement of rice quality in other areas.
Keywords: Arduino; Kingview; living environment

1. Introduction

Our country since the implementation of the South-to-North Water Diversion Project for the national people brought huge benefits, while the northern black land contains vitality by scientific methods can also be transferred to the south, looking for an implementation of the scheme has become a major topic of agricultural workers. With the improvement of people's living standard, people's taste and quality of rice is getting higher and higher, so the quality and yield of rice is of great significance to our country. With the development of computer network and agricultural technology, it is possible to establish and realize the rice production environment in the whole country. This paper is based on the Arduino as the control core, establish data acquisition monitoring system with Kingview PC, to achieve real-time acquisition of data using sensor technology [1]. The primary task of this design is to collect the data of the black land, as well as the temperature and humidity, oxygen, carbon dioxide and organic matter content, and ultimately to improve the environmental conditions of the South loess.

2. The establishment of black soil environmental monitoring system

The system consists of Arduino control board, Kingview monitoring system, pH sensor, humidity sensor, CO_2 sensor, O_2 sensor and microbial sensor. Among them, Arduino microcontroller is the control center of the system, is responsible for the collection of data, and sent it to the host computer. All kinds of sensors are used to detect soil environmental factorsp [2]. The detection of oxygen and carbon dioxide also provides a reference for the detection of microbial populations. The system framework is shown in Figure 1.

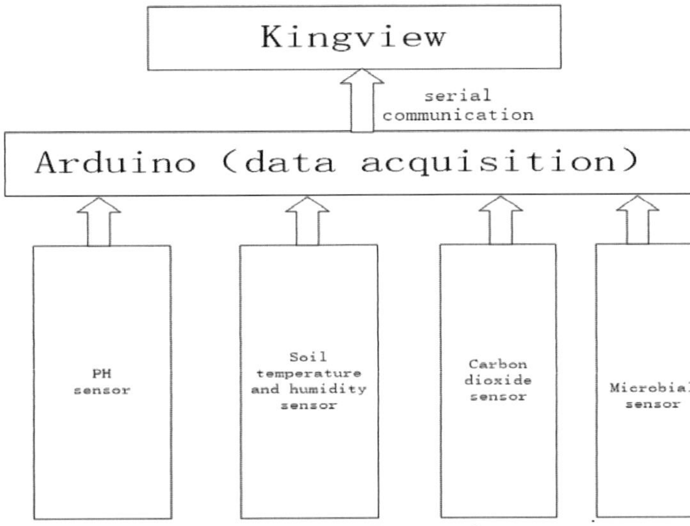

Figure 1. Structure diagram of system.

3. Hardware implementation of Arduino control board and sensor

Arduino is an open source hardware control platform, with its own integrated development environment (IDE).This system uses the Arduino unit of Arduino UNO R3, which is the latest version of Arduino USB series interface. The core processor is ATmega328, with 14 digital input/output (6 as PWM output), 6 analog input, support I2C and SPI communication protocol [3]. A variety of Arduino peripHeral functional circuits can be selected, including motor drive, wireless communications, music playback and a variety of sensors (humidity, temperature, speed, etc.).

3.1 pH value sensor

pH value measurement is simple, convenient and practical, designed specifically for the design of Arduino analog pH meter. When it works, only the pH sensor is connected to the onboard BNC, the onboard pH1.0 is connected to the analog acquisition port of the Arduino controller, and the Arduino open source test code can be used to measure the pH value of the solution. The software design includes the Arduino as the core of the signal processing module, data processing module and PC Ardul module, which mainly realizes the receiving of sensor signal conversion, receive, read. Using the built-in C/C++ based on Arduino programming language that has a huge repository function, this module to write simple programs related to the initialization, and then read the calculation of standard voltage, and then through the Aiduino signal data acquisition, PC USB serial voltage acquisition, and then the related data processing, the USB transmits the data to the host computer, display the measured data, the specific circuit is shown in Figure 2.

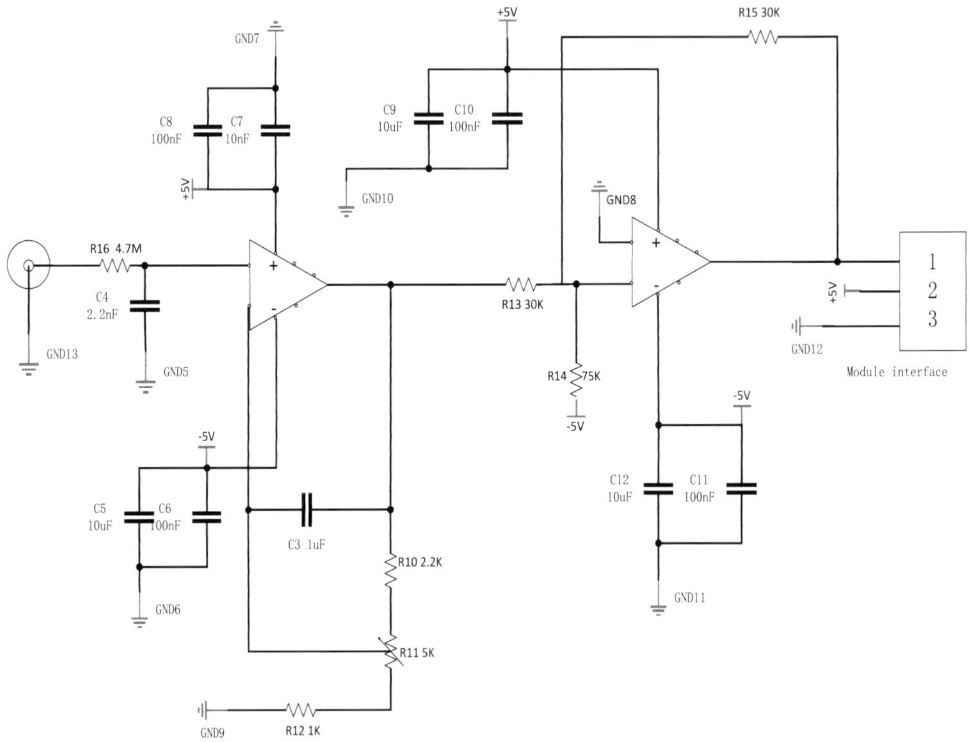

Figure 2. The schematic diagram of pH measuring circuit.

3.2 Soil temperature and humidity sensor

DHT11 can simultaneously measure the temperature and humidity, and humidity measurement range is 20~90%RH with a measuring accuracy of 5%RH. Due to DHT11 in the temperature measuring range is small (0~50 C), and the precision is relatively low [4]. It should be noted that the simultaneous measurement of the temperature and humidity will present a lager error. The specific circuit is shown in Figure 3.

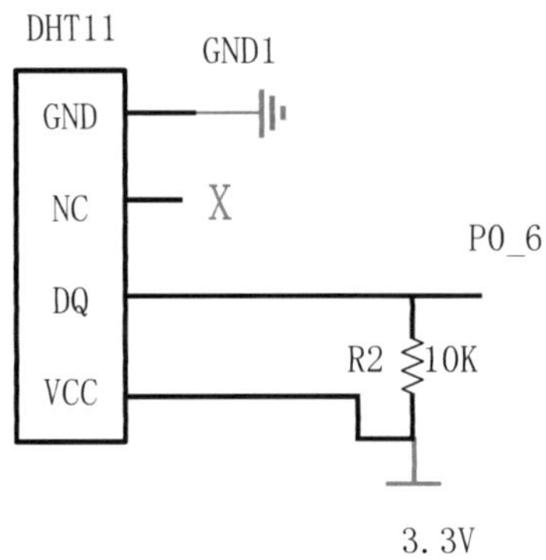

Figure 3. Humidity sensor.

Additionally, we have found that a higher accuracy could be obtained if a additional temperature sensor was used. DS18B20 is a digital temperature sensor, only one data line can be achieved with the microprocessor communication and a temperature range of -55-125 degrees centigrade. The specific circuit is shown in Figure 4.

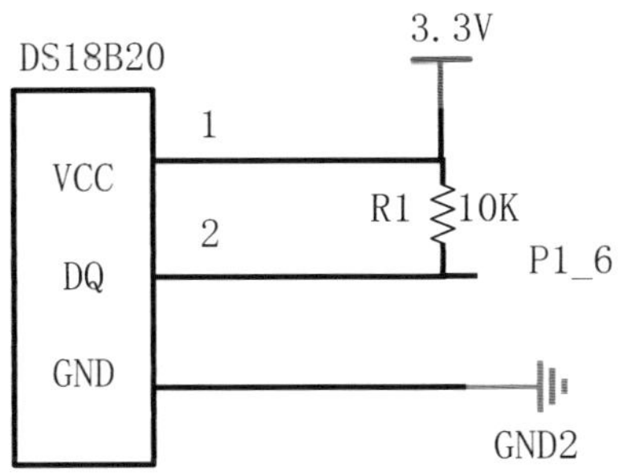

Figure 4. Temperature sensor.

3.3 Carbon dioxide sensor

TGS4160 carbon dioxide sensor is a kind of solid state electrochemical gas sensitive original [5]. The element is suitable for electricity at room temperature for long time. Its output signal voltage of CO2 concentration is 0-3V, which is equivalent to the concentration of 0-3000PPM. There are relay transfer control ports, which can output high and low threshold signals for external control. The specific circuit is shown in Figure 5.

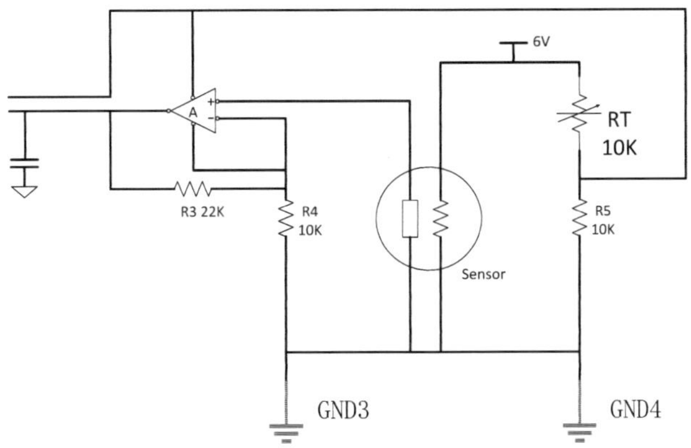

Figure 5. Carbon dioxide sensor.

3.4 Microbial sensor

The microbial sensor is composed of molecular recognition of the original (sensor) and the signal converter with a (transducer) composition analysis tool or system. It is senstive to the change of biochemical signals, and employed for sensing a measurable electrical signal, then realize connection detection and control board [6].

A BOD sensor (the sample dissolved in pure water) that used to detect polymer immobilized microorganism preparation of biological particles was selected, too. In a specific environment, it can fully contact with the measured samples. It works as the following process. Firstly, the decomposition of organic matter in the sample process will consume dissolved oxygen, then with the obtained dissolved oxygen concentration, the electrode will change detection record values of dissolved oxygen, and then it was compared with the standard value, thus get the value of organic matter by calculating. The current signal measured by the dissolved oxygen probe is connected in parallel with a large resistance to a voltage signal, and then the signal is amplified and input to the control panel to obtain the measured dissolved oxygen concentration.The specific circuit is shown in Figure 6.

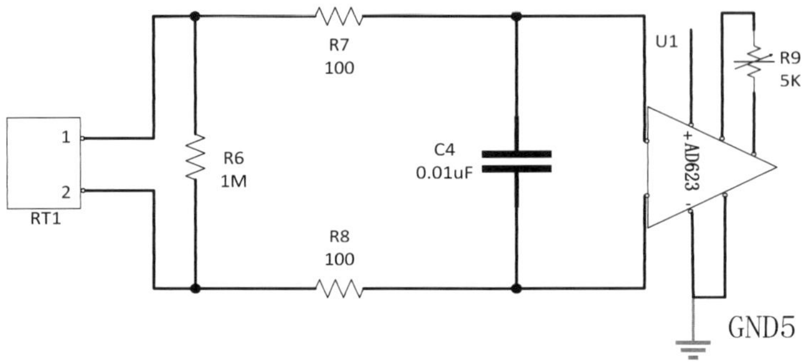

Figure 6. Microbial sensor.

4. Host computer

PC interface is developed by Kingview software design, the software supports more communication protocols, which has the advantageous of easy development, powerful and developed interface [7]. It can be very intuitive response system work process. Kingview built-in communication protocol, only need to configure serial port baud rate can be achieved and the communication of information.The system interface is shown in Figure 7.

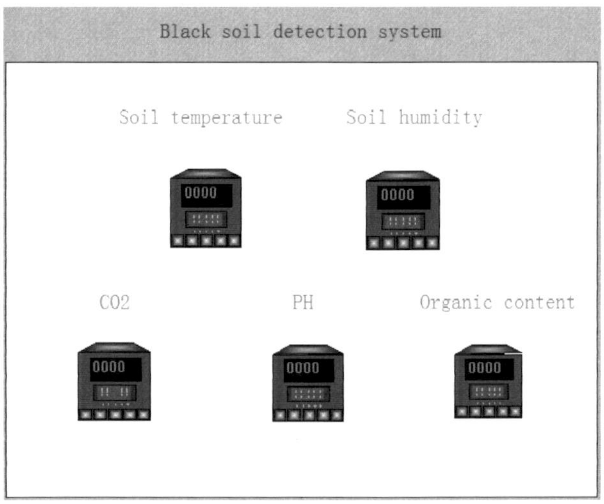

Figure 7. System interface.

5. Conclusion

The system of Arduino and Kingview analysis of black soil environment, including sensor module, data acquisition module, PDA display module and data storage etc. has been designed in current study. From the front of the sensor rod contact with the soil, soil pH, temperature and humidity can be collected. The specific environment of statistical analysis on soil organic matter is assisted by the data acquisition module, grapHics display. Based on data collected from the monitoring data to the host computer of the system configuration save, the rice planting environment has been presented. The system has the characteristics of high precision, simple operation and strong anti-interference ability, and has high popularization and application value. It will show a great potential in environment monitoring of rice planting. The current design can be improved by adding a automation control module.

6. References

[1] Zuo X G, Liu Y C and Wang J P 2016 *J. Agricul. Mechan. Res.* **2** pp 213

IOP Conf. Series: Earth and Environmental Science **67** (2017) 012008 doi:10.1088/1755-1315/67/1/012008

[2] Yao S F, Feng C G and JIA Y Y 2011 *J. Agricul. Mechan.Res.* **7** pp 190
[3] Fu J Q 2015 *J. Packaging Engineering.* **10** pp 76
[4] Liang N X 2016 *J. Science & Technology Ecnony Makket.* **3** pp 21
[5] Zhang H H, Zhao X S, Yu X H, Cheng L Y and Ma S J 2016 *J. Journal of Shenyang Normal University(Natural Science Edition).* **4** pp 454
[6] Zhang L, Liu J, Guo H and Li Y H 2017 *J. Chinese Journal of Public Health Engineering.* **1** pp 36
[7] Sun H H, Mo Y P, Ma R and Qian K 2016 *J. Industrial Instrumentation & Automation.* **4** pp 102

Integrating environmental management into food safety and food packaging in Malaysia: review of the food regulation 1985

N H Nordin[1*], H Hara[1], N Kaida[2]

[1] Department of Environment Engineering and Green Technology, Malaysia-Japan International Institute of Technology (MJIIT), Universiti Teknologi Malaysia
[2] Faculty of Engineering, Information and Systems, University of Tsukuba, Japan
*Corresponding Author: norulhajarnordin@gmail.com

Abstract. Food safety is an important issue that is related to public safety to prevent the toxicity threats of the food. Management through legal approach has been used in Malaysia as one of the predominant approaches to manage the environment. In this regard, the Food Regulation 1985 has been one of the mechanisms of environmental management through legal approach in controlling the safety of packaged food in food packaging industry in Malaysia.The present study aims to analyse and to explain the implementation of the Food Regulation 1985 in controlling the safety of packaged food in Malaysia and to integrate the concept of environmental management into the food safety issue. Qualitative analysis on the regulation document revealed that there are two main themes, general and specific, while their seven sub themes are included harmful packages, safety packages, reuse packages, polyvinyl chloride (PVC), alcoholic bottle, toys, money and others and iron powder. The implementation of the Food Regulation 1985 in controlling the safety of packaged food should not be regarded solely for regulation purposes but should be further developed for a broader sense of food safety from overcoming the food poisoning.

1. Introduction

1.1. Food packaging matters for food safety and environmental management.

Food safety is an important public safety issue to prevent the toxic threats of the food. According to the World Health Organization [1], safe food is defined as a food confidence degree where it does not cause disease or harm to the consumer when it is prepared, served or eaten according to their usefulness. Safe food means food prepared on clean and sanitized surfaces with utensils and dishes that also are cleaned and sanitized. Moreover, from consumer's perception, safe food are more practical like food that does not make a person ill, purchasing fresh chicken and not having the package leak or drip juice, making them wonder about the integrity of the initial seal [2]. Food safety problems can be caused by chemical, microbiological and physical food contaminants. According to Wilcock [3] and Kher et al. [4], microbial contamination of food is also an important issue for public health. Potentially harmful microorganisms proliferate rapidly in the human body once ingested. Changes in eating habits and increasing in the distribution of pre-prepared meals have also augmented the threat of microbial hazards. Moreover, chemicals such as heavy metals, agrochemical residues, food additives used during food processing, traces of antibiotics in animal products and veterinary chemicals all represent potential risks to public health [4].

Eating food that has been contaminated with dirt and germs will involve Food Water Borne Diseases (FWBD) in human bodies. Among various types of the diseases including food poisoning, cholera, typhoid, dysentery and viral hepatitis A, food poisoning is the most popular in FWBD in Malaysia in terms of a number of cases (Table 1). Food poisoning, which can be defined as a situation where taking a contaminated food or ingestion of contaminated food affected by bacteria or toxin bacteria [5] affect human bodies, can cause various symptoms such as diarrhea, vomiting, nausea, discomfort, headache, dizziness and abdominal pain [6]; [7]. These facts indicate that food poisoning is one of the important challenges to be tackled for food safety and particularly against FWBD in the country. In this regard, food packaging is considered as one of the effective approaches to overcome and reduce the toxicity of food that lead to food poisoning.

Environmental management was a stand points to the issue of controlling the safety of packaged food in food packaging. Food packaging helps consumers to take foods hygienically and safely. According to Hawkins [8], materials that are used for packaging food are sometimes considered as a source of risk to the food, can be eventually wastes after use if not reused or recycled, and they can be a source of water and air pollution if packages as wastes are not properly treated. These general concerns also explain the situations of Malaysia. The issue of food packaging therefore needs to be considered from multiple perspectives: food safety and environmental management.

1.2. Aims

This article aims to analyze and explain the legal developments of food safety in Malaysia with a focus on food packaging, taking a particular case of the Food Regulation 1985 as a key development in the country, and to integrate environmental management standpoints to the issue of controlling the safety of packaged food in food packaging industry in Malaysia.

Table 1. Number of cases and incident rates every 100,000 population (KI) for FWBD from 2002 to 2010

	Food poisoning		Typhoid		Cholera		Dysentery		Viral hepatitis A	
Year	Case	(KI)	Case	(KI)	Case	(KI)	Case	(KI)	Case	(KI)
2002	7,023	28.6	853	3.5	365	1.5	292	1.2	295	11.0
2003	6,624	25.4	785	3.0	135	0.5	310	1.2	-	-
2004	5,957	23.3	484	1.9	89	0.4	356	1.4	107	0.4
2005	4,641	17.8	1,072	4.1	386	1.5	141	0.5	44	0.2
2006	6,938	26.0	204	0.8	237	0.9	105	0.4	64	0.2
2007	14,455	53.2	325	1.2	133	0.5	146	0.5	94	0.4
2008	17,322	62.5	201	0.7	93	0.3	92	0.3	36	0.1
2009	10,238	36.2	303	1.1	276	1.0	154	0.5	40	0.1
2010	12,519	44.2	210	0.7	443	1.6	104	0.4	39	0.1

2. Food packaging: developments as technology and legal definition in Malaysia

Food packaging is defined as wrappers or containers used to protect food or other products from dirt, germs and damaged. The use of various plastics has developed as new materials for packaging in 1950, when there was a rapid rise in the number of food poisoning cases that reported to the authorities in the United Kingdom [7]. Nowadays, packaging manufacturing uses a variety of materials and labeling methods to protect and promote a product but some of the product may have chemicals material that may affect the health. Therefore, it is important to identify environmental management approaches in dealing with the safety control of food packaging to overcome food poisoning and other diseases that related with FWBD.

IOP Conf. Series: Earth and Environmental Science **67** (2017) 012009 doi:10.1088/1755-1315/67/1/012009

Packaging in Malaysia is under certification schemes related to food safety assurance under the Ministry of Health Malaysia, namely Good Manufacturing Practice (GMP) Certification Scheme. Section 2 in the Malaysia Food Act 1983 indicates that packages can include anything in which or any means by which food is wholly or partly cased, covered, enclosed, contained, placed or otherwise packed in any way whatsoever and include any basket, pail, tray or receptacle of any kind whether opened or closed.

3. Environmental management as concept

There can be many strategies under environmental management that are useful for ensuring a food safety such as review and update the legislation, strengthening the infrastructure, enhancing collaboration between the government, corporate bodies, non-governmental and international organizations. Environmental management in food safety also covers to develop and train human resources and deal a strategic plan on educating consumers to raise their awareness about the issues.

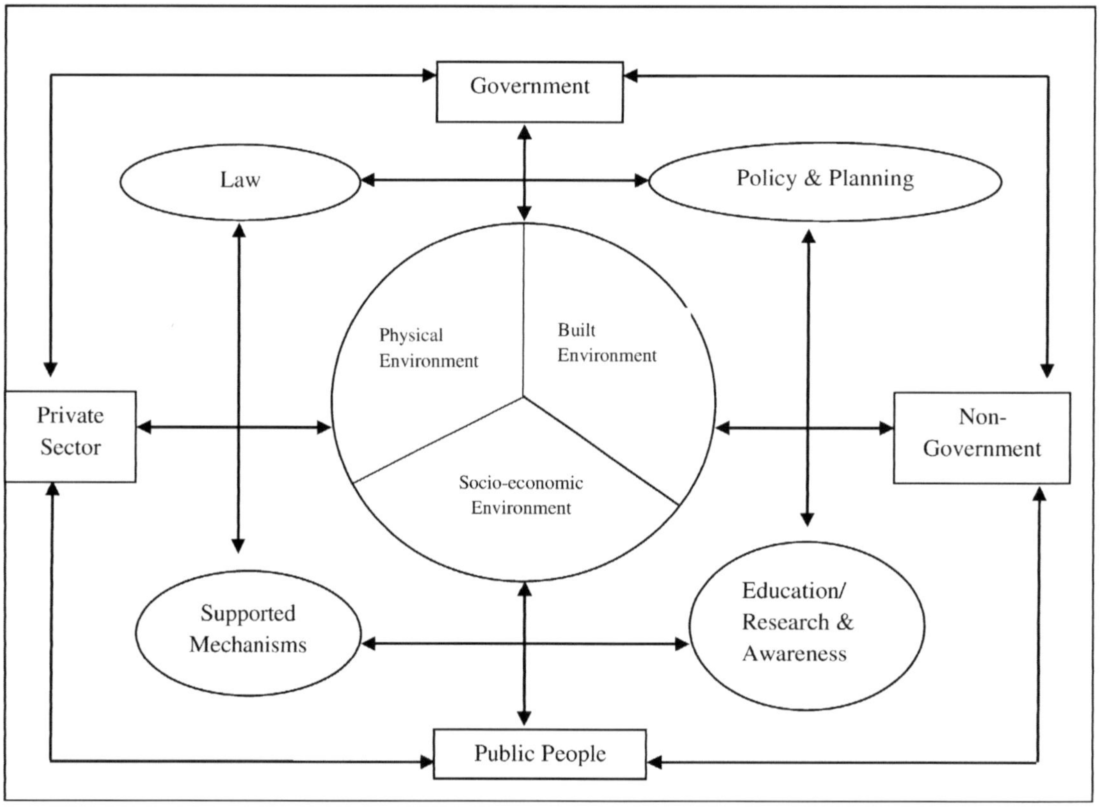

Figure 1. Environmental management process

Each individual in this world active to some extent environmental management, but the refer scope here is through a conscious and systematic measure by individual or groups of people acting together to produce a useful physical environment in terms of the economy, which has aesthetic values, ensures the health, comfort and safety [11]. The environment can be divided into two, namely the physical environment and human environment. The physical environment is the "site" to all forms of human activity that also known as a natural process of large systems. The environment is not merely made up from physical components, but also the social, economic and culture that shaped and carved by humans [11]. Environmental management as concept is to optimize the balance among different components of the environment through controlling their dynamic interactions.

For effective management, an understanding of the components and the relationship between them is very important and cooperation also required for all fields, science, engineering and social sciences. A combination of environment information systems acts as a focal point of cooperation between stakeholders and mechanisms of environmental management [10] (Figure 1). Government, non-government, private sector and public involve in cooperation between stakeholders together with law, policy and planning, supported mechanisms, and also education, research and awareness. Malaysia's international involvements have influenced the environmental management in the country. Malaysia has responded and taken in from its experience at the international level to accept the knowledge into the strategy and policy on the environment in the country [11].
According to Muhammad Rizal & Jamaluddin [12], environmental management can be divided into two approaches namely environmental management through legal approach and environmental management through non-legal approach. Management through legal approach has been used in Malaysia as one of the mechanism to manage the environment. According to this understanding, the Food Regulation 1985 is one of mechanism of environmental management through legal approach in controlling the safety of packaged food in food packaging industry in Malaysia.

4. The Food Regulation 1985

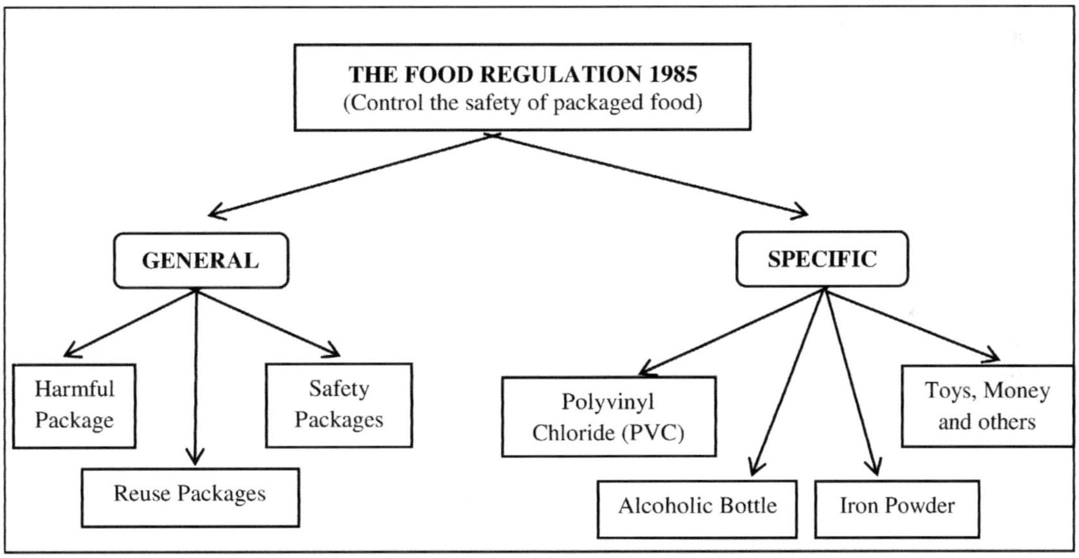

Figure 2. Themes and sub themes of the Food Regulation 1985 related to the safety of packaged food.

The Food Regulation 1985 was developed under the Malaysia Ministry of Health and enacted in 1985. The Food Regulations 1985 has created and aims to protect consumer from harmful food poisoning beside to ensure the safety of food sold in Malaysia. This Regulation consists of 10 Sections. The Food Regulation 1985 includes two main themes and seven sub themes that are focused on controlling the safety of packaged food as shown in Figure 2. The themes are divided into *general* and *specific*, while their sub themes are divided into harmful packages, safety packages, reuse packages, polyvinyl chloride (PVC), alcoholic bottle, toys, money and others and iron powder.

4.1. General theme

Food packaging with unsaid name, type and detail on package is categorised as *general*. This includes harmful packages, package security and usage packages again. Regulation 27 is about dangerous usage prohibition. Individuals are not allowed to import, manufacture, or advertise.

> "Except as to how stated others in this Regulation, no one else could import, manufacture, advertise for sales or sell, or use or causes or allow use in

> preparation, packaging, storage, past dispatcher or food exposure for sales, any package, tool, former or vessel which produces or can put out in the content any material that is poisonous, damage or deform, or which helps in the food attenuation"[Regulation 27]

While packaging purpose is to protect food products from source of pollution which caused the food product's perishable, it is also to facilitate operation and manufacturer, distributor, seller and consumer [13]. These can cause loss, damages and could threaten the safety and altogether can cause diseases whether in man or environment because it has poison content or toxic poison content or toxic. Harmful incidents sometimes occur due to chemical substances which include in packer when it recycled or utilized for different uses.

Safety of food packaging is a way to avoid the package to be dangerous and hazardous materials, which mostly cause unsafe package.

> "No person shall import, manufacture, advertise for sale or sell any package, appliance, container or vessel made of enamel or glis porcelain intended to be used in the preparation, packaging, storage,submit and post or exposure of food for sale whether it can transfer lead, antimony, arsenic, cadmium, or any other poisonous substances to the food prepared, packed, stored, delivered or exposed in it, or do not have resistance to acid unless the package, appliance, container or vessel satisfies test described in the Thirteenth Schedule " [Regulation 28]

There are many categories of distribution and reuse package related to food safety. It is stated in Regulation 32(1).

> "Subject to subregulation (2), no person shall use, or cause or permit to be used, in the preparation, packaging, storage, or exposure for sale, submit and post" [Regulation 32(1)]

4.2. Specific theme

Packages stated a name or information that is related to the specific packaged food are included within a *specific* theme. There are packages made from from polyvinyl chloride (PVC), alcoholic bottle, toys, money and others and iron powder. PVC which contains excess vinyl chloride monomer (VCM) is prohibited from using, importing, manufacturing food packages.

> "There is no even one could import, manufacture or advertise for sales or sell or are using in provision, packaging, storage, or past dispatcher to food exposure for sales, any package, tool, former or rigid vessel or not solid made from polyvinyl chloride which contains more than 1 mg / kg vinyl chloride monomer" [Regulation 29]

PVC is one of major thermoplastic substance because it is a polymer that is relatively cheap and versatile. In 1976 more eight million tons of PVC was invoked throughout the world [14].

> "No person shall import, manufacture or advertise for sale or use in the preparation, packaging, storage, submit and post or exposure of food for sale, any package, appliance, container or receptacle rigid or se,i-rigid made of polyvinyl chloride containing more than 1mg/kg of vinyl chloride monomer"[Regulation 29]

Toys, money and others are strictly prohibited from placing in food packaging mainly for sale. It may risk and attracted a child. Maybe they will eat together with the food.

> "Nothing toys, money or other things can be put in food for sale, or in the food packaging"[Regulation 36(1)]

They however may be put into the food packaging if stated in accordance with the following conditions.

> "Nothing in sub regulation (1) may prohibit placed in food or in food packaging,"
>
> a) A body to measure the quantity of food that is recommended for consumption, provided that the thing was sterile;

b) The label referred in sub regulation (6) of Regulation 12, or
c) Any reduced iron powder packets for the purpose of absorbing oxygen"
[Regulation36 (2a), (2b), (2c)]

Iron powder can be put into food packaging, only if it is packed in oxygen-absorbing packets that are not contaminated, damage or to enter the food. Labels on iron powder itself must not be contaminated.

"The reduced iron powder specified in paragraph (c) of sub regulation (2) Regulation 36 must be packed in packets in a manner that does not contaminate the oxygen absorber, spoil or get into the food"
[Regulation 36A (1)]

5. Conclusion

Management through legal approach has been used in Malaysia as one of the predominant approaches to manage the environment. In this regard, the Food Regulation 1985 has been one of the mechanisms of environmental management through legal approach in controlling the safety of packaged food in food packaging industry in Malaysia. The qualitative analysis of the present study revealed the Food Regulation 1985 provides two main themes related to the safety of packaged food, namely general and specific. It also identified that their sub themes consists of harmful packages, package security, used packaged, PVC, alcoholic bottle, toys, money and others and iron powder.

The implementation of the Food Regulation 1985 in controlling the safety of packaged food should not be regarded solely for regulation purposes but should be further developed for a broader sense of food safety from overcoming the food poisoning, FWBD and other health-related issues to sound environmental management. The present study suggests that the concept of environmental management may facilitate the integration of the food safety regulations and policies and human and environmental health through interactions between various stakeholders and components.

6. References

[1] WHO 2003 *Food Safety*. Geneva: World Health Organization.

[2] Schmidt Ronald H and Rodrick Gary E 2003 *Food Safety Handbook* John Wiley and Sons Publication Canada

[3] Wilcock A, Pun M, Khanona J, and Aung M 2004 Consumer attitudes, knowledge and behavior: a review of food safety issues *Journal of Trends in Food Science and Technology* **15**:56-66

[4] Kher S V, Jonge J D, Wentholt M T A, Deliza R, Andrade J C d, Cnossen H J, Lujickx N B and Frewer L J 2013 Consumer Perceptions of Risks of Chemical and Microbiological Contaminants Associated with Food Chains: A Cross National Study *International Journal of Consumer Studies* **37**: 73-83

[5] Fatan Hamamah Yahaya 2005 Emerging Infectious Diseases and the Impact to Man in Malaysia(Kemunculan Penyakit Berjangkit dan Kesannya Terhadap Manusia di Malaysia) *SARI 23* **(10)**: 153-168

[6] Norhayati Ismail 2000 *Food Poisoning (Keracunan makanan)* Kuala Lumpur Utusan Publications

[7] Crosby N T 1981 *Food Packaging Materials: Aspects of Analysis and Migration of Contaminants* Ripple Road Barking Essex England Applied Science Publishers Ltd.

[8] Hawkins G 2012 The performativity of food packaging: market devices, waste crisis and recycling *The Sociological Review* **60**: 66-83

[9] Malaysia 2011 *Annual Report 2010* Kuala Lumpur Department of Statistic Malaysia

[10] Fauza Abdul Ghaffar & Khairulmaini Osman Salleh *2004 Sustainable in Malaysia: Challenge and Accessibility Mampan di Malaysia: Cabaran dan Ketercapaian)* Mohd Razali Agus & Yazid Salleh (ed.) Pengurusan Perbandaran 31-61 Kuala Lumpur Dewan Bahasa dan Pustaka

[11] Jamaluddin Md Jahi 2010 From Stockholm to Rio de Janeiro and beyond: Environmental Mangement in Malaysia (Dari Stockholm ke Rio de Janeiro dan seterusnya: Pengurusan alam sekitar di Malaysia) Bangi Penerbit Universiti Kebangsaan Malaysia

[12] Muhammad Rizal Razman and Jamaluddin Md Jahi 2010 *Environment and Health Legal Perspective in Brazil: Focus on Particular Tort Low to Negligence Law (Alam Sekitar dan*

kesihatan dari perspektif undang-undang di Malaysia: tumpuan kepada Undang-Undang Tort khususnya kepada Undang-Unadang Kecuaian) 2nd National Conference on Environment and Health 2010: USM Kampus Kesihatan, Kubang Kerian, Kelantan 216-225

[13] Faridah Mohd Som 2003 *Packaging: The Impact on Quality and Food Safety(Pembungkusan:Kesannya terhadap mutu dan keselamatan Makanan)* MARDI Serdang

[14] Whelan A and Craft J L 1977 *Developments in PVC production and Processing* England: Applied Science Publishers LTD

IOP Conf. Series: Earth and Environmental Science **67** (2017) 012010 doi:10.1088/1755-1315/67/1/012010

Experimental study of the influence of varying ceiling height on the heat release rate of a pool fire

Jiahao Liu[1,2], Jian Wang[1], Richard Yuen[2]
[1]State Key Laboratory of Fire Science, University of Science and Technology of China, Hefei, China
[2]Department of Architecture and Civil Engineering, City University of Hong Kong, Hong Kong, China

Abstract. To investigate the influence of ceiling height on the combustion process of a pool fire whose flame impinges the ceiling, a sequence of pool fires with varying ceiling heights was performed using a scaled-down cone calorimeter. N-heptane and jet-A were employed as fuels to conducted the tests. Experimental findings reveal that with the decreasing ceiling height, the maximum and average heat release rates will initially increase due to the enhanced heat feedback, and then decrease as a result of the restriction of air entrainment caused by the extremely small ceiling height. In addition, the dimensionless ceiling height is found to have a linear relationship with the logarithm value of the dimensionless averaged heat release rate for the two given fuels with the similar slope of -2/3.

1. Introduction

When a flame impinges on an unconfined ceiling, the unburnt gas fuels will spread out radically and entrain air for combustion, and a circular flame will be established beneath the ceiling. This phenomenon has received comprehensive scientific attentions because the fire development under the ceiling is often the trigger for the occurrence of the flashover for a compartment fire [1-2]. The parameters concerned in previous studies mainly include the flame extension length and maximum excess temperature distribution beneath the ceiling, which are essential aspects in characterizing the danger of the flame impingement.

You and Faeth [3] initially proposed the correlation for the flame extension length along the ceiling based on the experimental data in the range of $(H_f\text{-}H)/D$=0~5.8, expressed as

$$r_f / D = 0.502\left[(H_f - H)/D\right]^{0.957} \tag{1}$$

wherer_f is the flame extension length, D is the fuel dimension,H_f and H are the flame height and ceiling height, respectively. This empirical equation has been validated by Zhang et al. [4], and extend its application to a wider range of 0~14.5.With respect to the maximum excess temperature distribution beneath an unconfined ceiling, Heskestad and Delichatsios [5] indicated that the non-dimensional excess temperature $\Delta T^* = (\Delta T / T_\infty)/Q^{*2/3}$ can be applied to partitionthe temperature profile of the turbulent impinging flame into two parts. For $r/H \le 0.2$, ΔT^* is independent of the horizontal distance r with constant value of 6.3, while for $0.2 < r/H < 4.0$, ΔT^* can be correlated with *r/H* as

$$\Delta T^* = (0.188 + 0.313 r / H)^{-4/3} \tag{2}$$

where$Q^*$is the non-dimensional heat release rate expressed as,

$$Q^* = Q/(\rho_\infty T_\infty c_p \sqrt{g} H^{5/2}) \tag{3}$$

Here, $Q$is the heat release rate (HRR) of fire source, T_∞ and ρ_∞ are the ambient temperature and density of air,c_p is the specific heat of air, g is the gravitational acceleration.

These classical correlations have been verified by many previous researchers [4,6-7]. However, these equations were mostly summarized from the experiments with constant HRR, where relatively steady flames were presented. Scarce research focused on the influence of presence of a ceiling on the combustion process of a fire with buoyancy-controlled turbulent jet diffusion flame, such as a pool fire. Karlsson and Quintiere [1] indicated that the boundaries including ceilings and walls couldsignificantly enhance the burning intensity as they were close enough to a fire. Other studies concerning the effect of presence of ceilings or sidewalls on the combustion behaviors of pool fires also demonstrated the enhancement effects [8-9].

To quantitatively understand the effect of varying ceiling height on the HRRs of pool fires, a sequence of pool fires under a ceiling with adjustable ceiling heights was performed in a cone calorimeter in current study. The HRRs for all the tests were measured and analyzed carefully. The results may be helpful for evaluating the risks of fire scenarios where the flame impingement occurs.

2. Experiments

2.1. Test platform

To cater for the current need, a scaled-down calorimeter with dimensions of 40% of that in ISO 9705 was built, as shown in Figure. 1. The dimensions of the combustion chamber were 1.2 ×1.2×1.2 m, and a gap of 0.15 m was reserved to guarantee the supplement of fresh air. A volumetric air flow rate of 0.18 m^3/s was sustained to remove the combustion products by the fan, which ensured the over-ventilation environment for all the tests. The exhaust gases including oxygen, carbon monoxide and carbon dioxide were continuously recorded by the Servomex 4100 analyzer, which was eventually calculated in the PC to give the real-time HRR. The specified introduction to this platform can be found in [10]. Before conducting all the experiments, a standard gas fire originating from the propane fuel with purity above 99.9% was employed to calibrate the test platform.

2.2. Experimetal setup and configurations

A 60×60 cm square mica smooth plate with excellent heat resistance performance was supported by the four iron stands at a distance of 70 cm away from the floor to serve as the ceiling. The circular steel pan with diameter of 14 cm and depth of 2 cm was placed on the lifting table with a 20×20 cm square insulation board between them to shield from the elevated temperature. These devices were positioned on an electronic scale with resolution of 0.1 g to measure the mass loss of the fuel during the test. The ceiling height, defined as the vertical distance between pan bottom and ceiling, was designed as 5, 15, 25, 35 cm for each fuel. Two kinds of fuels, i.e. n-heptane and jet-A, were employed to conduct the tests. The tests without the ceiling were also involved in current study to investigate the effect of the ceiling. A summary of experimental configurations is shown in Table 1. The ambient temperature and relative humidity were recorded for the tests, which were 25±3 ºC and 45±5%, respectively. The typical cases were repeated for three times and the results exhibited good repeatability with discrepancy less than 5%.

Table1. Experimental configurations for current tests

Fuel	Ceiling height (cm)
N-heptane	35,25,15,5, case without the ceiling
Jet-A	35,25,15,5, case without the ceiling

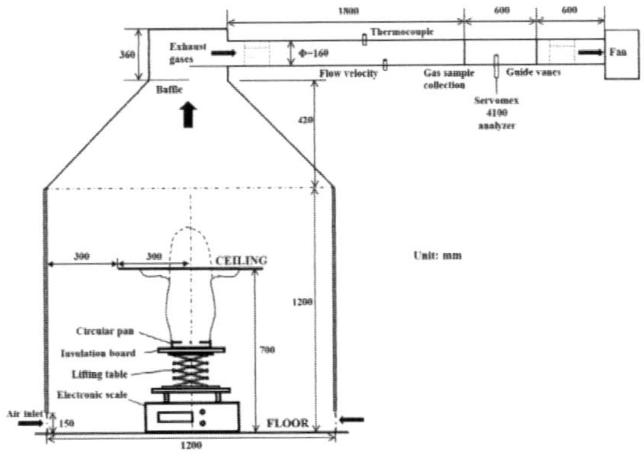

Figure 1. Schematic diagram of the test rig.

3. Results and discussion

3.1. General results

The HRRs for all the configurations are averaged among the repeated tests, and the ultimate results together with the cases without the ceiling are plotted in Figure. 2. It can be seen that the HRRs for cases with varying ceiling height exhibit quite different combustion process. The ceiling board significantly affects the buoyancy-driven surrounding flow and then the burning behaviours. For the cases without the ceiling, the combustion process shows the typical characteristics of the thin-layer pool fire, where four stages can be clearly identified, i.e. pre-burning stage, quasi-steady burning stage, boiling stage and decay stage [11-12]. The cases with the ceiling height of 35 cm also experience the similar burning process because the flame height is comparable to or slightly higher than the ceiling height, and thus the burning process will not be influenced by the ceiling markedly. Meanwhile, the presence of the ceiling definitely imposes restriction on the entrainment in the test, leading to a prolonged burning duration.

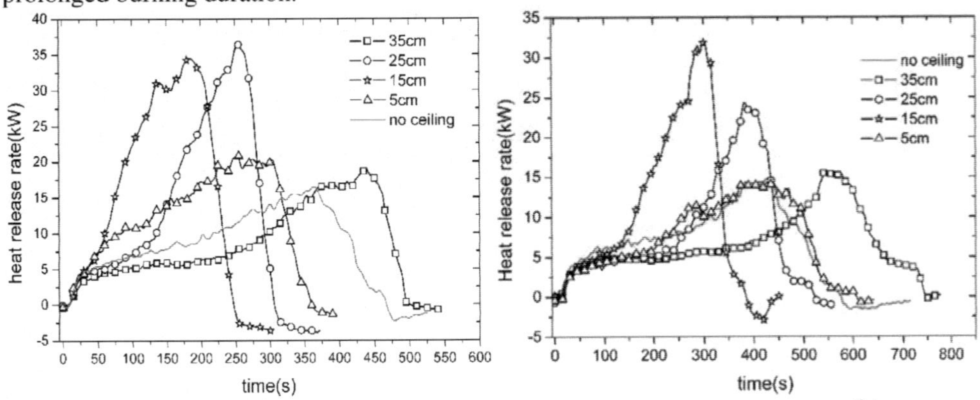

(a) (b)

Figure 2. HRRs of all the configurations: (a) N-heptane; (b) Jet-A

With the decreasing ceiling height, the quasi-steady burning stage for the thin-layer pool fire gradually disappears, and the HRRs sharply increase from the beginning of the combustion, as observed for the curves of cases with ceiling height of 25 and 15 cm. As a result of the flame extension beneath the ceiling, the heated ceiling will become hotter and thus generate a stronger radiation to the fuel [9]. Consequently, the boiling phenomenon occurs soon after the ignition. Previous researchers [13] indicated that a lower value of latent heat of gasification should be expected when the fuel boiled. Thus, the continuously increasing HRRs can be observed. It is worthwhile to note that the maximum HRR for the cases with ceiling height of 15 cm appear earlier than that of 25 cm, implying a higher fire risk. However, as the ceiling height decreases to 5 cm, the burning intensity seems to be weakened, and no sharp increase can be seen for the two fuels. This may be attributed to the effect of ceiling on the air entrainment when the fire source is extremely close to it.

3.2. Analysis on the results

The averaged HRR for a fixed configuration is defined as the average value over its whole combustion duration, expressed as

$$\bar{\dot{Q}} = \left(\int_0^t \dot{Q} dt \right) / t \tag{4}$$

The mass losses for all the configurations can be obtained by the electronic scale, and similarly, the averaged mass loss rate (MLR) can be calculated. The maximum and averaged HRRs of each test together with their corresponding MLRs are listed in Table 2, where AVE and MAX represent the average and maximum values, respectively. The combustion efficiencies η are also calculated by the following equation and listed in the table.

$$\eta = Q/(m \cdot \Delta H_c) \tag{5}$$

where ΔHc is the combustion heat of the fuel. Here, ΔHc=48 kJ/g for n-heptane [14], and ΔHc=43.5 kJ/g for Jet-A [15]. From the Table 2, the quantitative results can well coincide with the discussion in Section 3.1. In addition, the combustion efficiencies for maximum and average results of different configurations vary slightly, which also testifies the well-ventilation experimental condition in current study. The combustion efficiencies of n-heptane are always higher than that of jet-A, which may be explained by the different sooting levels of the two fuel. N-heptane is assigned to be moderately-sooting fuel, while jet-A is a highly-sooting fuel with lower combustion efficiency [16].
Based on the experimental findings, the ceiling height will undoubtedly impact the average HRR of a pool fire. To well understand the relationship between them, two dimensionless parameters are employed, i.e. dimensionless ceiling height *H/D* and a different dimensionless average HRR Q^*_{DH} concerning the effect of ceiling height and pool dimension [17].

Table 2. Experiment values for all the configurations

Fuel	Ceiling height (cm)	MAX-HRR (kW)	MAX-MLR (g/s)	η	AVE-HRR (kW)	AVE-MLR (g/s)	η
N-heptane	No ceiling	16.88	0.45	0.78	9.83	0.24	0.83
	35	19.16	0.47	0.84	9.14	0.24	0.81
	25	37.58	0.86	0.91	14.01	0.33	0.88
	15	35.23	0.82	0.89	20.37	0.45	0.92
	5	20.87	0.53	0.82	12.9	0.32	0.85
Jet-A	No ceiling	15.44	0.47	0.76	8.16	0.26	0.72
	35	15.79	0.44	0.82	6.84	0.21	0.75
	25	24.39	0.70	0.80	8.73	0.28	0.72
	15	31.84	0.86	0.85	12.88	0.45	0.66
	5	14.26	0.47	0.70	8.02	0.27	0.68

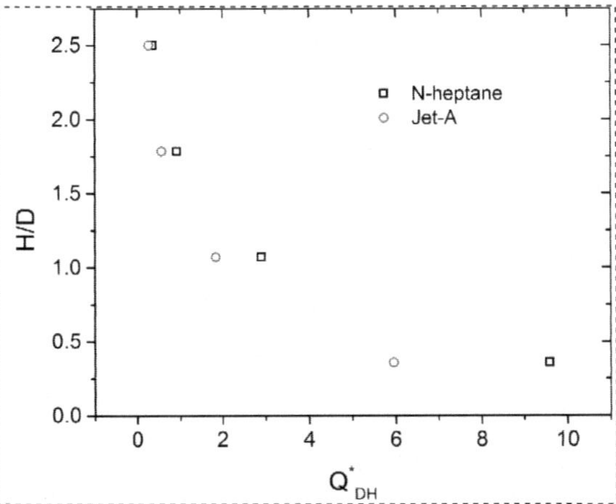

Figure 3. The relationship between H/D and Q^*_{DH}.

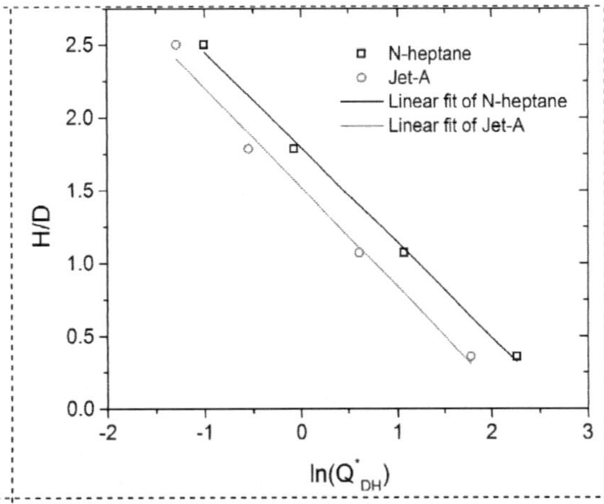

Figure 4. The relationship between H/D and $\ln(Q^*_{DH})$.

$$Q^*_{DH} = \bar{\dot{Q}} / \left(\rho_\infty T_\infty c_p \sqrt{g} D H^{3/2}\right) \tag{6}$$

Figure. 3 plots *H/D*versus Q^*_{DH}, where an exponential relationship can be observed. Furthermore, considering the correlation of *H/D* vs $\ln(Q^*_{DH})$, the relevance of them is presented in Figure 4. The linear fittings for the two fuels are successfully achieved with R^2= 0.995 and R^2= 0.985 for n-heptane and jet-A, respectively.

Moreover, as observed in Figure. 4, the slopes for the two fuels are quite similar, i.e. -0.65 for n-heptane and -0.68 for jet-A, which are approximate to -2/3. Therefore, the relationship between *H/D* and Q^*_{DH} can be summarized as

$$H/D = -\frac{2}{3}\ln\left(Q^*_{DH}\right) + C_1 \tag{7}$$

This expression can be further rearranged as

$$Q^{*2/3}_{DH} \cdot \exp(H/D) = C_2 \tag{8}$$

where C_1 and C_2 are the constants depending on the fuel, and C_2 equals to 6.0 and 4.6 for n-heptane and jet-A in current study, respectively. The data in this research are a little limited, and more experiments in such scale or larger scale with different pool dimensions and fuels are necessary to validate the equation (8) in future research.

4. Conclusions

The effect of ceiling height on the combustion behavior of a pool fire whose flame height is comparable or larger than the ceiling height was experimentally examined in current study by means of a scaled-down cone calorimeter. The HRR and corresponding MLR were quantitatively analyzed. The major results are summarized as follows:

1) The quasi-steady burning stage disappears with the reduced ceiling height, and instead, the HRR sharply increases from the beginning of the combustion, which may be attributed to the

enhanced heat feedback from the ceiling. As the ceiling height decreases to 5 cm, the HRR is mitigated due to the restriction of air entrainment.

2) The dimensionless ceiling height has a linear relationship with the logarithm value of the dimensionless averaged HRR for the two given fuels with the similar slope of -2/3. The expression of $Q_{DH}^{*2/3} \cdot \exp(H/D)$ is found to equal to a constant for a given fuel.

5. References

[1] B. Karlsson and J. G. Quintiere, *Enclosure fire dynamics*. (Boca Raton, USA: CRC Press), 1999.

[2] W. W. Yuen and W. K. Chow, The role of thermal radiation on the initiation of flashover in a compartment fire,*International Journal of Heat and Mass Transfer*, **47**, Sep. 2004, pp. 4265-76, doi:10.1016/j.ijheatmasstransfer.2004.05.017.

[3] H. Z. You and G. M. Faeth, Ceiling heat transfer during fire plume and fire impingement,*Fire and Materials*,**3**, Sep. 1979, pp. 140-7, doi:10.1002/fam.810030305.

[4] X. C. Zhang, L. H. Hu, W. Zhu, X. L. Zhang and L. Z. Yang, Flame extension length andtemperature profile in thermal impinging flow of buoyant round jet upon a horizontal plate,AppliedThermal Engineering, 73, Dec. 2014, pp.1522,doi:10.1016/j.applthermaleng.

[5] G. Heskestad and M. A. Delichatsios, The initial convective flow in fire,*Symposium (International) on Combustion, Elsevier*, 1979, pp. 1113-23, doi:10.1016/S0082-0784(79)80106-X.

[6] D. Gross, Measurement of flame lengths under ceilings,*Fire Safety Journal*, **15**, Jan. 1989, pp. 31-44, doi:10.1016/0379-7112(89)90046-5.

[7] Y. Z. Li, B. Lei and H. Ingason,The maximum temperature of buoyancy-driven smoke flow beneath the ceiling in tunnel fires,*Fire Safety Journal*, **46**, May 2011, pp. 204-10, doi:10.1016/j.firesaf.2011.02.002.

[8] Z. H. Zhou, W. Yao, H. H. Li, C. H. Lin, T. Wu and J. Wang, Experimental study on Jet-A pool fire at high altitude,*Fire Safety Science*,**11**, 2014, pp. 510-6, doi:10.3801/IAFSS.FSS.11-510.

[9] J. Ji, C. G. Fan, Y. Z. Li, H. Ingason and J. H. Sun, Experimental study of non-monotonous sidewall effect on flame characteristics and burning rate of n-heptane pool fires,*Fuel*, **145**, Apr. 2015, pp. 228-33, doi:10.1016/j.fuel.2014.12.085.

[10] M. Y. Chen, D. C. Zhou, X. Chen, W. X. Zhang, J. H. Liu, R. Yuen R and J. Wang, Investigation on the thermal hazards of 18650 lithium ion batteries by fire calorimeter,*Journal of Thermal Analysis and Calorimetry*,**122**, Nov. 2015, pp. 755-63, doi:10.1007/s10973-015-4751-5.

[11] J. H. Liu, Y. P. He, Z. H. Zhou, W. Yao, R. Yuen and J. Wang, The burning behaviors of pool fire flames under low pressure,*Fire and Materials*, **40**, Mar. 2016, pp. 318-34, doi:10.1002/fam.2289.

[12] J. H. Liu, Y. P. He, Z. H. Zhou, R. Yuen and J. Wang, Investigation of enclosure effect of pressure chamber on the burning behavior of a hydrocarbon fuel,*Applied Thermal Engineering*, in press, doi:10.1016/j.applthermaleng.2016.02.123.

[13] J. Q. Zhang, S. X. Lu, Q. Li, R. Yuen, M. Yuan and C. H. Li, Impacts of elevation on pool fire behavior in a closed compartment: a study based upon a distinct stratification phenomenon,*Journal of Fire Sciences*, **31**, Mar. 2013, pp. 178-93, doi:10.1177/0734904112460203.

[14] X. K. Hu, Y. P. He, Z. H. Li and J. Wang, Combustion characteristics of n-heptane at high altitudes,*Proceedings of the Combustion Institute*, **33**, 2011, pp. 2607-15, doi:10.1016/j.proci.2010.07.025.

[15] J. G. Quintiere, *Fundamentals of fire phenomena*, (England: John Wiley), 2006.

[16] J. H. Liu, P. Li, M. Y. Chen, X. Chen, R. Yuen and J. Wang, Experimental study on burning behaviors of liquid fuels with different sooting levels at high altitude,*Thermal Science*, in press, doi:10.2298/TSCI151218100L.
[17] H. W. Ding and J. G. Quintiere, An integral model for turbulent flame radial lengths under a ceiling,*Fire safety journal*, **52**, Aug. 2012, pp. 25-33, doi:10.1016/j.firesaf.2012.03.008.

Acknowledgments
This research was supported by the Major Projects of Civil Aviation of China (No. MHRD20130103) and the grant from the Research Grant Council of the Hong Kong Special Administrative Region, China (contract grant number CityU 11301015). The authors deeply appreciate that.

Pre-chlorination contact time and the removal and control of *Microcystis aeroginosa* in coagulation

T Lapsongpon[1], S Leungprasert[1] and C Yoshimura[2]

[1] Department of Environmental Engineering, Kasetsart Umiversity, Bangkok, Thailand
[2] Department of Civil Engineering, Tokyo institute of technology, Tokyo, Japan

Abstract. The use of pre-oxidation is known to improve algae removal by coagulation and control the growth of algae. The contact time between oxidants and algae in pre-oxidation stage has been found as important parameter. This study investigated the effect of pre-chlorination contact time on the control and removal of cyanobacteria *Microcystis aeruginosa* by coagulation. The results showed that when the alum dose was sufficient, increasing contact time showed an improvement in algae removal by coagulation in case of high chlorine dose. The algae removal ratio at high chlorine dose, 3 mg L-1 increased when contact time increased and it decreased after 30 minutes of contact time. In contrast, the result from chlorine dose, 2 mg L-1, showed an unclear trend when contact time increased. Adding 2 mg L-1 of pre-chlorination with 10 minutes of contact time was enough to control the regrowth of *M. aeruginosa*. In addition, dissolve organic carbon (DOC) and UV absorbance at 254 nm, which particulary indicates aromatic compounds, tended to increase when the contact time increased. The increased of DOC and UV 254 indicated the release of intracellular organic matter (IOM) from *M. aeruginosa*. High level of DOC, 0.68 mg L-1 in this study showed negative effect on *M. aeruginosa* removal by coagulation and could not be removed by coagulation process.

1. Introduction

Eutrophication has caused algae-related problems in many water treatment facilities. The major algae-related problems in water treatment plant are unpleasant tastes, odors and filter clogging [1, 2]. Filter clogging by algae has been reported as a technical problem by many water treatment plants (WTPs). For example, Cheong Ju water treatment plant in South Korea faced the decrease of filter run time from 20 hours to below 5 hours by the occurrence of diatom, Synedra [3]. Moreover, the presence of filamentous diatom, *Melosira* sp., at Wahnbach Reservoir, Germany, resulted in reduction of the filter run time from 30 to 8 h, which was further reduced to 4 h as a result of a simultaneous influx of smaller cyanobacteria *C. naegelianum* Unger [4]. The bloom of blue green algae, particularly *Anabaena* sp. and *Microcystis* sp. seriously impacted on filtration process in Morton Jaffray (MJ) water as evidenced by the increased backwashing frequency reported to be at every 4–8 h [5].

General methods to remove algae are physical processes such as filtration, membrane filtration, sonication and adsorption by activated carbon, biological processes such as activated sludge and chemical processes such coagulation and chlorination. Among these methods, the chemical treatment process has been considered to be cost-effective way because many chemical agents are inexpensive and they usually do not require plants to significantly change work-flow and structure [6, 7]. A number of studies have reported that pre-treatment using oxidants such as ozone, chlorine, potassium permanganate and potassium ferrate can improve algae removal by coagulation [4, 5, 8]. Oxidants has the ability to enhance micro-flocculation and to reduce conventional coagulant dosages [9]. It was found that the change of algae cell surface or cell stability and the lysis of algae cell caused by pre-oxidation are factors leading to more effective coagulation [8]. In addition, cellular component released from algae caused by oxidation was found as coagulant aids that can promote coagulation [10, 11].

Contact time between oxidants and algae in pre-oxidation stage has been found as important parameter in algae removal by coagulation. Hoko and Makado (2011) [5] conducted experiments to optimize the removal of algae, mainly *Anabaena* sp. by using jar test and copper sulfate as an oxidant, found that removal efficiency increased with increasing contact time and the optimum contact time was around 30 minute. As pre-oxidation has been found to be associated with the change of algae cell architecture and the release of algogenic organic matter (AOM), which consists of intracellular organic matter (IOM) and extracellular organic matters (EOM), serving as coagulant aid [4, 10-12], this could be the reason why the above study found that the removal efficiency increased with increasing contact time [5]. According to the above mechanisms, the inappropriate contact time can severely damage cell or cause cell lysis releasing intracellular organic matter (IOM) and toxins in case of toxins producing algae such as cyanobacteria and then increase the levels of dissolved organic matter (DOM) [12, 13] which might be as precursor of disinfection by products. The large amount of intracellular organic matter (IOM) released by chlorination such as protein can show positive effect or negative effect depending on its concentration, composition and characteristics such as molecular weight (MW) distribution. The negative effect such as forming protein–coagulant complexes which inhibit algae removal and increase alum demand [12].

The effect of contact time of pre-chlorination, which is a common oxidant, on algae removal by coagulation is still not well study and it is practical to know the optimum contact time of pre-chlorination. This article describes the effect of pre-chlorination contact time and concentration on the removal and control of *Microcystis aeruginosa* (Kützing) Lemmermann. In addition, we also determined the effect of the contact time on dissolved organic carbon (DOC) concentration and UV absorbance at 254 nm, which specifically indicates aromatic compounds.

2. Methodology

Microcystis aeruginosa (Kützing) Lemmermann, NIES-843 was purchased from The National Institute For Environmental Studies (NIES) and cultured in an incubator with following condition: 25 C, light intensity: 20-30 µmol photons/m^2/sec and Light(L)/Dark(D) cycle: 10L:14 D. The density of algae was counted under an optical microscope and harvested during the exponential growth phase. The algae was centrifuged at rate 700 G 5 minutes to separate culture medium and algae. Then, only algae was transfered to synthetic medium. The final algae density subjected to the experiments was around 80×10^6 cell L^{-1}.

Purified water was used as sample water. Turbidity, alkalinity and pH of sample were adjusted to average values during algae bloom in Chao phraya river, which were around 30-35 NTU, 100 mgL^{-1} and 7.5, respectively. Alkalinity and turbidity were adjusted by adding sodium bicarbonate ($NaHCO_3$) and kaolinite clay, respectively. Hydrochloric acid (HCl) and sodium hydroxide (NaOH) were used for pH adjustment. Chlorine solution for pre-chlorination was made by dissolving calcium hypochlorite ($Ca(ClO)_2$). Stirrer was used to perform pre-chlorination and coagulation experiment. Chlorine solution was added prior to coagulation process and mixed at 150 rpm with different contact time, including 0, 5, 10, 20, 30 and 60 minutes. The zero contact time in this study meant adding chlorine and alum together at the same time. After pre-chlorination, sample was collected immediately at the depth 5 cm below water surface for DOC and UV measurement. Then, alum solution was added and mixed at 200 rpm for 1 minutes and 40 rpm for 15 minutes. After 10 minutes of sedimentation, water was sampled at the same depth, 5 cm below surface to measure DOC, UV, turbidity and algae density. The amount of algae was counted under an optical microscope. The dose of alum, 30 mg L^{-1} was decided based on the pre-liminary result of coagulation experiment because it was found as an optimum dose for turbidity removal. To investigate the algae re-growth experiment, chlorine solution,

2 and 3 mg L^{-1} were added and mixed at 150 rpm with 10 and 30 minutes of contact time. After that, sample was collected and then centrifuged to remove chlorine solution. After removing chlorine solution and washing algae, algae was transferred to culture medium and cultured in incubator. The algae's density was periodically checked.

pH was measured by electromatic method using HM-25R pH meter, DKK-TOA Corp., Tokyo, Japan. Turbidity was measured by nephelometric method using turbidity meter, 2100Q - HACH. DOC and UV254 were measured by UV spectrometer and TOC analyzers (A non-purgeable dissolved organic carbon method was employed) [14].

All analyses were conducted in triplicate. Removal ratio results were tested with oneway ANOVA to determine the significant effect of contact time.

3. Results and discussions

3.1. *Effect of pre-chlorination dose and contact time on the removal of M. aeruginosa and turbidity*

Pre-chlorination with chlorine 2 and 3 mg L^{-1} did not significantly improve M. aeruginosa removal by coagulation (p-value from ANOVA was higher than 0.05 in both concentration and at every contact time). However, Removal ratio of algae at chlorine 3 mg L^{-1} showed a significant change (p-value from ANOVA was 0.002) when contact time increased while removal ratio at chlorine 2 mg L^{-1} did not show a significant change with contact time (p-value from ANOVA was 0.821). Removal ratio at chlorine 3 mg L^{-1} increased with increasing contact time and reached the maximum 95.05% at contact time 30 minutes and then slightly decreased into 92.11% at 60 minutes of contact time (Figure 1). The increase and decrease of removal ratio in Figure 1 may result from the released intracellular organic matter (IOM) which has been found to show positive and negative effect on coagulation [12]. Black line in Figure 1 represents the removal ratio, which is 90.7 percent and has 3.51 of SD value, from coagulation with alum 30 mg L^{-1}.

The result of turbidity after coagulation in pre-chlorination 3 mg L^{-1} experiment significantly changed with contact time (p-value from ANOVA was 0.002). It showed a reduction trend during 0 to 30 minutes of contact time and then sharply increased at 60 minutes of contact time (Figure 1). This turbidity result conformed to algae removal ratio of pre-chlorination 3 mg L^{-1} experiment shown in Figure 1. In case of pre-chlorination 2 mg L^{-1} experiment, turbidity did not significantly change when contact time increased (p-value was higher than 0.05).

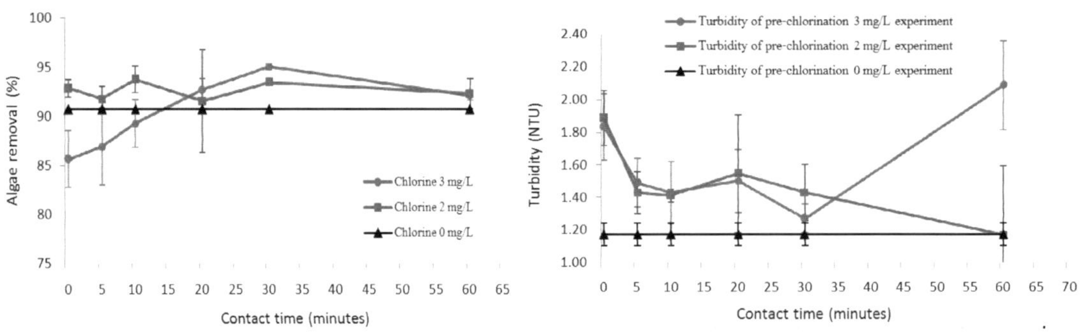

Figure 1. Relation of pre-chlorination contact time and algae removal ratio (Left) and turbidity (Light) after coagulation (at 30 mg L^{-1} of alum dose). The results are shown with average, standard deviations from triplicate experiments.

3.2. *Effect of pre-chlorination dose and contact time on the regrowth of M. aeruginosa*

IOP Conf. Series: Earth and Environmental Science **67** (2017) 012011 doi:10.1088/1755-1315/67/1/012011

Adding pre-chlorination 2 mg L^{-1} with 10 minutes of contact time was sufficient to control the growth of M. aeruginosa. As shown in Figure. 2, the density of algae remained stable after being reacted with chlorine 2 and 3 mg L^{-1}, 10 and 30 minutes of contact time while the density of algae that did not expose to chlorine increased obviously. This means chlorine damaged the algae cell resulting to algae growth inhibition and the increase of DOC concentration.

3.3. *Effect of pre-chlorination dose and contact time on DOC and UV 254*

DOC concentration before coagulation of chlorine 3 mg L^{-1} was higher than those 2 mg L^{-1} at every contact time (Figure 3). DOC concentration before coagulation of chlorine 3 mg L^{-1} increased gradually during 5 to 20 minutes of contact time and then largely increased at 30 minutes of contact time. At 60 minutes of contact time, DOC concentration remained same as contact time 30 minutes. In case of the lower chlorine dose, 2 mg L^{-1}, DOC concentration did not show important change within 30 minutes of contact time, but it showed sharp increase at 60 minutes. The increase of DOC concentration indicated the release of AOM, especially IOM as pre-chlorination has been found to damage algae cell wall and release IOM of algae [12, 13].
Absorbtion of UV at 254 has been used as organic compounds monitoring parameter, especially aromatic compounds throughout water treatment plants. In this study, UV 254 showed large increase in case of chlorine 3 mg L^{-1}, particularly at 30 minutes of contact time and then decreased at 60 minutes of contact time (Figure 3). The decrease of UV 254 at 60 minutes might be resulted from the degradation of high MW organics to lower MW organics by chlorine [12, 13]. In addition, UV 254 value of chlorine 2 mg L^{-1} showed no large different when contact time increases. Since UV 254 specially indicates aromatic compounds, so not only conatact time but the dose of chlorine also affects characteristics of organic matter.

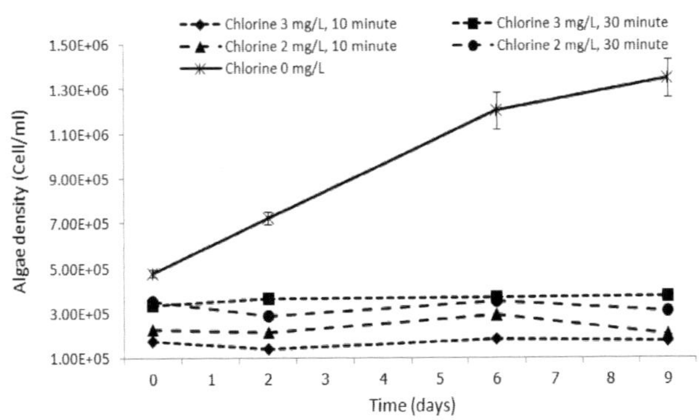

Figure 2. The growth of algae after exposing to chlorine. The results are shown with average and standard deviations from triplicate experiments.

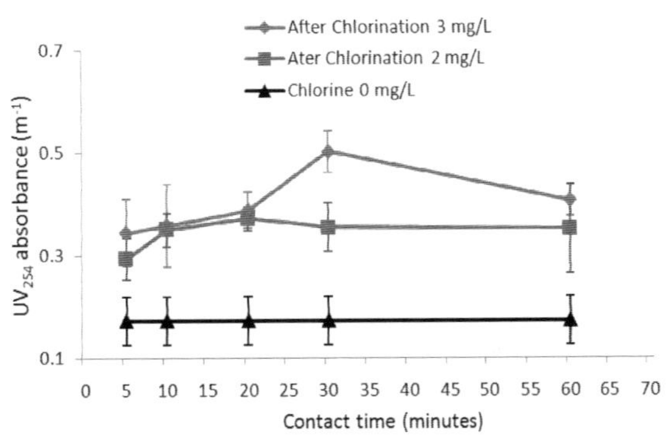

IOP Conf. Series: Earth and Environmental Science **67** (2017) 012011 doi:10.1088/1755-1315/67/1/012011

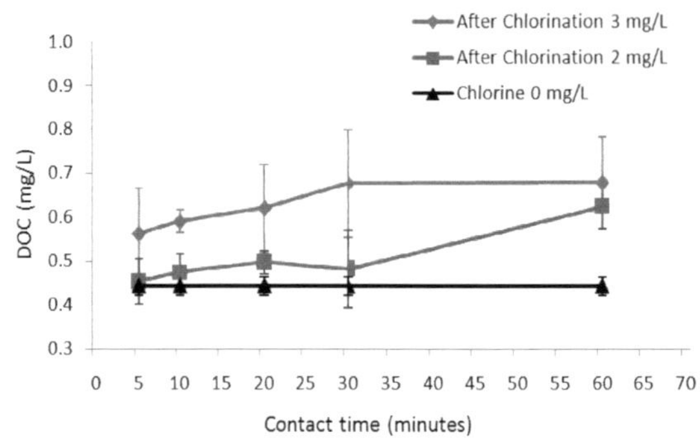

Figure 3. Relation of pre-chlorination contact time and DOC concentration (Left) and UV 254 (Right) before coagulation (at 30 mg L^{-1} of alum dose). The results are shown with average and standard deviations from triplicate experiments.

According to the result of DOC concentration, the trend of DOC and UV were similar, especially at 30 minutes of contact time of chlorine 3 mg L^{-1} which shows obviously increased in both DOC and UV 254.

The intracellular organic matter (IOM) has been found to affect coagulation process and may influence coagulation performance for removing algae in this study as shown in Figure 4. The removal ratio and DOC increased when contact time increase d umtil 30 minute s of contact time. After that , removal ratio show ed decreasing while DOC remained stable. The increase and decrease trend of algae removal ratio could be influenced by DOC concentration as it has been found that the low level of DOM benifited coagulation while high level of DOM inhibited the aggregation of algae cells and increased the alum dose [12, 15]. In addition, the low ratio of coagulant/DOC was found to decrease coagulation performance by causing charge stabilization [16]. Base on the above result, the high level of DOC in this study showed negative effect on algae removal.

The results of DOC concentration after coagulation in Figure 5 shows that low level of DOC, produced in chlorine 2 mg L^{-1} experiment, was removed by coagulation process to the same level of blank experiment.

In contrast, high level of DOC in chlorine 3 mg L^{-1} experiment remained high after coagulation process (Figure 5). This means 30 mg L^{-1} of alum dose in this experiment was not sufficient to remove high level of released organic matter.

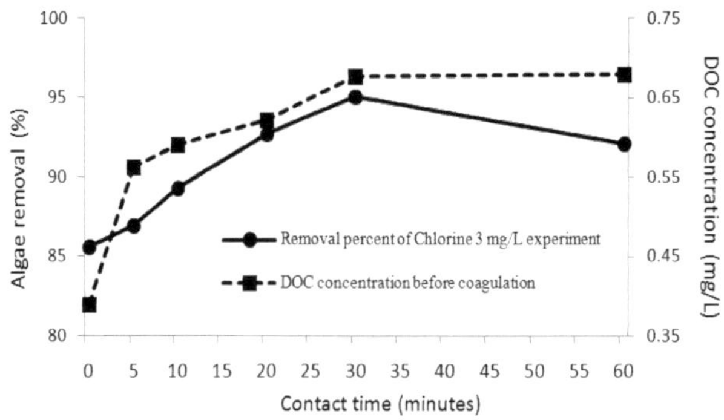

Figure 4. Relation between pre-chlorination contact time, DOC concentration (before coagulation) and algae removal (at 30 mg L^{-1} of alum dose). DOC value of 0 contact time is value of water containing algae.

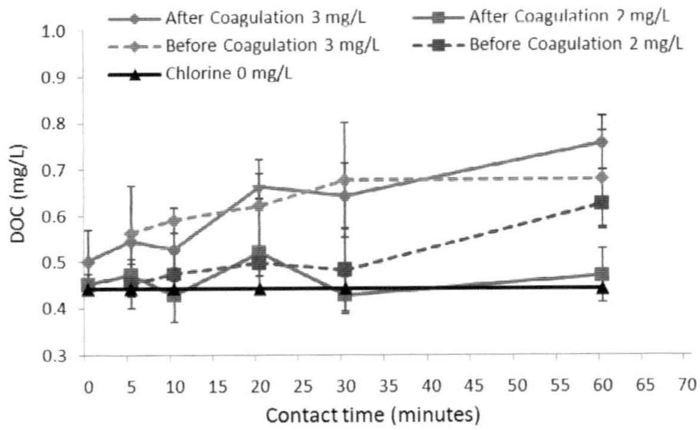

Figure 5. Relation of pre-chlorination contact time and DOC concentration (before and after coagulation at 30 mg L^{-1} of alum dose). The results are shown with average and standard deviations from triplicate experiments. (There is no result of before coagulation at zero contact time because that contact time is adding alum and chlorine at the same time).

4. Conclusion

Adding 2 and 3 mg L^{-1} of pre-chlorination did not show a significant improvement in *M. aeruginosa* removal by coagulation. However, pre-chlorination contact time showed a significant effect on *M. aeruginosa* removal ratio at chlorine 3 mg L^{-1}. The success in algae growth control and the increase of DOC concentration supported that chlorine damaged algae cell and released AOM, especially IOM. According to results from DOC, and UV 254, pre-chlorination contact time in this study affected removal ratio of *M. aeruginosa* by increasing organic level, especially 3 mg L^{-1} of chlorine at 20–60 minutes of contact time. The increase of organic concentration increased and decreased the removal ratio as shown in Figure 6. In addition, 30 mg L^{-1} of alum was sufficient for removing algae but not sufficient to remove organic compounds produced at high chlorine dose. Moreover, due to the increase of disinfection by product precursor found in this study, the pre-chlorination contact time should less than 30 minutes when high chlorine dose is used.

5. References

[1] Knappe D. Water Resources Research Institute of the University of North, and A.R. Foundation, *Algae detection and removal strategies for drinking water treatment plants*. 2004, Denver, CO: AWWA Research Foundation : American Water Works Association. xlviii, 466 p.

[2] Westrick JA, Szlag DC, Southwell BJ and Sinclair J. A review of cyanobacteria and cyanotoxins removal/inactivation in drinking water treatment. *Anal Bioanal Chem*, 2010. **397**(5): p. 1705-14.

[3] Jun H, Lee Y, Lee B and Knappe D. *E*ffectiveness of coagulants and coagulant aids for the removal of filter-clogging Synedra. *Aqua*, 2001. **50**: p. 135-148.

[4] Henderson R, Chips M, Cornwell N, Hitchins P, Holden B, Hurley S, et al. Experiences of algae in UK waters: a treatment perspective. *Water and Environment Journal*, 2008. **22**(3): p. 184-192.

[5] Hoko Z and Makado PK. Optimization of algal removal process at Morton Jaffray water works, Harare, Zimbabwe. *Physics and Chemistry of the Earth*, Parts A/B/C, 2011. **36**(14–15): p. 1141-1150.

[6] Shen Q, Zhu J, Cheng L, Zhang J, Zhang Z and Xu X. Enhanced algae removal by drinking water treatment of chlorination coupled with coagulation. *Desalination*, 2011. **271**(1-3): p. 236-240.

[7] Wu Z, Shen H, Ondruschka B, Zhang Y, Wang W and Bremner DH. Removal of blue-green algae using the hybrid method of hydrodynamic cavitation and ozonation. *J Hazard Mater*, 2012. **235-236**: p. 152-8.

[8] Plummer JD and Edzwald JK. Effects of chlorine and ozone on algal cell properties and removal of algae by coagulation. *Journal of Water Supply: Research and Technology - Aqua*, 2002. **51**(6): p. 307-318.

[9] Shehata S, Badr S and Wahba S. Drinking Water Treatment Options for Eliminating Freshwater Algae. *International Journal of Environmental Studies*, 2002. **59**(6): p. 679-688.

[10] Ma, J. and W. Liu. Effectiveness and mechanism of potassium ferrate (VI) preoxidation for algae removal by coagulation. *Water Research*, 2002. **36**(4): p. 871-878.

[11] Petrusevski vB, Van Breemen A and Alaerts G. Effect of permanganate pre-treatment and coagulation with dual coagulants on algae removal in direct filtration. *Aqua*, 1996. **45**(6): p. 316-326.

[12] Ma M, Liu R, Liu H, Qu J and Jefferson W. Effects and mechanisms of pre-chlorination on Microcystis aeruginosa removal by alum coagulation: Significance of the released intracellular organic matter. *Separation and Purification Technology*, 2012. **86**: p. 19-25.

[13] Ma M, Liu R, Liu H and Qu J. *Chlorination of* Microcystis aeruginosa suspension: cell lysis, toxin release and degradation. *J Hazard Mater*, 2012. **217-218**: p. 279-85.

[14] Association, A.P.H., et al., *Standard methods for the examination of water and wastewater*. Vol. 2. 1915: American Public Health Association.

[15] Bernhardt H, Hoyer O, Schell H and Lüsse B. Reaction mechanisms involved in the influence of algogenic organic matter on flocculation. *Zeitschrift für Wasser-und Abwasser-Forschung*, 1985. **18**(1): p. 18-30.

[16] Pivokonsky M, Safarikova J, Bubakova P and Pivokonska L. Coagulation of peptides and proteins produced by Microcystis aeruginosa: Interaction mechanisms and the effect of Fe–peptide/protein complexes formation. *Water research*, 2012. **46**(17): p. 5583-5590.

Acknowledgment

This work was supported by JSPS Core-to-Core Program, B. Asia-Africa Science Platforms. The authors would like to thank senior members in Water Environemental Lab, Tokyo Institue of Technology, Kornravee SAIPECH, Mohamed Ateia IBRAHIM, NATSUIKE Masafumi and others members who always support this study.

IOP Conf. Series: Earth and Environmental Science **67** (2017) 012012 doi:10.1088/1755-1315/67/1/012012

Hazardous waste minimization challenge in autocomponent industry, West Java, Indonesia

L Handayani[1] and S S Moersidik[2]
1.School of Environmental Science, University of Indonesia, Salemba, Jakarta, Indonesia
2.Environmental Engineering Program, Departemen of Civil Engineering, Faculty of Engineering. University of Indonesia, Depok, West Java, Indonesia
E-mail: 1. lidyahandayani008@gmail.com, 2. ssarwanto@eng.ui.ac.id

Abstract. Modern industries have managed their hazardous waste through hazardous waste management with End of Pipe approach. As part of the most robust industry, autocomponent industry have to manage their hazardous waste from production process. To meet sustainable manufacturing, waste minimization is required. Hazardous waste minimization in practice is relatively difficult to implemented. This paper explore hazardous waste management and waste minimization activity in one of autocomponent industry in Indonesia. Hazardous waste minimization regulation also explain in this paper. Regarding waste minimization implementation there were some obstacle such as lack of awareness and knowledge, lack of time and economic factor

1. Introduction

The Automotive sector is one of the most robust industries in Indonesia and the country ranks 15th in world in term of sales. The industry has steady domestic growth averaging 15% in 2010 as is projected to continue to growth till the end of decade [1]. This industry growth is parallel autocomponent industry development. Autocomponent industry in its production plant process uses raw material and chemicals component that generated hazardous waste. Hazardous waste burden give impact to water and soil that potentially harmful to human health. Modern industries have managed their hazardous waste through hazardous waste management with *End of Pipe* approaching at end of the process stage. This approach tends to ignore various kinds of environmental problem that appeared during operational process. On the other hand, this methode is not effectively implemented by the manufacturing industry [2].

To meet the sustainable manufacturing, waste minimization is one of the best technique in the industry through efficiency of raw material, energy and natural resource in process. Some automotive industries has waste minimization program but there were some industries more focused in delivery, quality and cost.According to the Environment Protection Agency waste minimization refers to strategies that are aiming to prevent waste at source through upstreaminterventions[3]. Waste minimization program in the industry affected by government regulation, industrial behavior and strategy related with awareness of the benefits from waste minimization program [4].

The concept of waste minimization look easy, but in practice is relatively difficult due to factors not directly related such as government regulation, quality human resources from various disciplines and the most important is the strong commitment from top management [5]. Availability of technology and facility for waste minimization can affect the social behavior as well as pro environmental atitudes [6]. Nevertheless waste management behavior might have been motivated to avoid the cost for waste

disposal, not due to the proenvironmental attitude. Assem and Karima revealed that indirect causes of waste were found to be the lack of legal and contractual incentives [7]. The ability of operative contribute to a waste reduction in industry as good as level of management support offered[8].
Some research of waste minimization in Indonesia has been conducted in Pharmaceutal industry. Waste minimization effort could be implemented by reduction of waste at source, modification of raw material and process, and water saving during production process [9]. Waste minimization in paint industry applied by housekeeping, reused of used water and segregation of waste contaminated and uncontaminated[10].
In this paper, we discuss current status of hazardous waste management and implementation of waste minimization in autocomponent industry, in one of autocomponent Industry in Indonesia. We summarize the role of government regulation in hazardous waste management in autocompanent industry. We also represent barrier of waste minimization implementation in industry.

2. Methode

Hazardous waste minimization data in Autocomponent Industry obtained through observation in industry. The hazardous waste management implementation reviewed based on Indonesia government regulation. Information about barrier of hazardous waste minimization come from structured questionnaire collected from employees in indutry and semi dept interview toward middle management in industry.

3. Discussion

3.1 Minimization hazardous waste regulation

Indonesia has Government Regulation number 101 year 2014 regarding Hazardous Waste Management. Based on article 1 in this regulation definition of hazardous waste is residue of a business activity that contain hazardous and toxic substance. In article 3, any person who generates hazardous waste will be required to manage the hazardous waste generated. In other hand, waste reduction is obligated for every person who generates hazardous waste. Waste reduction conducted through material substitution, process modification, and environmental friendly technology. Material substitution implemented by eliminated toxic material. For process modification set up through implementation of more efficiency production process. This hazardous waste reduction mechanism revealed in article 10 about Hazardous waste reduction
Hazardous waste that generated collect in hazardous storage. Guideline for storage and collection of hazardous waste revealed in this regulation, however detail explaination about this guidance exist in Decree of Head of BAPEDAL: 01/BAPEDAL/09/1995. Duration of hazardous storage in this store not more than 90 days. For treating hazardous waste, in case every body or organization industry being incapable of treating hazardous waste individually the treatment shall be entrusted to other party. Treatment of hazardous waste shall secure license from Environemental Minister. Hazardous waste transport from waste producer to treatment using manifest. Regarding manifest system of hazardous waste, the regulation noted in Decree of Head of BAPEDAL:2/BAPEDAL/09/1995. The transporter of hazardous waste needed licence to transfer hazardous waste from its source to treatment and disposal facilities.

3.2 Hazardous waste management in autocomponent industry

Autocomponent industry located in West Java Province, Indonesia. Industry has been implemented Environmental Management System (ISO 14001) and has Environmental Management and Monitoring Document license from Local Environmental Authority. Based on production activity industry generated various hazardous waste such as Scale quenching, Shotpeen Ash, Painting Sludge, Saw Contaminated, WWTP Sludge.
Hazardous waste from process collected using drum 200 litre (52 gallon). Hazardous waste then store in safe storage including in this store is fire extinguishers, secondary contaiment, eye wash, hazardous waste management standart. Duration of hazardous waste storage is 90 days. There is operator who

has task for storage cotainer and storage are inspection. Description of hazardous waste from autocomponent industry could be seen in Table 1

Table 1. Hazardous Waste from Autocomponet Industry

Type	Description	Hazard Category	Waste Code*
Scale Quenching	Steel material heated in tempering process then dipped in quenching oil	Explosive	A 323-3
Shotpeen Ash	Residue from shotpeen process,	Flammable	B323-1
Painting Sludge	Residue from paint spray mixed with water wall	Flammable	B323-2
Saw Contaminated	Absorbent of oil spill, chemical spill	Explosive	A108d
WWTP Sludge	Residue from Wastewater Treatment Plant process	Flammable	B323-5

*Categories and coding based on Government Regulation of The Republic Indonesia Numb. 101 year 2014 on Hazardous and Toxic Waste Management

Hazardous waste transportation and treatment facilities conducted off site using manifest system. Lack of inspection did by autocomponent industry during transport, treatment, and disposal due to it belief that private companies has permit that they must be complying with regulation from Environmental Ministry. Limited time and personil become barrier in hazardous waste transport, treatment and control. This uncontrol activity lead to release of hazardous waste dumping in environment. However number of hazardous waste in a year could be seen in Table 2.

Table 2. Number of Hazardous Waste from Autocomponent Industry in A year (2014-2015)

Type	Sceal Quenching	Shotpeen Ash	Painting Sludge	Chemical absorben	WWTP Sludge
Volume (tonnase)	69,84	10,8	44,95	0,475	12,8

Based on Table 2. Total all of hazardous waste generated from Autocomponent industry were 138,865 Ton in a year.The most dominant hazardous waste generated was scale quenching with a composition of 50%, then followed by 33% of painting sludge volume

3.3 Waste Minimization Implementation

Waste minimization includes source reduction and recycling. Source reduction is defined as any activity that reduces or eliminates the generation of waste at the source, usually within a process. Recycling is defined not only as recovery and or reuse of what would otherwise be a waste material but also quality improvements [11],[12].

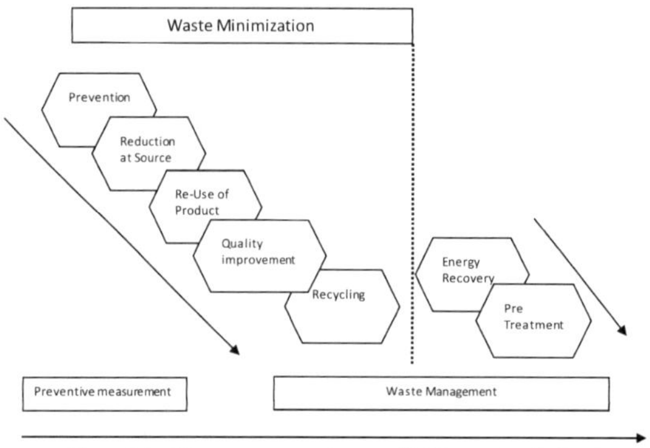

Figure 1. Graphic Representation of Waste Minimization

In production process, waste minimization strategy focuses on optimization of raw material, energy use and reduction chemical material usage. In general, primary route of waste reduction from its source are [13]:

1. Decreasing or eliminating the amount or toxicity of material used in the manufacture and packaging of product
2. Redesigning products for increased life span, reusability, and repairability
3. Changing purchasing decisions to favor those products that have minimized residual toxicity and waste associated with them
4. Modifying patterns of consumption and material use in a way that reduces the amount and toxicity of waste generated

Hazardous waste reduction from source has been implemented on production process with various approaches. Hazardous waste minimization implementation in autocomponent industry could be seen in Table 3

Table 3. Hazardous Waste Type and Its Waste Minimization Approach

Methode	Waste Type				
	Chemical Absorbant	Painting sludge	Sceale Quenching	Shotpeen Ash	WWTP Sludge
Raw Material Management	X	X	X	X	X
Setting Process Condition	X	O	O	O	X
Technology Modification	X	X	X	X	X
Good Housekeeping	O	X	X	X	X
Off site Treatment	O	O	O	O	O

X : Not Implemented O : Implemented

Based on hazardous waste minimization observation in industry, there were some technique with process approach such as setting process condition for painting sludge, sceal quenching, and shotpeen ash and good housekeeping for saw contaminated. However, this waste minimization strategy not measure effectively. Based on interview with production manager, methode in setting process and good housekeeping only generated minimizing hazardous waste about 10% from total. Minimum result from waste minimization implementation due to setting process modification conducted based on improvement suggestion from employees who were not expert waste minimization knowledge. Good house keeping also set up based on routine activity, there was no Plan Do Check Action in this

program. All hazardous waste such as saw contaminated, painting sludge, scale quenching, shotpeen ash, and WWTP sludge after collected on-site and stored in a segregation area based on their characteristic. Then transported off-site by a licensed regulated waste transpoter, to a licensed facility for treatment and disposal.
Autocomponent industry has been implemented Environmental Management System (EMS), it can be verified that EMS System contributes to made environmental management in industry better however hazardous waste still be managed by the end stage of production process. This condition relevant with research conducted by Cervelini and Souza that Environmental Management System certified does not establish conditions to minimize the use of *End of Pipe* technique [14]. In research ISO 14001 and waste minimization in metallurgy industry state that implementing EMS clearly brings positive changes in waste treatment, however it is insufficient to bring forth waste prevention [15].

3.4 Waste minimization obstacle
Previous studies have been conducted regarding waste minimization implemention issue, there are common issue highlight among industrial sectors such as lack of awareness and knowledge, lack of time, and economic factor.

3.4.1 Lack of awareness and knowledge
Employees awareness in waste minimization implementation are still lack due to respondents stated that waste minimization can run if carried out by certain employees who has task control the waste. This perception is consistent with research in Constraction Industry that most workers assumed that the reduction of waste are not part of their regular duties and responsibilities. Based on management statement, waste reduction activity has not been implemented properly and was not specific due to limited knowledge of employees [16]. This result in line with research measurement of waste preventive that calculation of waste minimization is complex process due to needed quality of human resources from multidiciplines and require strong commitment from top management [12].

3.4.2 Lack of time
Waste minimization implementation has not been effective due to unoptimal attention from management. Most of the industry's attention more focused on development of projects for new products based on consumer demands. Through interviews with the management, information was obtained that problems were found in the implementation of waste minimization activities were cost, time, labor constraints in an effort to conduct improvement ideas, and lack of support from the top management. This is consistent with studies that have been done in the implementation of waste minimization in various sectors of the industry that found a similar problem are the lack of awareness, not understand how to implemented it, no time [17][18].

3.4.3 Economic factor
Great motivation of Autocomponent Industry in waste reduction are cost savings however there was no budget allocation to run this program. Waste management efforts based on the cost savings also in line with the idea of waste management and recycling research in Korea in which the waste management activities over the background by the motivation to avoid the cost of elimination disposal [6]. In addition, research on some industrial contractors in the United Arab Emirates that the main benefits obtained by the industry in the application of waste minimization is to increase profits while benefiting environmental protection finished last [7].

4. Conclusion
Based on study result it can be concluded hazardous waste minimization well not developed, however regulation for industry to reduce hazardous waste generation exist. Government regulation abou

hazardous waste management required industry to eliminated toxicity material and modify production process to made activity more efficient. There were some barriers in hazardous waste implementation such as lack of awareness and knowledge, lack of time, and economic factor.

5. References

[1] Euromoney Institutional Investor Company. (2014). Automotive Sector Indonesia, December 2013

[2] Hasibuan, S. (2000). KarakteristikDukunganIndustriTerhadapUpayaImplementasiProduksiBersih. *JurnalTeknologiLingkungan, Vol.1, No.1.Januari 2000:54-62*

[3] Environmental Protection Agency (EPA). 2014. Guide for Industrial Waste Management. http://www.epa.gov accessed in May 26th, 2015

[4] Poonprasit, M., Philips, P S., Smith, A., Wirojanagud, W.,Naseby, D.(2005). *The Application of Waste Minimisation to Business Management to Improve Environmental Performance in the Food and Drink Industry*

[5] Panggabean, S.M. (2000). MinimisasiLimbahpadaPusatPengembanganPengelolaanLimbahRadioaktif. *BuletinLimbah Vol. 5 No.1* PusatPengembanganPengelolaanLimbahRadioaktif, BATAN

[6] Lee, S and Paik, HS. (2011). Korean household waste management and recycling behavior. Building and Environment 46(2011) 1159-1166. Elsevier

[7] Assem, A.H danKarima, H. (2011). Material Waste in the UAE Construction Industry Main Causes and Minimisation Practices. Heriot-Watt University

[8] Saunders, J., Construction, H., Wynn, P. (2002). Attitudes Towards Waste Minimization AmongsLabour Only Sub Contractors. Departement of Built Environment, Anglia Polytechnic University, Chelmsford, UK

[9] Zakarya, EN. (2004). MinimasiLimbahPadaIndustriFarmasi. Program StudiIlmuLingkungan. Program Pascasarjana. Universitas Indonesia. Jakarta

[10] Hernadewita., Rahman, M.N.A.,Deros, B.M. (2007). PenangananLimbahIndustri Cat DitinjaudariSisi Clean Technology dalamManajemenIndustri. JurnalTeknikMesin Vol.4, No.2, Desember 2007. ISSN 1829-8958

[11] Begum, RA., Chamhuri, Siwar., Pereira, JJ. (2009). Attitude and behavior factors in waste management in the construction industry of Malaysia. *Journal of Resources, Conservation, and Recycling* 53 (2009) 321-328

[12] Zorpas, AA danLasaridi, Katia. (2013). Measuring waste preventive. *Journal Waste Management 33 (2013) 1047-1056*

[13] Tchobanoglus, G danKreith, F. (2002). *Handbook of Solid Waste Management*. New York City. McGraw-Hill

[14] Cervellini, F. M., dan Souza, M. T. S. (2009). *A contribution of the Cleaner Production Program to the ISO 14001 Management System: A case study in the metal-mechanic sector*. Paper present at the Flagship Research Journal of International Conference of the Production and Operation Management Society

[15] Kvist, S., Pongracz, E., Keiski, R.L. (2005). *ISO 14001 and waste minimization in metallurgy industry*. Paper present at Seminar "Waste Minimization and utilization in Oulu Region". Oulu University Press, Oulu. P. 61-70.

[16] Kareem,W.A.,Asa, O.A., Lawal, M.O (2014) Resources Conservation and Waste Management Practices in Costruction Industry

[17] Doniec, A. (1995).Obstacles in Application of Cleaner Production n the Polish Industry. Faculty of Process and Environment Engineering. Technical University of Lodz. Poland

[18] Mallak, SK.,Ishak,MB.,Mohamed, AF. (2014). Waste Minimization Benefit and Obstacles for Solid Industrial Waste in Malaysia. IOSR Journal of Environmental Science, Toxicoogy and Food Technology

IOP Conf. Series: Earth and Environmental Science **67** (2017) 012013 doi:10.1088/1755-1315/67/1/012013

Evaluation of Mismatch Losses due to Shunts in industrial Silicon Photovoltaic Modules

P Somasundaran[1], M Shilpi[2] and R Gupta[1]

[1] Department of Energy Science and Engineering, Indian Institute of Technology, Bombay, Mumbai, 400076, India.
[2] Department of Electrical and Electronics Engineering, Government Engineering College, Wayanad, Mananthavady, 670644, India.
Email: somasundarankerala@gmail.com

Abstract. In order to achieve higher efficiencies in photovoltaic module technology, it is important to characterize the shunts and other defects which degrade the performance of cells and modules as well as decrease their efficiency. These shunts also affect the reliability of cells and modules. It is important to understand how much fill factor and power loss is caused by the presence of shunts in the module. Shunts not only reduce the module power output, but also affect the *I-V* characteristics of the cell and hence the characteristics of the shunted cells are different from those of the shunt-free cells connected in the module leading to the mismatch effect. This is an interesting effect which has been systematically investigated in the present work. Moreover, the flow of increased shunt current will give rise to increased temperature in the region of shunt, which will affect the cell and hence module performance. In the present study, the distributed diode model has been extended to the module level and applied to evaluate the electrical mismatch losses and thermal mismatch losses due to shunts in industrial Silicon PV modules.

1. Introduction

Photovoltaic cells often suffer from localized shunts, which are internal short circuits causing degradation in the cell performance and overall efficiency [1, 2]. The presence of a shunted cell within a module can severely influence the efficiency and reliability of the entire module [3, 4]. Further, the reliability of Silicon-based photovoltaic cells can be limited by thermal issues occurring in regions with high localized self-heating or hot spots born from the shunts, which are often observed to induce a strong degradation of the photovoltaic cell and of the module itself leading to a drastic reduction of the module life time and even to hazardous operating conditions [5]. The reliability issues cited above can be aggravated when a PV cell operates in reverse bias condition, which might take place when one PV cell in a module is shaded and others are still generating current [5]. In this case, a significant reverse current can flow through the shaded cell and this may lead to a premature break down and permanent degradation [6, 7], because the current is observed to flow through the shunt resistance. In the presence of a low shunt resistance, breakdown often occurs at localized sites [3].

It would be interesting to understand the possible mismatch effects arising due to the presence of a shunted cell or a number of shunted cells on the PV module performance. Since the shunts can have a dramatic influence on the *I-V* curve of the PV cell, degrading the cell's output power, open circuit voltage and fill factor, it is natural to expect a critical influence on the performance of PV module as well.

IOP Conf. Series: Earth and Environmental Science **67** (2017) 012013 doi:10.1088/1755-1315/67/1/012013

The aim of the present work is to present a systematic study of the effect of an ohmic shunt on the performance of an industrial Silicon photovoltaic module. Distributed diode model [8-11] developed based on experimentally measured parameters has been extended to simulate the PV module performance. A series connected module with 36 cells has been considered in the present work.

2. Approach and methodology

Since the shunts essentially affect the *I-V* characteristics of the solar cells, it is expected that they would cause mismatch losses in the PV module performance. The quantum of the mismatch losses due to shunts in a PV module has not been studied thoroughly as yet. Hence an approach has been devised to study the mismatch losses due to shunts. The focus will be on the part of the losses which could be possibly caused by the fact that the particular cell or cells which due to shunting operates at a lesser voltage compared to the other cells in the series connection. To take into account the fact that generally the modules are series connected and to simplify the approach, only the series connection of the module has been considered presently.

Photovoltaic modules have a large area and hence, there are a lot of parameters that possess spatial inhomogeneities like material defects, illumination intensity etc. Distributed diode model offers freedom in employing the inhomogeneities in one or all parameters and estimating the output from the PV module. Many effects can be studied and their influence on module performance can be analyzed.

Distributed diode model developed in the present work has been exploited to find the shunt currents in the shunted and non-shunted regions of the shunted cell or cells connected in the PV module. Presently, an improved 3-D model has been utilized in order to find the temperature rise. Since the most important factor which is principally affected by temperature is the reverse saturation current, its value has been found by the 3-D model and the new value is used to find the module I-V after temperature rise due to shunts.

In the distributed diode model, each PV cell is divided into smaller elementary areas and the model has been utilized to study thermal mismatch losses by computing the shunt currents flowing for various cases of shunting at the shunt locations.

The presence of shunts causes increase in leakage current. Through the simulation of distributed diode model, the currents flowing through each shunt resistance in the elementary area can be evaluated.

These shunt currents flowing in the shunt locations generates heat, which further causes an increase in temperature in that region. This temperature rise can be estimated by using an electrical equivalent circuit of thermal model, which has been explained in section 3.

3. Model

3.1. Distributed Diode Model

In the distributed diode model [10, 11], as shown in figure 1, each PV cell is divided into smaller elementary areas and each elementary area is modelled by Shockley's one diode model. These smaller areas are assumed to be homogeneous and are replaced by lumped single diode model circuits. The smaller area circuits are interconnected to make the entire model of solar cell. Each cell in a module is connected in series, which is interconnected with bus bar resistors. In this way, the homogeneity can be localized to small regions and the effects of individual region on the entire module performance can be analyzed.

For simulation of a PV cell and module using distributed diode model, each cell area is divided into 1421 elements. The parameters of the division are calculated according to the dimensions of the elementary area of the distributed diode model provided in the table 1. The simulation was carried out using PSpice circuit simulator.

3.2. 3D Thermal Model

3.2.1. Analogy between electrical and thermal quantities. The analogy between electrical and thermal quantities, from table 2 is utilized to build an electrical R-C network model that simulates the thermal response. The heat generated at the shunt locations is used in this model to find out the temperature rise at that location. The quantities are explained in table 3.

3.2.2. Electrical Equivalence of thermal model. Any material, during the heat transfer phenomenon, absorbs a part of the heat and the rest of the heat gets diffused through the material and transmits according to Fourier's law. If the material is located in the atmosphere, part of the energy is lost through convection to the surrounding air. The whole process can be modelled by a RC-network model using the analogy between electrical and thermal quantities. The thermal resistance and heat capacity of the material in figure 2 are transformed into electrical resistance and capacitance accordingly as in figure 3. The electrical parameters corresponding to the thermal quantities are given in table 4.

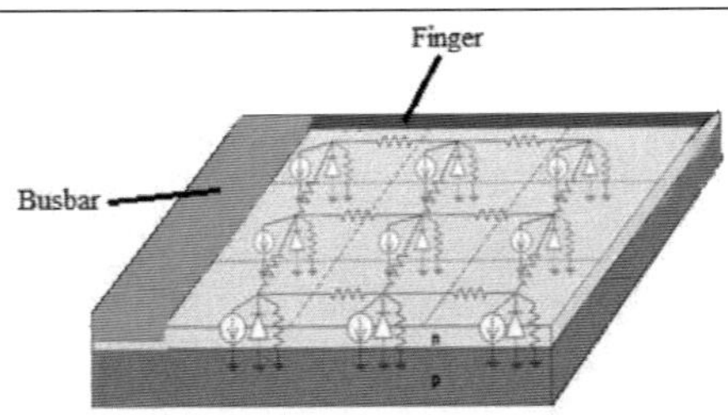

Figure 1. Distributed diode model of the solar cell showing only 3x3 elementary areas and wherein each elementary area is modelled by Shockley's one diode model.

Table 1. Simulation parameters.

Parameters	Value
Number of busbars	1
Number of fingers	14
Cell length	6.2 cm
Cell breadth	3.7 cm
Number of elementary areas	1421
Number of rows	29
Number of columns	49
Photo-generated current	4.715E-04 A
Reverse saturation current	7.83E-10 A
Ideality factor	1.7
Shunt resistance (shunt-free area)	7.815E05Ω/sq.
Sheet resistance	40 Ω/sq.

Table 2. Electrical and thermal quantities analogy.

Equations	Analogy
$I = \frac{\Delta V}{R_e}$; $q = \frac{\Delta T}{R_t}$	$V \leftrightarrow T$
$R_t = \frac{l}{kA}$	$R_e \leftrightarrow R_t$
	$I \leftrightarrow q$
$I = C\frac{dV}{dt}$; $q = m\,c_p\frac{dT}{dt}$	$C \leftrightarrow m\,c_p$

The heat that is conducted through the layers can be seen as the current flowing through the resistors. During this process the layers get heated up and some heat is stored in the layers. This storing of heat is similar to storage of charge by a capacitor when a current flows through it. The temperature rise at any location can be found from the electrical equivalence of thermal model as the voltage at that point. Steady state response is taken into consideration in the methodology adopted for solution.

3.2.3. Effect of Temperature. Rise in temperature causes change in the saturation current of the semiconductor. Even if the temperature rise at a single elementary area is small, the effect of change in saturation current, which result in considerable drop in performance, becomes significant on considering the large number of elementary areas involved in the case of an extended shunt.

The saturation current is affected by many parameters as shown in equation (1), of which the intrinsic carrier concentration is the most significant. The intrinsic carrier concentration is dependent on band gap and the temperature, as seen in equation (2). Their concentration increases as the band gap reduces, which is caused by the temperature rise. Equation (3) reveals the critical influence of temperature on the saturation current.

$$I_o = qA\frac{Dn_i^2}{LN_D} \tag{1}$$

$$n_i^2 = 4\left(\frac{2\pi kT^3}{h^2}\right)(m_c^* m_h^*)^{3/2}\, exp\left(\frac{-E_{GO}}{kT}\right) \tag{2}$$

$$I_o = B'T^3 exp\left(\frac{-E_{GO}}{kT}\right) \tag{3}$$

where I_0 is the reverse saturation current, q is the electron charge, A is area of the elementary area, D is the diffusivity of minority carriers, N_D is doping, k is Boltzmann constant, h is Planck's constant, m_c^* is the effective mass of electron, m_h^* is the effective mass of hole, n_i is the intrinsic carrier concentration, L is the diffusion length, T is temperature, B' is a constant, and E_{GO} is band gap linearly extrapolated to absolute zero.

3.2.4. Material Specifications of the PV Module. The simulation model has been formulated based on a 10W PV module whose parameters are summarized in table 1. Material specifications of the PV module modelled in the study are detailed in table 5.

Accuracy in determination of temperature rise has been improved by 3D thermal model. Distributed diode model has been exploited thereafter to determine thermal mismatch losses due to shunt currents flowing in some defined regions in the cells connected in series in the module. Thermal mismatch has effect on power, fill factor, module open circuit voltage, and operating point of the module. The last mentioned effect is felt on the loss in voltage and current at the MPP (Maximum Power Point).

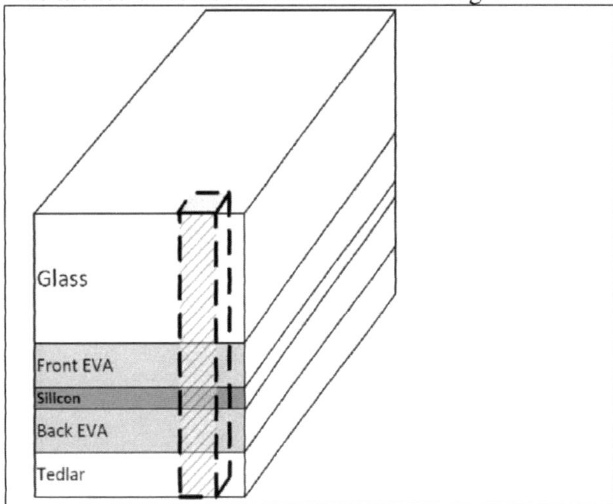

Figure 2. Cross-section used for thermal model simulation.

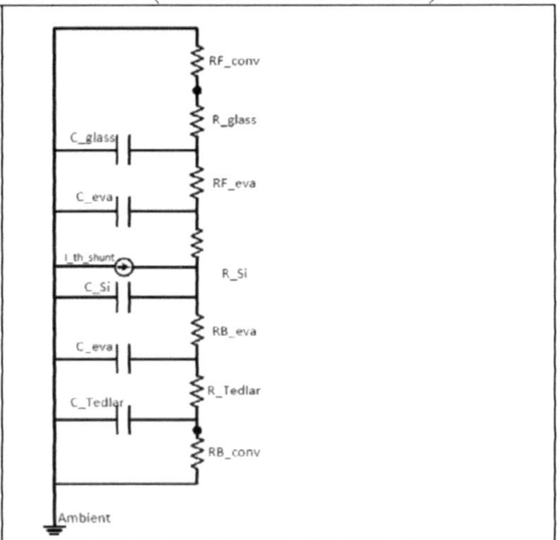

Figure 3. Electrical equivalence of thermal model.

Each of these has been studied and results are shown in bar charts for more clarity and for comparison. Both thermal and electrical mismatch effects have been combined in order to show explicitly the effect of temperature rise due to the shunt currents.

Table 3. Description of quantities.

Symbol	Quantity
I	Current
V	Voltage
Re	Electrical resistance
C	Capacitance
q	Heat energy
Rt	Thermal resistance
l	Thickness of layer
k	Thermal conductivity
A	Area of the elementary area
m	Mass of the elementary area
Cp	Specific heat capacity of material

Table 4. RC network elements corresponding to thermal quantities.

Thermal Resistance	Electrical Resistance
Top convection	RF_conv
Glass	R_glass
Front EVA	RF_eva
Silicon	R_Si
Back EVA	RB_eva
Tedlar	R_Tedlar
Bottom convection	RB_conv

Table 5. PV module material specifications. [12]

Parameter	Thickness (m)	Thermal Conductivity (W/m-K)	Specific Heat (J/kg-K)	Density (kg/m^3)
Glass	3.20E-03	1.8	5.00E+02	3.00E+03
Front EVA	4.00E-04	0.35	2.09E+03	9.60E+02
Silicon	2.25E-04	148	6.77E+02	2.33E+03
Back EVA	4.00E-04	0.35	2.09E+03	9.60E+02
Tedlar	1.25E-04	0.2	1250	1200

4. Results and discussion

The intensity of illumination falling on the different solar cells connected in the module determines primarily the amount of current generated in each cell. Hence it becomes an important parameter to be considered in the mismatch situation. However, the present study assumes that all cells in the module are uniformly illuminated.

4.1. Comparison of performance of shunted cell in the module and non-shunted cell

The most important parameter determining the quantum of the effect due to shunting is the shunt magnitude or to put it more simply the value of the shunt resistance offered by the cell. The value of the shunt resistance of the cell is in turn determined primarily by the shunt resistance of the shunted area of the cell and area of the shunted region. To have a realistic picture of the shunting severity and the resulting mismatch, an overall shunt magnitude of 10 ohm has been chosen for the shunted region, which value has been found acceptable from the published literature.

From the Lock-in Infrared Thermography (LIT) images, it has been observed that the edge shunt is normally occurring in many of the cells connected in the module. The edge shunt has been simulated by taking an area of 90 nodes, whereas the total cell area is represented by 1421 nodes as shown in Table 1. An overall shunt magnitude of 10 ohm has been simulated over the shunted area. The severity of the shunting due to this shunted area can be gauged from the graph shown in figure 4, which shows the power curve of the shunted and non-shunted cell connected in series in the module. The curve has been plotted with respect to the module voltage instead of corresponding cell voltages for the purpose of comparison.

The shunted cell power curve goes even into the third quadrant and becomes negative for some voltage range which means effectively it becomes a power dissipater rather than a power generator. The power generated by the shunted cell in each situation has been estimated from the cell current and the corresponding cell voltage. To have a common reference for comparison, the module voltage in the situation which has only one shunted cell has been taken on x-axis.

Due to mismatch, the shunted cell has been forced into reverse bias by the remaining non-shunted cells operating at a forward voltage, for module voltage ranging from 0V (short circuit condition) up to 15 V. This means that if the operating point of the module shifts to the region corresponding to the said voltage range, the effect of mismatch losses would impact the module output power adversely. For the particular case of module shown in the figure 4, the shunted cell was dissipating a power of 6.15 mW at near short circuit or no load conditions.

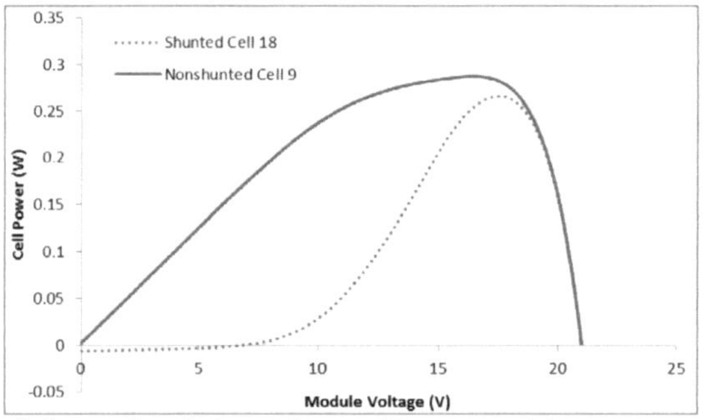

Figure 4. Variation of shunted and shunt-free cell power with respect to the module voltage.

4.2. Mismatch losses when all cells are equally illuminated

The module, normally a series connection of 36 PV cells, is simulated using distributed diode model where each cell is an interconnected network of one diode model circuits. The shunt currents are used to find the heat generated followed by investigation of rise in temperature at the shunt location by the simulation of electrical equivalent thermal model which utilizes the analogy between electrical and thermal quantities. The change in saturation current due to temperature rise is calculated. Modified saturation current is used at the shunt locations to simulate the module performance again using the distributed diode model. The performance of shunted module before and after the temperature rise is compared with that of the non-shunted module under study. The comparison between the electrical and thermal degradation effects for different spatial locations of shunting has been studied.

4.2.1. Effect of Mismatch on Module Output Power. figure 5 shows the effect of thermal mismatch on module relative power loss as a percentage of module power when all cells are non-shunted. . If only one cell is shunted by a magnitude of 10 ohms, the mismatch loss in power will be a small fraction of total power. However, as number of shunted cells increases, the effect becomes significant.

4.2.2. Effect of Mismatch on Module Fill Factor. The loss in Fill Factor (FF) is depicted in figure 6. Comparing the effects on power and Fill Factor leads to the following conclusion that, the FF is degraded more than power, but because of combination of these two effects there is a notable degradation due to electrical and thermal mismatch in the module performance.

4.2.3. Effect of *Mismatch on Module Open Circuit Voltage.* An interesting insight has been gained from the study that, even though only 10 ohm magnitude shunts have been considered, there is a notable change in the open circuit voltage of the module due to mismatch effect. figure 7 shows the variation of loss in module open circuit voltage (Voc) before and after temperature rise due to shunts for different number of shunted cells. The loss in Voc is increased considerably due to thermal effects. The increase in temperature clearly degrades Voc rather than Isc, the short circuit current, possibly due to the change in band gap of semiconductor with temperature.

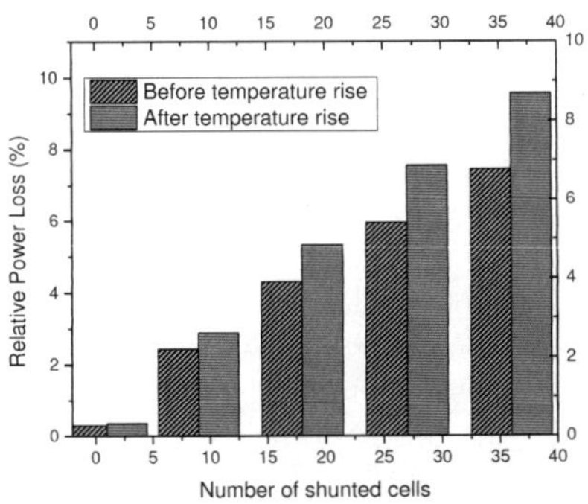

Figure 5. Variation of relative power loss before and after temperature rise due to shunts for different number of shunted cells.

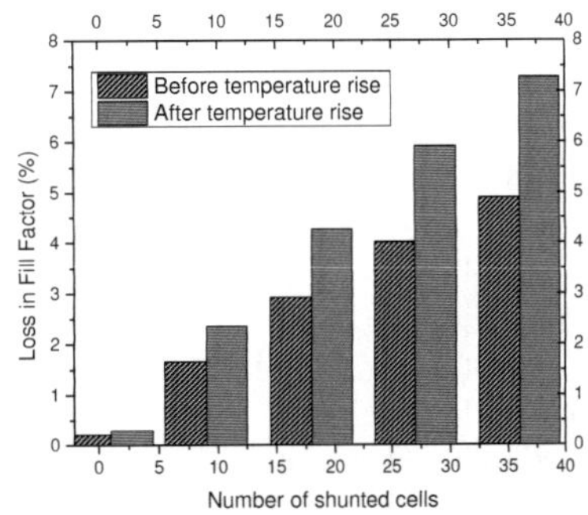

Figure 6. Variation of loss in fill factor before and after temperature rise due to shunts for different number of shunted cells.

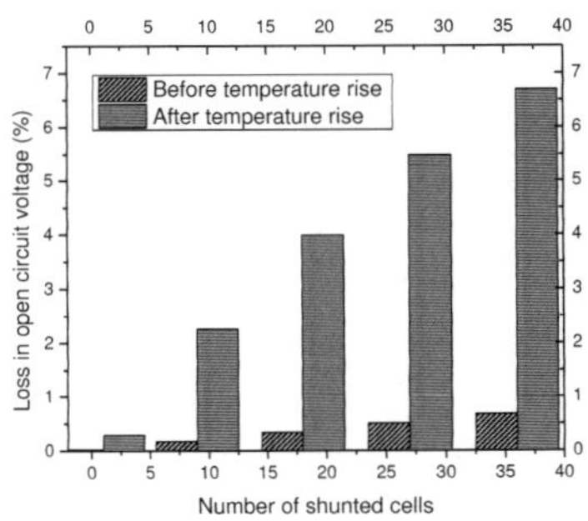

Figure 7. Variation of loss in module open circuit voltage before and after temperature rise due to shunts for different number of shunted cells.

5. Conclusions

A detailed study on the electrical and thermal mismatch effects of shunts on the performance of PV modules, have been conducted using the distributed diode model in combination with the 3-D thermal model. The investigation has yielded very interesting results, on how the shunts may cause performance loss due to mismatch and presented a novel approach to evaluate the actual power loss and fill factor loss as well as loss in open circuit voltage caused by the shunts and its temperature effects. The temperature rise due to shunt currents has been studied with a 3D thermal model so that the lateral heat flow also is considered in the model.

6. References

[1] Somasundaran P and Gupta R 2016 Influence of local shunting on the electrical performance in industrial Silicon solar cells *Sol. Energy* **139**.581-590

[2] Somasundaran P and Gupta R 2016 Evaluation of shunt losses in industrial Silicon solar cells *International Journal of Photoenergy* **2016** http://dx.doi.org/10.1155/2016/8029608.

[3] Barbato M Meneghini M Cester A Mura G Zanoni E and Meneghesso G 2014 Influence of shunt resistance on the performance of an illuminated string of solar cells: theory, simulation, and experimental analysis *IEEE Trans. Device and Materials Reliability* **14(4)** 942-950

[4] Dongaonkar S Loser S Sheets E J Zaunbrecher K Agrawal R Marks T J and Alam MA 2013 Universal statistics of parasitic shunt formation in solar cells, and its implications for cell to module efficiency gap *Energy Environ. Sci.* **6** 782-787

[5] Compagnin A Meneghini M Barbato M Giliberto V Cester A Vanzi M Mura G Zanoni E and Meneghesso G 2013 Thermal and electrical investigation of the reverse bias degradation of silicon solar cells *Microelectronics Reliability* **53** 1809–1813

[6] Johansson A and Gottschalg R Infield D G 2003 Modelling shading on amorphous silicon single and double junction modules *Proc. 3rd World conf. on photovoltaic energy conversion* vol **2** 1934-1937

[7] Woyte A Nijs J and Belmans R 2003 Partial shadowing of photovoltaic arrays with different system configurations : literature review and field test results *Sol. Energy* **74** 217-233

[8] Galiana B Algora C Rey-Stolle I Vara I G 2005 A 3-D Model for concentrator solar cells based on distributed circuit units *IEEE Trans. Electronic Dev.* **52 (12)** 2552-2558

[9] Giaffreda D Magnone P Meneghini M Barbato M Meneghesso G Zanoni E Sangiorgi E and Fiegna C 2014 Local shunting in multicrystalline Silicon solar cells: distributed electrical simulations and experiments *IEEE J. Photovoltaics* **4(1)** 40-47

[10] Gupta R Somasundaran P and Nandi D K 2012 Electrical simulation and characterization of shunts in solar cells *J. Appl. Mechanics Mater.* **110-116** 2453-2457

[11] Somasundaran P Sinha A and Gupta R 2012 Simulation and characterization of spatial variation of shunts in industrial solar cells by PSpice and dark lock-in infrared thermography *Proc. 27th European Photovoltaic Solar Energy Conf. (Frankfurt)* 744-748

[12] Armstrong S and Hurley WG 2010 A thermal model for photovoltaic panels under varying atmospheric conditions *Applied Thermal Engineering* **30** 1488-1495

IOP Conf. Series: Earth and Environmental Science **67** (2017) 012014 doi:10.1088/1755-1315/67/1/012014

Performance and economic analysis of a plug and play regenerative brake for improving energy efficiency for traction elevators

Chuttchaval Jeraputra and Supun Tiptipakorn

Department of Electrical Engineering, Mahidol University, Salaya, Nakonpathom 73170, Thailand

Email: supun.tip@mahidol.ac.th; chuttchaval.jer@mahidol.ac.th

Abstract. This paper presents performance and economic analysis of a plug and play regenerative brake for improving energy efficiency for traction elevators. The proposed regenerative brake recycles the energy loss of a dynamic brake and feeds into the grid while an elevator inverter is operating in the braking mode. According to field measurement of energy consumption, it reveals that the efficiency can be improved as much as 18%. The prototype of a regenerative brake 12 kW, 400V, 3ϕ is developed and tested on an elevator simulator. It is shown that it can transfer energy out of a DC capacitor before the dynamic brake kicks in. Further, an economic analysis is provided to carry out the payback period and the present worth equivalent to confirm economic feasibility.

1. Introduction

Traction elevators driven by electric motors are typically composed of several components, such as a car, sling, elevator machine, electric motor, inverter, control equipment, counter weight, hoist way, guide rail, and pit as shown in figure 1 [1]. In the present, induction motors fed by pulse width modulated (PWM) inverters are commonly used to control speed and position of elevators. They are capable of controlling speed from zero to the rated speed accurately regardless of load variations. For instance, the elevator under study employs a three-phase induction motor rated at 12 kW /230V /50Hz/ 4 Poles driven by a three-phase PWM inverter.

When elevator inverters operate in the braking mode, the kinetic energy accumulated in the elevator car is transformed to electrical energy and stored in the inverter's DC capacitors. If voltage across the inverter's DC capacitor becomes higher than the upper limit, the stored energy will be dissipated in the form of heat by the dynamic brake, which is a group of power resistors connected in series with an IGBT switch. When this switch turns on, the DC capacitor voltage will drop rapidly. It will turn off if the DC capacitor voltage is less than the lower limit. This method has a drawback that the dissipated heat will cause temperature rise in a machine room and shorten lifetime of electronic control devices. In some cases, air conditioners are installed to bring down the room temperature, which results in a loss of electrical energy.

In the past, recovering the energy loss had not gained interest due to high costs of regenerative inverters [2]. At present, advances in power semiconductor technologies and digital microcontrollers have significantly contributed to economic costs of the regenerative inverters so that they can be made commercially available [3-6]. However, their performance and economic feasibility have not been reported.

IOP Conf. Series: Earth and Environmental Science **67** (2017) 012014 doi:10.1088/1755-1315/67/1/012014

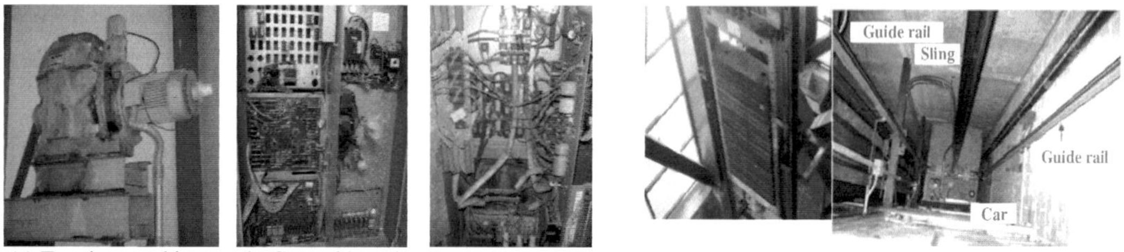

(a) Traction machine (b) Control board (c) Inverter (d) Hoist way (e) Counter weight

Figure 1. Typical components of traction elevators

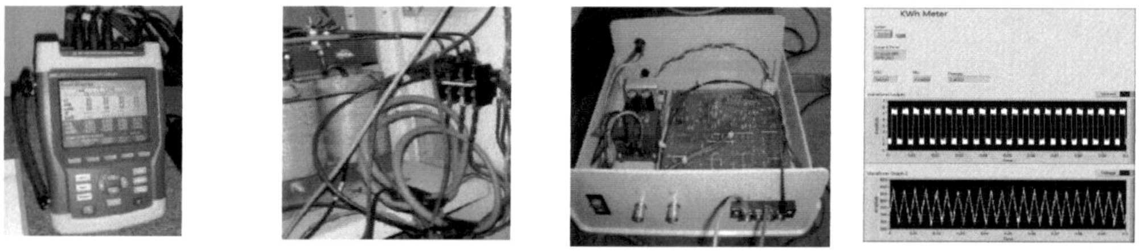

(a) Power analyzer (Fluke 435) (b) Current clamp (Flex 3000) (c) Signal conditioning circuit (d) Energy calculation software

Figure 2. Electrical energy measurement

This research proposes a plug and play regenerative brake which is placed in parallel to the existing elevator inverter without altering and re-commissioning the original control configuration. The energy consumption of the elevator under study is measured at various loads. The prototype of the regenerative brake is developed and tested on the elevator simulator. The experimental results confirm that the proposed regenerative brake can substitute the original dynamic brake and can supply electrical energy back to the grid safely. The economic analysis of cost/benefit of the device is also presented.

2. Electrical energy measurement

The electrical energy of the elevator is measured at various loads ranging from no-load to full-load capacity using the power quality analyzer, Fluke 435 in figure 2(a), (b). The electrical energy loss in a dynamic brake is measured by the electrical energy measurement device which is developed specifically for the load having discontinuous voltage and/or current in figure 2(c), (d). The measuring device comprises a differential high voltage probe with bandwidth of 50 MHz and voltage gain of 10mV/V to measure the voltage across the dynamic brake resistor and a current clamp with bandwidth of 200 kHz and output voltage gain of 100 mV/A. That is, the maximum measured voltage of 1,000 V and the maximum measured current of 100 A yield the signal output of 10 V.

All measured signals must be filtered and amplified by signal conditioning circuits before they are supplied to a data acquisition card (PCI-6013). Labview 2013 software is used for configuring the simultaneous 50 kHz sampling rates and calculating the kWh electrical energy. The sampled voltage and current signals are multiplied to calculate the electrical power and the electrical energy can be obtained from integration of the calculated electrical power. In addition, the developed software is capable of recording data such as voltage, current, power, energy in a text file.

Figure 3 illustrates the consumed energy and electrical energy loss in a dynamic brake at various loads. At 500 kg load being the same mass as the counter weight, the elevator consumes the same amount of energy when it moves up or down between the 1st and 5th floor. figure 4 shows electrical energy loss in the dynamic brake's resistors. It reveals that the elevator can operate efficiently when weight of the load and the counter weight are equal. According to the analysis as shown in figure 5, if the energy loss in the dynamic brake's resistors can be recycled and fed into the grid, the energy efficiency can be improved by as much as 18%. Considerable friction loss in elevator machines, guide rails, and pulleys is a major cause that prohibits improving the energy efficiency much less than the theoretical limit (50%).

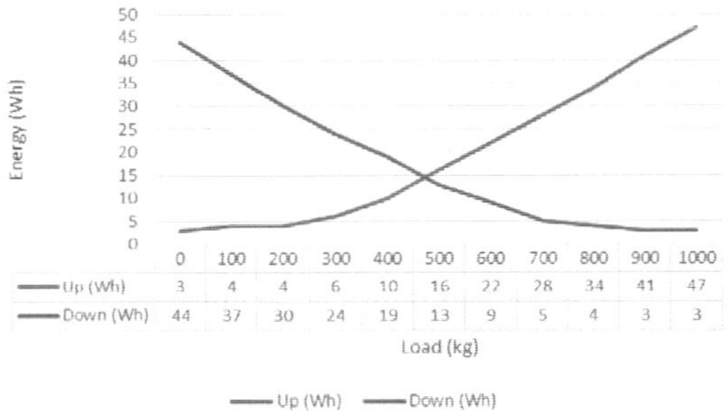

Figure 3. Electrical energy consumption

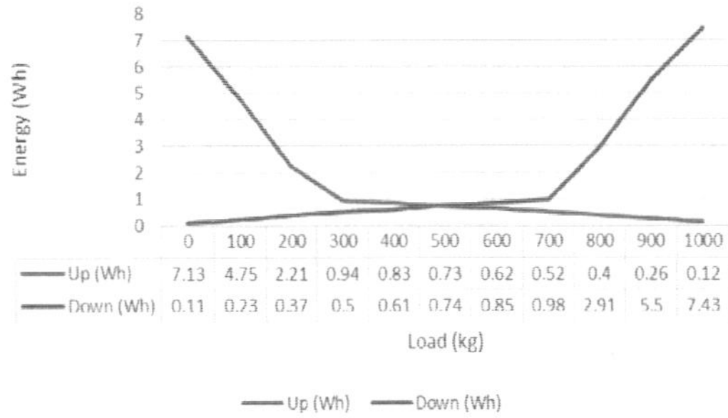

Figure 4. Electrical energy loss in the resistors

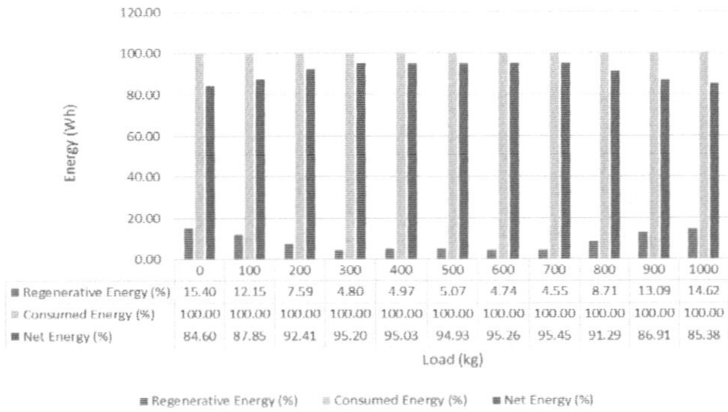

Figure 5. Analysis of electrical energy consumption

3. Plug and play regenerative brake prototype and testing

Figure 6 shows circuit configuration of the proposed plug and play regenerative brake. It is placed in parallel to the front-end diode rectifier of the inverter without altering the original control configuration. When the elevator's inverter operates in the braking mode. The mechanical energy stored in the elevator car will be removed and transformed into electrical energy stored in the DC capacitor. Thus, it causes the capacitor voltage to rise abruptly. Then, the regenerative brake must transfer energy stored in the DC capacitor and feed it into the grid before the dynamic brake does. A blocking bridge diode is employed to control energy to flow only in one direction. Just in case, a galvanic isolation transformer is recommended in practice to prevent malfunction of the elevator control board or the regenerative inverter.

Figure 7 shows the experimental setup. The operation of a traction elevator under study is emulated by two induction motor drives (ABB ACSM204-1). The first induction motor drive is programmed to emulate operation of the actual elevator's motor drive. The rotational speed is regulated to follow the speed profile of the elevator under study. The second induction motor drive exerts the load torque against the other that is considered as it were a passenger load. Both motor drives must work in synchronism by using a programmable logic controller (PLC S7-300). This method enables the developed elevator simulator to have similar speed and load torque characteristics to that of the actual elevator system.

Figure 8 shows the block diagram and speed/torque characteristics of the simulator. Figure 9 shows the input current of the regenerative brake while the elevator emulator is operating in the braking mode. The plug and play regenerative brake succeeds in recycling energy stored in the DC capacitor to the grid before the dynamic brake does. The DC capacitor voltage is well regulated at a constant 650V.

4. Economic analysis

The economic analysis and payback period is carried out by considering cumulative present worth (CPW) of all activities from the incoming and outgoing cash flows within an estimated lifetime of 15 years (180 months) and comparing it with the minimum attractive rate of return (*MARR*). The present worth factor is calculated by the expression (1).

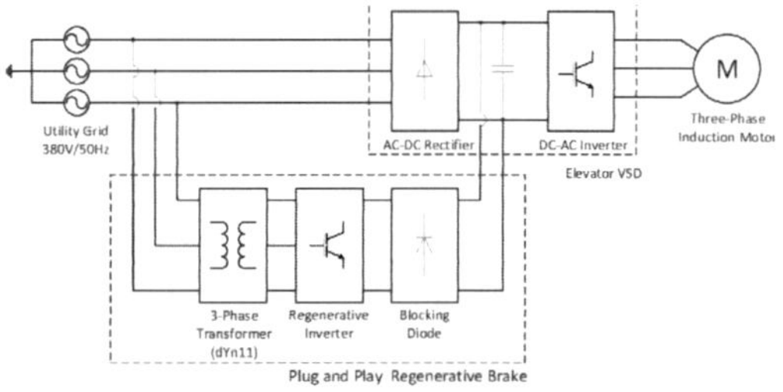

Figure 6. Plug and play regenerative brake

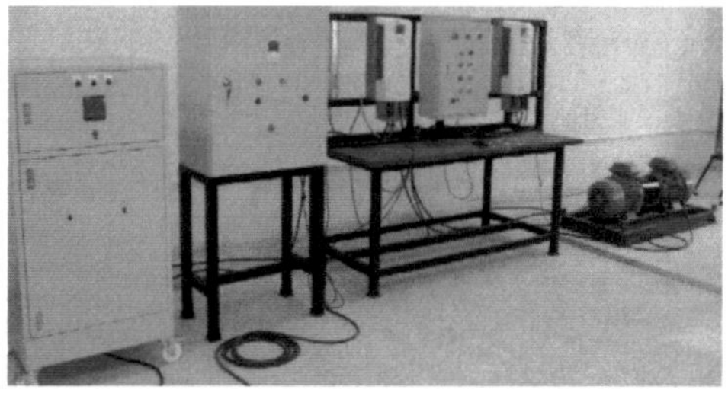

Figure 7. Experimental setup

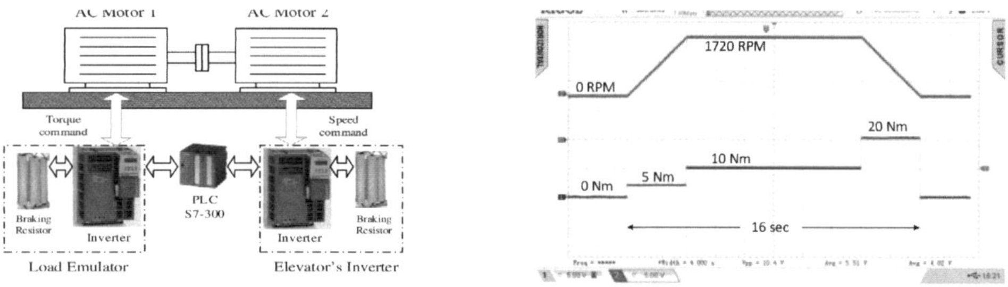

Figure 8. Block diagram and speed/torque characteristics of the simulator

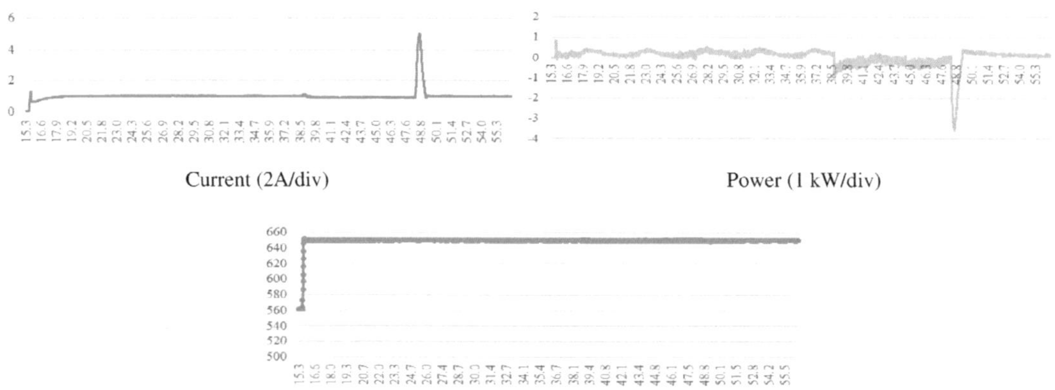

Figure 9. Test results

$$\mathrm{PWF} = \left(\frac{1}{1 + \mathrm{MARR}}\right)^{m} \tag{1}$$

where the number of year is denoted by *m* and the MARR is the minimum rate of investment return, in which an investor makes a decision whether it is worth investing.

If an investor borrows money for this investment, the MARR should be greater than the lending interest rate. If the investor uses his own cash instead, then the MARR should be greater than the deposit interest rate. This analysis chooses the MARR of 0.5% per month based on general inflation rate of 0.38%.in June 2016 from the Bank of Thailand. It is assumed that at the end of the 15th year, the salvage value will be zero, the maintenance cost approximates 1,000 Baht/year with additional 100 Baht increasing each year. Then, the installation of the regenerative brake will improve the overall efficiency of the elevator by 15%. If the device were installed with the existing elevator rated 1,000 Kg load at Building 1 of the Faculty of Engineering, Mahidol University, the electricity costs will be reduced approximately by 1,200 Baht/month. The electricity price per kWh increases 5% each year. According to the information, the cumulative sum of present worth equivalent can be calculated and compared with the investment in low risk 15 year bonds with 2.8% yearly interest rate and the discount yield rate of 2.38%.

Table I shows the calculation of the cumulative sum of present worth equivalent in the first 12 months. Starting at the 1st month, the cumulative sum is negative because of the outgoing cash flow of 125,000 Baht or the initial investment. Afterwards, the cumulative sum of present worth equivalent increases as a result of the electricity cost saving presented in the incoming cash flow. Figure 10 shows that present worth equivalent becomes positive in year 10. The cumulative sum of the present worth equivalent for the 15 year lifetime is 53,538 Baht.

In comparison, an investor could use his/her initial fund of 125,000 Baht to invest in low risk bonds for 15 years, such as "The 2016 3rd State Railway of Thailand Bond" with the fixed interest rate of 2.8% or 3,500 Baht yearly. Table II shows the cumulative sum of present worth equivalent when combining the interest with the repayment of principal at maturity with the adjusted discount rate, the present worth equivalent of this bond investment is 6,517.6 Baht.

From the economic analysis, it can be concluded that the installation of the regenerative brake device to improve the efficiency of the elevator is worthwhile, and the investment return is 8 times greater than the investment in low risk bonds at the same duration of 15 years.

Table 1. Cumulative sum of present worth equivalent in year 1 (first 12 months)

Year	Outgoing cash flow		Incoming cash flow	Total cash flow	Present worth factor	Present worth equivalent	Cumulative sum of the present worth equivalent
	Investment	Maintenance	Saving				
1	-125,000			-125,000	1	-125,000.00	-125,000.00
1			1,500	1,500	0.99502	1,492.53	-123,507.47
1			1,500	1,500	0.99007	1,485.11	-122,022.37
1			1,500	1,500	0.98515	1,477.73	-120,544.64
1			1,500	1,500	0.98025	1,470.38	-119,074.27
1			1,500	1,500	0.97537	1,463.06	-117,611.21
1			1,500	1,500	0.97052	1,455.78	-116,155.43
1			1,500	1,500	0.96569	1,448.54	-114,706.90
1			1,500	1,500	0.96089	1,441.34	-113,265.56
1			1,500	1,500	0.9561	1,434.15	-111,831.41
1			1,500	1,500	0.95135	1,427.03	-110,404.39
1			1,500	1,500	0.94661	1,419.92	-108,984.47
1		-1,100	1,500	400	0.94191	376.76	-108,607.71

Table 2. Cumulative sum of present worth equivalent of the 15 year bond investment

Year	Cash flow			Present worth factor	Present worth equivalent
	Outgoing cash flow	Incoming cash flow	Total		
0	-125,000		- 125,000.00	1.0000	-125,000.000
1		3,500.00	3,500.00	0.9767	3,418.553
2		3,500.00	3,500.00	0.9540	3,339.001
3		3,500.00	3,500.00	0.9318	3,261.300
4		3,500.00	3,500.00	0.9101	3,185.408
5		3,500.00	3,500.00	0.8889	3,111.281
6		3,500.00	3,500.00	0.8683	3,038.880
7		3,500.00	3,500.00	0.8480	2,968.163
8		3,500.00	3,500.00	0.8283	2,899.092
9		3,500.00	3,500.00	0.8090	2,831.628

10		3,500.00	3,500.00	0.7902	2,765.734
11		3,500.00	3,500.00	0.7718	2,701.374
12		3,500.00	3,500.00	0.7539	2,638.511
13		3,500.00	3,500.00	0.7363	2,577.111
14		3,500.00	3,500.00	0.7192	2,517.140
15		128,500.00	128,500.00	0.7024	90,264.448
					6,517.623

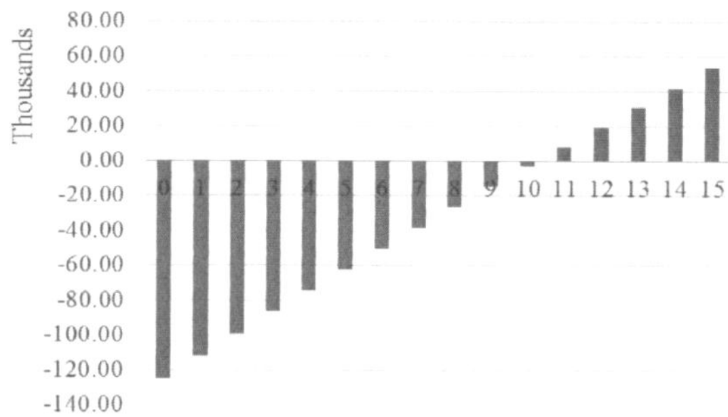

Figure 10. Cumulative sum of present worth equivalent within the lifetime of 15 years

5. Conclusion

This paper presented the performance and economic analysis of the regenerative brake to improve efficiency of traction elevators. The electric energy consumption and energy loss in the dynamic brake at various loads were measured and analyzed. If the energy loss in the dynamic brake can be recycled and fed back to the grid, the efficiency can be improved as much as 18% when the elevator carries no load or full load. The prototype of the regenerative brake was developed and tested with the elevator simulator. The test results confirmed viability of the method, and the economic analysis proved its investment worthiness. In comparison, its return from the energy saving is 8 times greater than the investment in low risk bonds with the same yield duration.

6. References

[1] G. R. Strakosch and R. S. Caporale, "The vertical transportation handbook," 4th ed. Hoboken, N.J. John Wiley & Sons, 2010.

[2] N. C. Linder. "Getting the power back: Line regenerating with variable speed A-C drives," Online available: ftp://ftp.unicauca.edu.co/Facultades/FIET/DEIC/Materias/Control%20de%20maquinas/bibliografia/variadores/Lixis%20Jairo%20toquita/Regeneraci%F3n%20a%20la%20red.pdf

[3] N. Mitrovic, M. Petronijevic, V. Kostic, and B. Jeftenic, "Electrical Drives for Crane Application," Online available: http://www.intechopen.com/books/mechanical-engineering/electrical-drives-for-crane-application

[4] Yasakawa, "Power regenerative unit (VS-656RC5)," Online available: www.yaskawa.com.cn/technology/download/manual/yea-toa-s656-3.pdf

[5] Schindler, :AC Regeneration Module," Online available: www.schindler.com/content/us/internet/en/modernization/energy-savings/_jcr_content/rightPar/downloadlist/downloadList/163_1336060675322.download.asset.163_1336060675322/ac-regeneration-module-sheet.pdf

[6] T. Ota, Y. Okui, N. Nakamura, and M. Takasugi, "Development of the regenerative power compensation device SANUPS K23A (R Type)," Online available: https://www.sanyodenki.com/archive/document/corporatedata/technicalreport/2012/TR33e_p24_K23A_2nd_en.pdf

[7] H. Sachs, "Advancing Elevator Energy Efficiency," Online available: http://kms.energyefficiencycentre.org/sites/default/files/a1501.pdf
[8] M. Fatih Adak, N. Durd, and H. T. Duru, "Elevator simulator design and estimating energy consumption of an elevator system," Energy and Buildings, ScienceDirect, 2013, pp. 272-280
[9] Energy performance of lifts, escalators and moving walks – Part I: Energy measurement and verification, ISO 25745-1, 2015
[10] Energy performance of lifts, escalators and moving walks – Part II: Energy calculation and classification for lifts, ISO/DIS 25745-2, 2015
[11] L. Blank and A. Tarquin, "Engineering economy," 7th ed. New York, NY: McGraw-Hill, 2011

Acknowledgments
The research is supported by the National Science and Technology Development Agency (NSTDA) and the Electricity Generating Authority of Thailand (EGAT).

IOP Conf. Series: Earth and Environmental Science **67** (2017) 012015 doi:10.1088/1755-1315/67/1/012015

Multi-objective optimal power flow for active distribution network considering the stochastic characteristic of photovoltaic

Bao-Rong Zhou[1], Si-Liang Liu[2], Yong-Jun Zhang[2], Ying-Qi Yi[2] and Xiao-Ming Lin[2]

[1]Technology Research Center of China Southern Power Grid Co.Ltd, Guangzhou, China
[2]School of Electrical Power South China University of Technology, Guangzhou, China
Email: zhoubr@csg.cn

Abstract. To mitigate the impact on the distribution networks caused by the stochastic characteristic and high penetration of photovoltaic, a multi-objective optimal power flow model is proposed in this paper. The regulation capability of capacitor, inverter of photovoltaic and energy storage system embedded in active distribution network are considered to minimize the expected value of active power the T loss and probability of voltage violation in this model. Firstly, a probabilistic power flow based on cumulant method is introduced to calculate the value of the objectives. Secondly, NSGA-II algorithm is adopted for optimization to obtain the Pareto optimal solutions. Finally, the best compromise solution can be achieved through fuzzy membership degree method. By the multi-objective optimization calculation of IEEE34-node distribution network, the results show that the model can effectively improve the voltage security and economy of the distribution network on different levels of photovoltaic penetration.

1. Introduction

As the problems of energy depletion and environmental pollution becoming increasingly serious, distributed generations (DGs) has got rapid development on account of less investment, flexible generating method and environmental compatibility [1-3]. However, large scale integration of DGs has an impact on security, economy and reliability of distribution networks (DNs). With the rapid development of active distribution network (ADN) technologies, regulating means and objectives for optimal optimization becomes more complex , and the traditional reactive power optimization is no longer applicable [4]. Considering the characteristic of the regulating equipments on ADN, the optimization is a nonlinear programming problem including continuous and discrete variables. Reference [5] proposed a new multi-agent immune algorithm based on the mechanism of antibody cluster and compete expansion by integrating the reactive power of different types of DGs and traditional voltage regulating equipments to minimize the power loss. Reference [6] considered the stochastic characteristic of photovoltaic (PV) station, took the minimum of power loss as the objective function and built mathematical model of reactive power optimization for DNs based on chance constraints, the simulation results verified the model could decrease the power loss on the premise that the voltage satisfied the chance constraints. The models proposed above can improve the economy of DNs, but they only consider a single objective for optimization. A voltage control model was proposed in [7] with active/reactive power integrated optimization to reduce voltage deviation, power loss and environmental pollution by using micro-turbine and fuel cell. Reference [2] proposed a two-stage reactive power planning model whose objective was to minimize the power loss, leaping transmission of reactive power and the investment of the compensators considering the uncertainty of power output of PV and load. They try to achieve the optimization on the aspect of both economy and security, but actually when it comes to multi-objective optimization, they change it into a single objective optimization by giving weight, or use layered optimization method to solve it, which ignores the contradictory between the sub-objectives, also fails to provide variety of decision schemes. In light of the problem of multi-objective optimization, a probabilistic model with the DGs for DNs was established in [8] which is based on the two-point estimation method for probabilistic power flow, and used the particle swarm optimization (PSO) algorithm to obtain the Pareto frontier, but it sets the voltage offset as the objective function which fails to quantify the weakness of the voltage security on DNs.

When it refers to regulating means for optimization, the model mentioned above all consider the reactive power regulation capability of DGs, some take advantage of the active power regulation capability of micro-turbine and fuel cell, but the intermittence of power output of DGs is outstanding, sometimes these regulating means are not very effective, even have no effect. In recent years, energy storage system (ESS) has attracted the interests from the researchers because it can quickly regulate power output and flexibly store excess energy from the power system [9, 10]. Therefore, integrated utilization of the reactive power of DGs and the active power of ESS is benefit for the optimal operation of the DNs.

Motivated by the observations above, this paper proposes a multi-objective optimal power flow model considering the regulation capability of capacitor, inverter of photovoltaic and energy storage system, and sets the expect value of active power loss and probability of voltage violation as its objective functions. A probabilistic power flow based on cumulant method is introduced to calculate the value of the objectives and NSGA-Ⅱ algorithm is adopted for optimization to obtain the Pareto frontier. Through fuzzy membership degree method, the best compromise solution can be achieved. Finally, the model is tested on IEEE34-node distribution network, the results show that compared with other different objective functions, the model can reduce loss more effectively under the same voltage security level; compared with other models considering different control variables, the model can keep the system operating in the optimal state on different levels of photovoltaic penetration.

2. Probabilistic model of load and PV

2.1. Probabilistic model of load

Error caused by the forecasting and measurement of load is unavoidable, so its active power P_L and reactive power Q_L should be described as random variables. Many researches demonstrate that these two random variables can be described by a normal distributed function.

2.2. Probabilistic model of PV

Some researches have shown that the solar irradiance r in a certain period of time can be described by a Beta distribution function as following:

$$f_R\left(\frac{r}{r_{\max}}\right)=\frac{\Gamma(\alpha+\beta)}{\Gamma(\alpha)\Gamma(\beta)}\left(\frac{r}{r_{\max}}\right)^{\alpha-1}\left(1-\frac{r}{r_{\max}}\right)^{\beta-1} \tag{1}$$

In (1), $f_R(\bullet)$ is the PDF of the random variable R, r and $r_{\max}$ represent the measured value and the maximum of the solar irradiance, $\Gamma(\bullet)$ is the Gamma function, α and β are two shape parameters of Beta distribution.

The power output of PV is mainly determined by r, and their mathematical relation can be expressed by the equation as following:

$$P_{\mathrm{PV}}=A\eta r \tag{2}$$

where A is the area of PV cells, η is the conversion efficiency. According to (1) and (2), the PDF of the active power of PV can be expressed by the equation as following:

$$f_{P_{\mathrm{PV}}}\left(p_{\mathrm{PV}}\right)=\frac{1}{A\eta r_{\max}}\bullet\frac{\Gamma(\alpha+\beta)}{\Gamma(\alpha)\Gamma(\beta)}\left(\frac{p_{\mathrm{PV}}}{p_{\mathrm{PV\,max}}}\right)^{\alpha-1}\left(1-\frac{p_{\mathrm{PV}}}{p_{\mathrm{PV\,max}}}\right)^{\beta-1} \tag{3}$$

Define λ_{PV} as the power factor of PV, so the PDF of the reactive power of PV can be expressed by

$$f_{Q_{\mathrm{PV}}}\left(q_{\mathrm{PV}}\right)=\frac{\lambda_{\mathrm{PV}}\dfrac{\Gamma(\alpha+\beta)}{\Gamma(\alpha)\Gamma(\beta)}\left(\dfrac{q_{\mathrm{PV}}}{q_{\mathrm{PV\,max}}}\right)^{\alpha-1}\left(1-\dfrac{q_{\mathrm{PV}}}{q_{\mathrm{PV\,max}}}\right)^{\beta-1}}{p_{\mathrm{PV\,max}}\left(1-\lambda_{\mathrm{PV}}^{2}\right)^{1/2}} \tag{4}$$

3. Multi-objective optimal power flow model

Multi-objective optimal power flow model mainly consists of objectives, equality constraints and inequality constraints. It uses the probabilistic method to quantify the economy and voltage security of the DNs and considers the regulation capability of capacitor, inverter of PV and ESS for optimization, which can ensure the accuracy and objectiveness of the optimization results.

3.1. Objectives

Expect value of the active power loss and the probability of voltage violation are set as the two objectives, mathematically

$$\begin{aligned} F_1&=\min\overline{P_{loss}}=\min\left(\sum_{i=1}^{N_{\mathrm{G}}}\overline{P_{\mathrm{G}i}}-\sum_{i=1}^{N_{\mathrm{L}}}\overline{P_{\mathrm{L}i}}\right)\\ F_2&=\min\left[\max\left(P_{\mathrm{V1\_violation}},P_{\mathrm{V2\_violation}},\cdots,P_{\mathrm{Vn\_violation}}\right)\right]\end{aligned} \tag{5}$$

In (5), $\overline{P_{\mathrm{G}i}}$ is expect value of the active power of generators at bus i, $\overline{P_{\mathrm{L}i}}$ is expect value of the active power of load at bus i, $P_{\mathrm{Vn\_violation}}$ is the probability of voltage violation at bus i.

3.2. Constraints

3.2.1. Equality Constraints.

The equation constraints of the active power and reactive power flow equations are given by

$$\begin{aligned} &P_{is}-V_i\sum_{j\in\Gamma}V_j\left(G_{ij}\cos\delta_{ij}+B_{ij}\sin\delta_{ij}\right)=0\\ &Q_{is}-V_i\sum_{j\in\Gamma}V_j\left(G_{ij}\cos\delta_{ij}+B_{ij}\sin\delta_{ij}\right)=0\end{aligned} \tag{6}$$

3.2.2. Inequality Constraints.

The variables in this model consists of control variables and state variables. The control variables include the reactive power outputs of capacitors, the active power output of the ESS and the power factor of PV. The inequality constraints of control variables are given as following:

$$\begin{cases} Q_{C(i)\min} \le Q_{C(i)} \le Q_{C(i)\max} \\ P_{ESS(j)\min} \le P_{ESS(j)} \le P_{ESS(j)\max} \\ \lambda_{PV(k)\min} \le \lambda_{PV(k)} \le \lambda_{PV(k)\max} \end{cases} \tag{7}$$

In (7), $Q_{C(i)}$ is the reactive power outputs of capacitors at bus i, $P_{ESS(j)}$ is the active power output of the ESS at bus j, $\lambda_{PV(k)}$ is the power factor of PV at bus k. $Q_{C(i)\min}$, $Q_{C(i)\max}$, $P_{ESS(j)\min}$, $P_{ESS(j)\max}$, $\lambda_{PV(k)\min}$ and $\lambda_{PV(k)\max}$ represent the lower bound and the upper bound of $Q_{C(i)}$, $P_{ESS(j)}$, and $\lambda_{PV(k)}$.
The state variables include the node voltage, the branch transmission power and the leaping transmission of reactive power, The inequality constraints of state variables are given as following:

$$\begin{cases} V_{i\min} \le V_i \le V_{i\max} \\ |I_j| \le |I_{j\max}| \\ Q_S \ge 0 \end{cases} \tag{8}$$

where V_i is the voltage amplitude at node i, I_j is the current of branch j, Q_s is the leaping transmission of reactive power. $V_{i\min}$, $V_{i\max}$ represent the minimum and the maximum of V_i, $I_{j\max}$ represent the maximum of I_j.

4. Solution

4.1. NSGA- II

NSGA-II, also known as non-dominated sorting genetic algorithm [11], has high efficiency, rapid speed of convergence and strong capability of global search. The specific steps are as follows:

4.1.1. Coding. The decimal system is used to code for each gene on the chromosome and each gene represents each control variable. When coding, make sure that the value of each control variable should be in the range set before.

4.1.2. Initialization of Population. Generate the value of each control variable randomly and calculate the value of objective functions. Then the first solution can be obtained, and so on.

4.1.3. Non-domination Rank and Crowding-distance. The non-domination rank of each solution can be calculated through the fast non-dominated sorting approach. If the value of objective functions is closer to the Pareto frontier, the non-domination rank of the individual will be higher. The crowding-distance of an individual is the average distance of two individuals on either side of this individual along each of the objectives.

4.1.4. Selection, Crossover and Mutation. The aim of selection is to select the individual with higher non-domination rank in different ranks, and the individual with bigger crowding-distance in the same non-domination rank. In order to prevent the prematurity and improve the diversity of the new population, also to avoid trapping into the local optimization, crossover and mutation will be operated to form children population after selection.

4.1.5. Elite Strategy. The elite strategy is to avoid superior individuals losing when optimizing. Firstly, combine the parent population P_t of size N and its children population Q_t of the same size N to form a new population R_t. Then calculate its non-domination rank and crowding-distance and select the top N individuals as the new parent population P_{t+1}. Finally, obtain the new children population C_{t+1} by the selection, crossover and mutation refer before.

4.1.6. Pareto Frontier and Best Compromise Solution. Suppose control vector $\boldsymbol{X}$ and objective function $F_i(\boldsymbol{X})$ ($i=1, 2, \ldots, n$), if a vector $\boldsymbol{X}'$ exists and it satisfies the following relation:

$$F_i\left(\boldsymbol{X}'\right) \le F_i\left(\boldsymbol{X}\right) \tag{9}$$

then we can draw a conclusion that $\boldsymbol{X}'$ is the Pareto optimal solution. The Pareto frontier is a set of optimal solutions like $\boldsymbol{X}'$ and the best compromise solution can be obtained through fuzzy membership degree method, see reference [12].

4.2. PPF based on cumulant method

In this paper, the probabilistic power flow (PPF) based on cumulant method is applied to calculate the expect value of active power loss and probability of voltage violation. The linear power flow model is adopted for power flow calculation, and Gram-Charlier expansion is used to gain the PDF and CDF of the state variables, such as node voltage, active line flow and so on. The specific mathematical procedure of the PPF based on cumulant method can see the reference [13, 14].

5. Case Study

This paper selects IEEE34-node distribution network as the simulation example as shown in Figure 1. In this case, the base capacity is 1MVA and the base voltage is 24.9kV (1.03p.u.). Line parameters, expected value of load refer to reference [16] and the standard deviation is the 10% of the expected value. Two groups of switchable shunt capacitors are connected to node 4, 10, 13, 23 and 28, the capacity of each group is 5kvar. A photovoltaic-energy storage system are connected to node 34, of which the maximum charge and discharge power of energy storage system is 200kW, and the power factor of PV inverter ranges from lagging 0.85 to leading 0.85. Some parameters for NSGA-II are set as following: size of population is 100, maximum of evolution generation is 50, rate of crossover is 0.9, and rate of mutation is 0.1. The voltage security range is set as [0.95, 1.05] for the probability of voltage violation calculation.

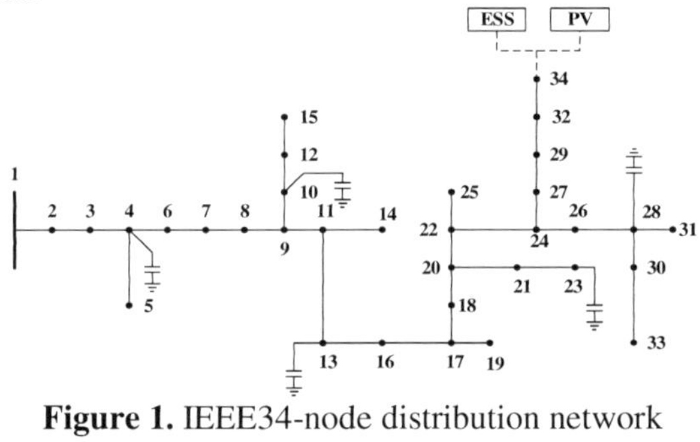

Figure 1. IEEE34-node distribution network

This paper uses HOMER software to obtain the data of solar irradiance in Guangzhou, China (23°6' N, 113°2' E). Assuming that the solar irradiance follows Beta distribution [15], then the shape parameter α is 1.54, β is 0.96, the maximum of solar irradiance rmax is 1.03kW/m2. The PV penetration is defined as the ratio of expected value of active power output of PV and expected value of active power of load on the distribution network, shown as following:

$$\rho_{\mathrm{PV}} = \frac{P_{\mathrm{PV}}^{\mu}}{P_{\mathrm{L}}^{\mu}} = \frac{A\eta r_{\max}\mu_{r/r_{\max}}}{P_{\mathrm{L}}^{\mu}} \tag{10}$$

And three levels of PV penetration are set as following: low penetration (10%), medium penetration (40%) and high penetration (80%), where the corresponding PV parameters are shown as following:

Table 1. PV parameters corespond to three levels of PV penetration

PV parameters	Low penetration (10%)	Medium penetration (40%)	High penetration (80%)
Area of PV cells (m^2)	500	2000	4000
Conversion efficiency	13%	13%	13%
Expected value of active power output of PV (kW)	41.26	165.05	330.09

In order to verify the validity of the proposed model, this paper carries out the following two-part simulation. The first part is to verify the validity of the objective functions, and four kinds of optimization model—Model A, Model B, Model C and Model D are set in this paper, of which the corresponding objectives and solving algorithms are shown in Table 1. F_{10}、F_1 represent expected value of active power loss before and after optimization respectively; F_{20}、F_2 represent probability of voltage violation before and after optimization respectively; F_{30}、F_3 represent voltage satisfaction before and after optimization; a、b represent the weight of the two objectives respectively, of which the value is 0.4 and 0.6 respectively. Model D is the proposed model in this paper. In addition, the function of voltage satisfaction is shown in equation (11):

$$V_{\text{satisfy}} = \sum_{i=1}^{N}\left(V_i - \frac{V_{i\max} + V_{i\min}}{2}\right)^2 \tag{11}$$

where N is the total number of system nodes; $V_{i\min}$, $V_{i\max}$ represent the minimum and the maximum of voltage in node I respectively, which are 0.95 and 1.05 respectively.

Table 2. The objective functions and solving algorithm for four kinds of optimization models

Model	objectives	solving algorithms
A	$\min\left\{a\frac{F_1}{F_{10}} + b\frac{F_3}{F_{30}}\right\}$	Cataclysmic GA
B	$\min\left\{a\frac{F_1}{F_{10}} + b\frac{F_2}{F_{20}}\right\}$	Cataclysmic GA
C	$\min\{F_1, F_3\}$	NSGA-II
D	$\min\{F_1, F_2\}$	NSGA-II

The second part is to verify the validity and adaptability of control method in the proposed model. Three control methods—Method A, Method B and Method C, are set in this part. The control variables of each method are

Method A: Q_C
Method B: Q_C and Q_{PV}
Method C: Q_C, Q_{PV} and P_{ESS}

And Method C is the control method of the proposed model.

5.1. Validity of objective functions

On the high level of PV penetration, the Pareto frontier of four models optimizing by the NSGA- II algorithm are drawn in Figure 2.

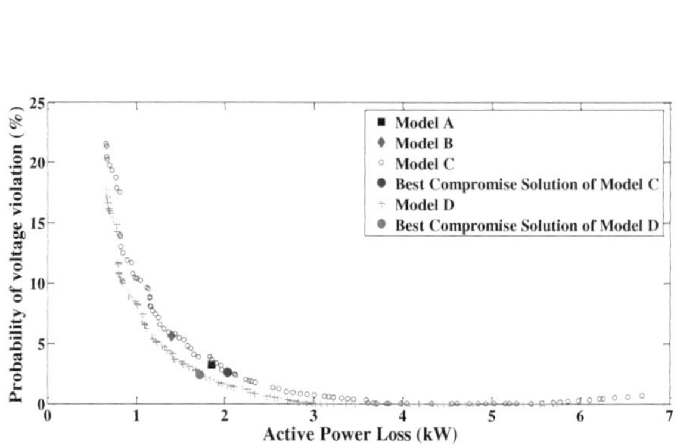

Figure 2. The comparison of the Pareto frontier of four models

According to Figure.2, we can draw conclusions as following:
(1) The optimal solution for Model A does not fall in the Pareto frontier, the same with Model B. It suggests that on the one hand the models can't always lead to the non-inferior solutions for multi-objective optimization because of the weight given without the basis, on the other hand they can't provide variety of solutions.
(2) As a whole, the Pareto frontier obtained from Model C is higher than that of Model D in the figure, showing that under the same probability of voltage violation, Model D gets the lower expected value of active power loss. This is mainly because the voltage satisfaction can' t quantify the actual risk of voltage violation, tending to conservative optimization results. The improvement of optimization results in Model D is shown in Table 3.

Table 3. Improvement of optimization results in Model D

Expected value of active power loss (kW)	Probability of voltage violation				
	1%	2%	3%	4%	5%
Model C	2.70	2.24	1.97	1.66	1.56
Model D	2.29	1.84	1.64	1.41	1.24
Improvement	15.2%	17.9%	16.8%	15.1%	20.5%

5.2. Validity of objective functions

The optimization results of best compromise solutions are compared in Figure 3 and Figure 4.According to Figure 3 and Figure 4, we can draw some conclusions as following:
(1) On the aspect of economy, the optimization effect of Method C is more effective than that of the other two methods on the low and medium levels of penetration. Compared with the other two methods, the rates of loss reduction reach 82.76% and 50.74% respectively. However, the loss of Method C increases slightly compared with that before optimization on the high level of penetration. This is mainly because the active power output of PV is high and the voltage generally exceeds the limits. In order to reduce the risk of voltage violation, ESS will absorb the excess active power to reduce the probability of voltage violation, which will increase the loss. In addition, on different levels of PV penetration, the variances of the optimization results reach up to 52.95 and 43.16 respectively in Method A and B, while it is only 0.61 in Method C, suggesting that Method C has better adaptability.
(2) On the aspect of voltage security, the probability of voltage violation is nearly zero on the low and medium levels of penetration, but on the high level of penetration, the probability of system voltage violation increases sharply. In this case, the optimization effect of Method C is more effective—the probability of voltage violation drops from 8.43% to 2.22% and the rate of improvement reaches 73.67%. In addition, Method C has better adaptability compared with the other two methods on different levels of PV penetration.

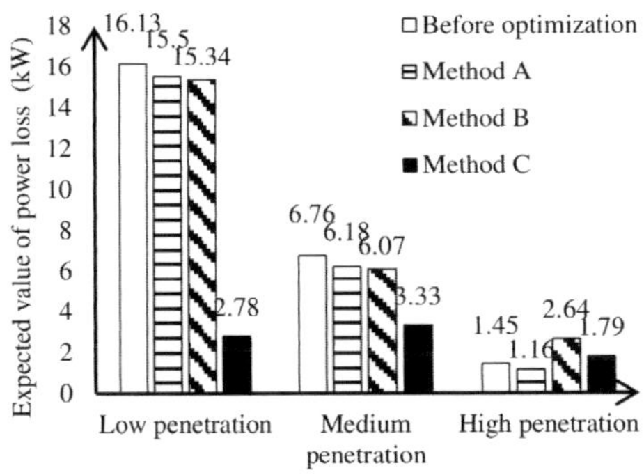

Figure 3. The comparison of expected value of active power loss

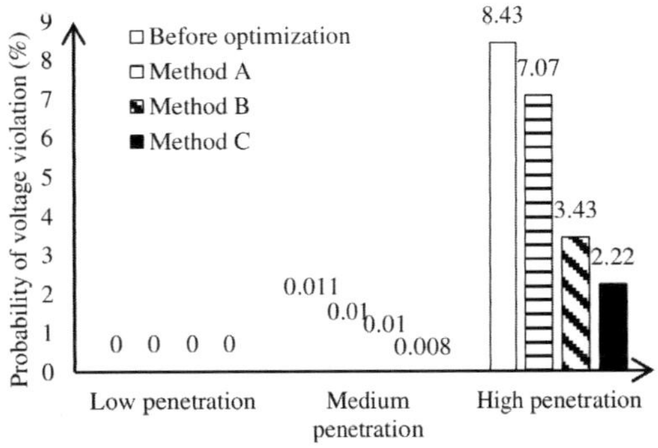

Figure 4. The comparison of probability of voltage violation

6. Conclusions

In this paper, a multi-objective optimal power flow model for ADN is proposed to minimize active power loss and probability of voltage violation. NSGA-II algorithm and PPF based on cumulant method are applied to find the Pareto optimal solutions. The results show that the model has evident advantage in reducing loss and ensuring voltage security, also it has better adaptability on different levels of PV penetration. The uncertainty of other types of DG and the impact of correlation between input variables will be considered in the future.

7. References

[1] Cheng B, Chen M Y and Huang Y C 2013 *J. Power System Protection and Control.* **41** 45.
[2] Xiao Y B, Cheng H Z, Yang W H, Wang X Y and Luan W J 2016 *J. Modern Electric Power.* **33** 38.
[3] You Y, Liu D, Yu W P, Chen F and Pan F 2012 *J. Automation of Electric Power Systems.* **36** 10.
[4] Ye L H, Yang Q and Yan W J 2015 *34th Chinese Control Conference (CCC).* IEEE pp. 9025-9030.
[5] Zhang L, Xu Y Q, Wang Z P, Li X D and Li P 2011 *J. Transactions of China Electrotechnical Society.* **26** 18.
[6] Guo K, Xu Y Q, Zhang L, Yue J F and Sun L F 2012 *J. Power System Protection and Control.* **40** 53.
[7] Lv Q J, Wang S and Liu T L 2012 *J. Power System Protection and Control.* **40** 71.
[8] Ding X Q, Zhang H, He J, Deng J X and Meng Z K 2014 *J. Water Resources and Power.* **32** 195.
[9] You Y, Liu D, Zhong Q and Xu N H 2014 *J. Automation of Electric Power Systems.* **38** 46.
[10] Zhu G L, Liu Z H, Liu W Z and Zhang Y J 2015 *J. Journal of South China University of Technology (Natural Science Edition).* **43** 49.
[11] Deb K, Pratap A, Agarwal S and Meyarivan T 2002 *J. IEEE Trans on Evolutionary Computation.* **6** 182.
[12] Liu W Y, Wen J Xie W, Wang W Z and Liang C 2015 *J. Proceedings of the CSEE.* **35** 1079.
[13] Fan M, Vittal V, Heydt T and Ayyanar R 2012 *J. IEEE Trans on Power Systems.* **27** 2251.
[14] Liu M B, Jian J Y and Dong P 2014 *J. Journal of South China University of Technology (Natural Science Edition).* **42** 13.
[15] Rahman S, Khallat M A and Salameh Z M 1988 *J. Energy.* **13** 63.
[16] Ciric R M, Padilha-Feltrin A and Denis I F E D 2004 *J. European Trans. On Electrical Power.* **14** 347.

Acknowledgment
This work was supported in part by 2015 Science and Technology Project of China Southern Power Grid (CSGTRC-K153010-002-XT).

IOP Conf. Series: Earth and Environmental Science **67** (2017) 012016 doi:10.1088/1755-1315/67/1/012016

The active control strategy on the output power for photovoltaic-storage systems based on extended PQ-QV-PV Node

Xu Chen[1], Bao-Rong Zhou[1], Jian-Wei Zhai[2], Yong-Jun Zhang[2] and Ying-Qi Yi[2*]

[1]Technology Research Center of China Southern Power Grid Co.Ltd, GuangZhou City , China

[2]School of Electrical Power South China University of Technology, GuangZhou city, China

Email: chenxu@csg.cn/1515017780@qq.com

Abstract. In order to solve the problem of voltage exceeding specified limits and improve the penetration of photovoltaic in distribution network, we can make full use of the active power regulation ability of energy storage(ES) and the reactive power regulation ability of grid-connected photovoltaic inverter to provide support of active power and reactive power for distribution network. A strategy of actively controlling the output power for photovoltaic-storage system based on extended PQ-QV-PV node by analyzing the voltage regulating mechanism of point of commom coupling(PCC) of photovoltaic with energy storage(PVES) by controlling photovoltaic inverter and energy storage. The strategy set a small wave range of voltage to every photovoltaic by making the type of PCC convert among PQ, PV and QV. The simulation results indicate that the active control method can provide a better solution to the problem of voltage exceeding specified limits when photovoltaic is connectted to electric distribution network.

1. Introduction

Zinc Distributed Photovoltaic Power Generation (DPPG) is a kind of clean energy, which can convert solar energy into electric energy directly. With DPPG's great development, more and more DPPG is connected to distribution network, which makes it convenient for the application of electricity, however, it brings great challenge to the safe operation of power system at the same time[1,2].

As to the problem that voltage exceeds specified limits when DPPG is connected to distribution network, the main solution is to regulate voltage of PCC by controlling power of photovoltaic[2]. A self-adaption control strategy with the limitation of active power and the compensation of reactive power is proposed to restraint over-voltage in [3], but the economic benefit is low while the strategy does not take active and reactive power loss of the network into consideration. Two conditions based on degree of voltage exceeding specified limits are discussed in [4]. One is normal condition and photovoltaic inverter is used to regulate voltage in this case, the other is emergency condition and reactive power of photovoltaic system in the same feeder is used to regulate voltage in this case. But, once there is communication failure between network nodes, the algorithm is invalid. Reactive power control strategy that based on the voltage of grid-connected node and photovoltaic active power output is put forward in [5]. The premise condition of this strategy is that the loss of active power and reactive power is positive correlation with the reactive power output of inverter when the active power output of photovoltaic system is fixed. But the active power outputs of photovoltaic system change over time. In [6], inverter's Automaticly adjusting voltage of distributed photovoltaic by controlling its power factor cos$\theta(U)$ is introduced. But in this strategy, the loss of network and the cost of inverter's adjustment aren't considered. In [7], a control method of over-voltage prevention based on active-reactive power collaborative optimization is established, which doesn't analyze complementary adjustment between energy storage and photovoltaic inverter.

Now the development and application of energy storage technology can increase the penetration of DPPG. The combination of DPPG and energy storage can effectively make up the randomness ,the intermittent and uncertainty of DPPG and limit it, it also plays an important role in improving the quality of DPPG, adjusting peak of power grid and stabilizing voltage fluctuation. In [8], energy storage is configured with goals of reliability and economy. A configured method of energy storage takes voltage constraints and voltage disturbance when photovoltaic is connected to distribition network into consideration in [9] .But the method doesn't allow for the inhibition effect of photovoltaic inverter itself to voltage fluctuation. In [10], grid-connected nodes of PVES are extended to QV node, and a day-ahead schedule of PVES based on voltage control method of PCC is put forward.All these works do research in areas of controlling the PVES, photovoltaic inverter and the voltage of PCC. But they didn't give well-directed research to the connotation of grid-connected node in the system of PVES. The combination of DPPG and ES in electric power distribution network and the active control strategy to voltage of PCC based on the transformation of node type of PCC haven't been found at home and abroad yet.

When we talk about voltage exceeding specified limits in distribution network with photovoltaic, it mainly refers to the voltage that is near PCC which is always caused by higher photovoltaic power output and lower absorb of load. The impedance ratio in the line is a little big,so photovoltaic inverter that is needed to restrain over-voltage maybe big. While the capacity of photovoltaic and the carrying capacity of lines both can lead to the invalidation of above-mentioned control strategy related to inverter's reactive power. This paper analyses the influence to voltage caused by photovoltaic inverter and energy storage of PCC.Then on the basis of detail study of QV node's power flow calculation, this paper gives a active control strategy on the output power for photovoltaic-storage systems based on extended PQ-QV-PV node, which provides optimal power reference value for every inverter and energy storage. The voltage can be effectively restrained within a certain range without communication between nodes by using this method.

2. The Mechanism of the restrain of voltage disturbance based on photovoltaic connected energy storage

In order to make full use of electric power generated by photovoltaic, unity power factor is used in DPPG when photovoltaic is connected to power grid. Energy storage can effectively regulate the voltage of photovoltaic PCC while it does well in balancing active power output. The reactive power output of photovoltaic inverter is mainly restrained by self capacity of DPPG. When the active power output of photovoltaic is lower than its rated capacity, the remaining capacity can be used by photovoltaic inverter to provide reactive power for power grid, which can regulate the voltage of PCC. Small DPPG is connected to electric power system by low voltage distribution network, and the impedance ratio of low voltage line is generally big. Fig.1 presents the simplified model of photovoltaic connected to low voltage distribution network. In the Fig.1, PV represents distributed photovoltaic source, R represents impedance of line, X represents reactance of line, V_{pcc} represents the voltage of PCC, V_s represents the equivalent voltage of low voltage distributed network, P_{pv} represents the active power of photovoltaic, Q_{pv} represents the reactive power of photovoltaic inverter, P_L represents the local active power of PCC, and Q_L represents the local reactive power of PCC.

V_{pcc} can be calculated by formula 1:

$$\dot{V}_{pcc}=V_s-\frac{(P_L-P_{pv})R+(Q_L-Q_{pv})X}{V_{pcc}}-j\frac{(P_L-P_{pv})X-(Q_L-Q_{pv})R}{V_{pcc}} \tag{1}$$

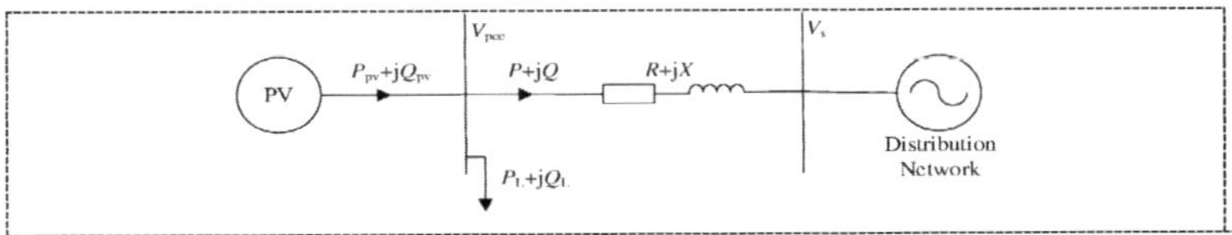

Figure 1. Equivalent circuit of grid-connected photovoltaic system

Solve formula 1, then we get the value of V_{pcc} as formula 2:

$$V_{pcc}=\sqrt{\frac{V_s-2a+\sqrt{V_s^4-4aV_s^2-4b^2}}{2}} \tag{2}$$

a and b in formula 2 can be calculated by formula 3 and formula 4:

$$a=(P_L-P_{pv})R+(Q_L-Q_{pv})X \tag{3}$$

$$b=(P_L-P_{pv})X-(Q_L-Q_{pv})R \tag{4}$$

This paper will discuss the question in two cases.
A) When the active power output of DPPG is less than local need of load (when $P_{pv}<P_L$), the power shortage will be met by power network. Obviously, the power which is given by power network flows into the PCC is P_L-P_{pv}. According to the formula(2), the voltage of PCC is got. If P_{pv} is small but the load is heavy, a is very big so V_{pcc} may be less than the lower limit. In such a situation, if there are energy storage devices in the PCC that provide real power or the photovoltaic inverter provides the system with inductive reactive power,the voltage is easier to be controlled within a qualified range and its quality is improved.
B) When the power output of DPPG is more than local load(When $P_{pv}>P_L$), the photovoltaic power output need to transport extra power to distribution network and meet local load. The extra power is P_{pv}- P_L. In calculating the voltage of PCC, P_L-P_{pv} is replaced by P_{pv}- P_L and Q_L-Q_{pv} is replaced by Q_{pv}-Q_L .Recalculating a, b and the voltage. In the situation, if P_{pv} is big while the load is small, V_{pcc} may be over the upper limit. If there are energy storage devices which absorb real power or making the photovoltaic inverter absorbs inductive reactive power to reduce voltage of PCC ,voltage amplitude will be controlled within a reasonable range and its quality is improved.

3. The active control strategy on output power for photovoltaic-storage systems based on extended pq-qv-pv node

A Power flow calculation with QV node

In conventional power flow calculation of power system, the nodes has three types that are PQ, PV and balanced nodes[9].With the increase of the penetration of DPPG, the distribution of power flow of power distribution networks has a profound change. Generally, in order to maximize the use of renewable energy, photovoltaic energy only generates real power with a power factor of 1.0.
Supplying DPPG may cause the voltage of PCC rising and even over the limitation. Therefore, under the premise of voltage constraints, it is unavoidable to face the problem that we need to find the real power output of the photovoltaic PCC in order to get the capacity of DPPG and the energy storage[10]. Thus, the calculation of power flow with QV node is derived. It means the reactive power Q and voltage magnitude V is given so the active power P and voltage phase angle δ are the unknowns. Generally pure active power sources to restrain voltage are QV nodes, such as the photovoltaic power station considering the constraint of voltage with energy storage in distribution power networks. The detailed steps solving power flow calculation including QV nodes based on Newton-Raphson algorithm are elaborated in [10].
The determination of voltage magnitude V is a key problem in treating photovoltaic-storage PCC as QV nodes. At present, the setting of V depends mainly on the historical operation data of photovoltaic system and experiences of experts. Then, the data of power flow of distribution power networks with

QV node based on a fixed value of V is not necessarily the optimal. When the penetration of DPPG in distribution power networks is high, the running cost of energy storage may be high if the PCC are treated as QV node.

A Active control strategy on the output power of photovoltaic-storage system based on PQ-QV-PV node

Configuring the energy storage device at photovoltaic PCC helps effectively to solve the over-limit voltage. Conventional photovoltaic-storage system outputs power in model of limiting the output power of the PVES. Taking into account of the uncertainty of DPPG output and the load fluctuation, the voltage fluctuation of the PCC need to be as small as possible to limit the voltage fluctuation of other nodes and decrease the probability of voltage over its limitation.

Based on this, from the point of voltage control with active transformation among PQ, PV and QV nodes of the node type of PCC, this paper proposes the method that controls the real power of the storage energy and the reactive power of photovoltaic Inverter.

The program block diagram of the active control method on the output power of photovoltaic-storage based on PQ-QV-PV node is shown in Fig.2.

The active control method on output power of photovoltaic-storage based on PQ-QV-PV node is listed:

a) Get data of the distribution network frame.

b) Get the forecasting data of load and photovoltaic power output next day .

c) Treat the PCC as PQ-PV-QV node and configure the upper and lower limit value: V_U, V_L.

d) Each node is treated as PQ node in distribution power system. The real power loss in distribution network is the objective function. Choosing the interior point methods of the optimal power flow to calculate voltage magnitude V_i and real power output P_{opti} of the storage energy and reactive power output O_{opti} of photovoltaic inverter.

f) To determine whether the V_i meets $V_i \in [V_L, V_U]$.If it is satisfied, active power output of energy storage of each PCC is that $P_{si}=P_{opti}$ and reactive power output of photovoltaic inverter is that $Q_{si}=O_{opti}$. If not, transform the node type whose V_i is over the upper(lower) limit into QV node. Then, reactive power Q is set to be 0,V is set to be V_U (V_L).Then, after calculating the power flow with QV node, real power of storage energy based on the QV node P_{qvi} is got.

4. An example analysis

A simulation model based on 10kV electric power distribution network is established in MATLAB. In order to emphasis the voltage change of key nodes, a complicated electric power distribution network is abstracted to a equivalent electric power distribution network. The structure figure of network flutter is given in Fig. 3. The node 1 is a balance node, and voltage V_1 =1.05pu.Node 6 and Node10 are connected with photovoltaic whitch capacities are 800kW and 1000 kW. Loads change over 24 hours according to their characteristics, and the outputs of photovoltaic change over a day according to its characteristics.

In order to test the effect of the active control method based on photovoltaic connected energy storage output power, comparisons are made among methods of photovoltaic grid-connected node without ES, a day-ahead schedule of photovoltaic connected energy storage based on the voltage control of grid-connected node. There are three cases.

Case 1: photovoltaic grid-connected node is not connected energy storage, node 6 and 10 are free outputs of photovoltaic power. For the convenience of observation, voltages of node 2,3,5,6,9,10 are focused on. The voltage curve of each node is calculated by power flow calculation, which is presented in Fig. 4.

The safe operation constraint of 10kV electric power distribution network is 0.93pu~1.07pu. The symbol "▲" represents that the voltage is over the upper limit.

Voltage fluctuation of each node is displayed in Fig. 4. Voltage fluctuation of node 6 and 10 is more violent because they are photovoltaic grid-connected nodes and their voltage fluctuation change along with the randomness and intermittent of photovoltaic outputs. Specially, voltage values of node 6 are

IOP Conf. Series: Earth and Environmental Science **67** (2017) 012016 doi:10.1088/1755-1315/67/1/012016

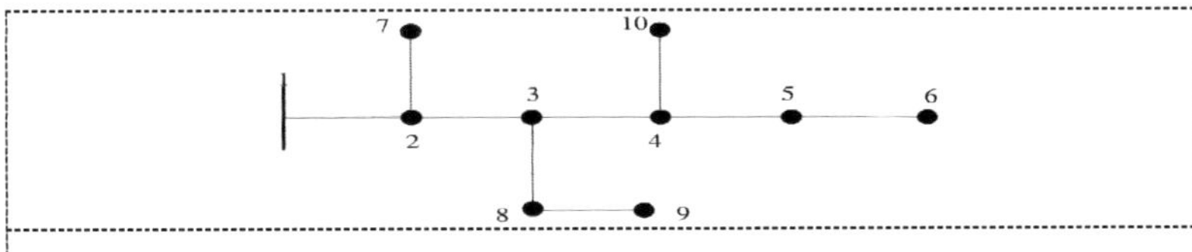

Figure 3. Topology of a 10kV distribution network this case

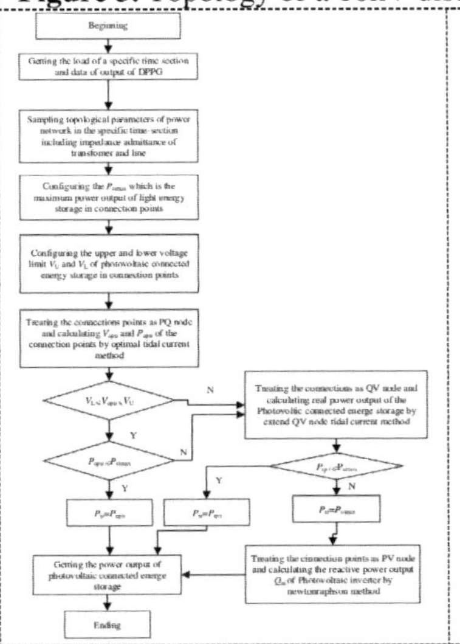

Figure 2. Program block diagram of The PQ-QV-PV node alternation.

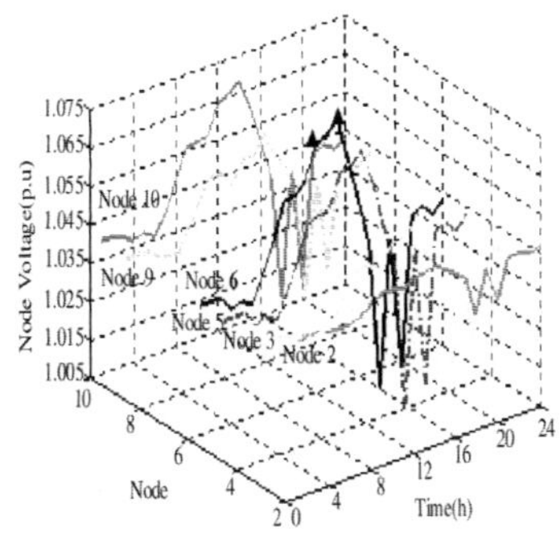

Figure 4. Voltage curves in case 1

over voltage upper limit at 13h and 14h. Generally, the amplitude fluctuation of each node is relatively large.

Case 2:Photovoltaic grid-connected node with energy storage, applying a day-ahead schedule of photovoltaic connected energy storage based on the voltage control method of grid-connected nodes. The power output of energy storage is controlled by the voltage of grid-connected node. The voltage of grid-connected node maintain constant which achieve by combining the active power output of energy storage with the active power output of photovoltaic. The voltage value of node 6 and 10 is supposed to be 1.050, that is to say, node 6(Q=0,U=1.050) and node 10(Q=0,U=1.050) are QV nodes, and the method including power flow calculation of QV node is used to analyze the control effect of the voltage of every node for case 2.

The maximum power of energy storage is confined in case 2, which is 0.7 times of the maximum output power of photovoltaic. The maximum output power of energy storage that is equipped to node 6 is 560kW,and node 10 that is 700kW. So when the value of output power of energy storage is over the limit, the node related should be transformed from QV node to PQ node, and the output of energy storage is the maximum output power of it. Under this circumstance, the voltage curve and the output power curve of photovoltaic connected energy storage are given in Fig. 5, Fig. 6 and Fig. 7.

Compared to Case 1, the voltage fluctuation of each node in case 2 is obviously gentler according to Fig. 5. Because of using the joint power output of photovoltaic connected energy storage, the voltage of photovoltaic grid-connected node is directly controlled. It seems like there are some nodes whose voltage is almost constant. So, this method could effectively solve the voltage fluctuation problem. But, the power output of energy storage in photovoltaic connection point is maximum as QV node is transformed into PQ node, the voltage is not constant. It will cause large voltage fluctuation and influence the capacity of energy storage to some extent.

Case 3: Active control strategy on the output power of photovoltaic-storage system based on PQ-QV-PV node

This case transforms the type of grid-connected node among PQ,QV and PV node in turn according to the voltage of the PCC. Based on this operation, voltage of connection point is stabilized with the real power output of energy storage and reactive power output of photovoltaic inverter. After configuring voltage limitation [1.047, 1.053](p.u) of connection points of node 6 and node 10, the output power curve and voltage curve of each node is shown in Fig. 8, Fig. 9 and Fig. 10.

The biggest power output of ES in case 3 is same as what in case 2. But voltage fluctuation is not only obvious gentler than what in case 2 but not beyond upper and lower limit. Case2 and case 3 all need enough ES to regulate voltage of photovoltaic connection point. In order to avoid excessive charging and discharging in ES, the state of charge of the energy storage battery is limited to 10%-90%. In the situation, the capacity of ES and real power loss of distribution network for case 2 and case 3 are compared in table 1. The real power loss of case 3 is less than 104.929KW and storage capacity is less than 10.061MW.h compared to case 2.

Based on the above analysis, the active control method on the output power of photovoltaic-storage systems based on PQ-QV-PV node makes a better performance in solving the problem that the node voltage is out of limit, improving the power flow of the distribution network and filling and shifting the peak load. Therefore, the capacity of ES could be reduced and voltage fluctuation could be controlled with the appropriate operation mode. Furthermore, the appropriate operation mode is helpful to reduce the investment of storage cell and voltage regulating device and the costs of the device need in its regulating action.

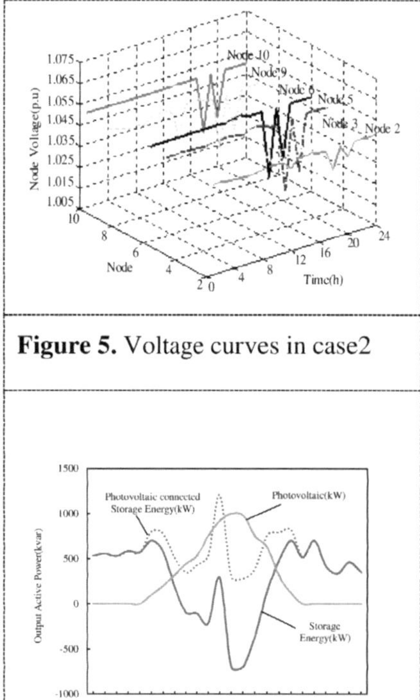

Figure 5. Voltage curves in case2

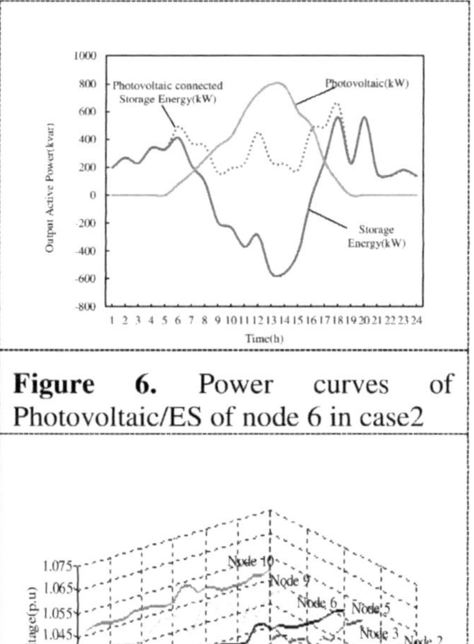

Figure 6. Power curves of Photovoltaic/ES of node 6 in case2

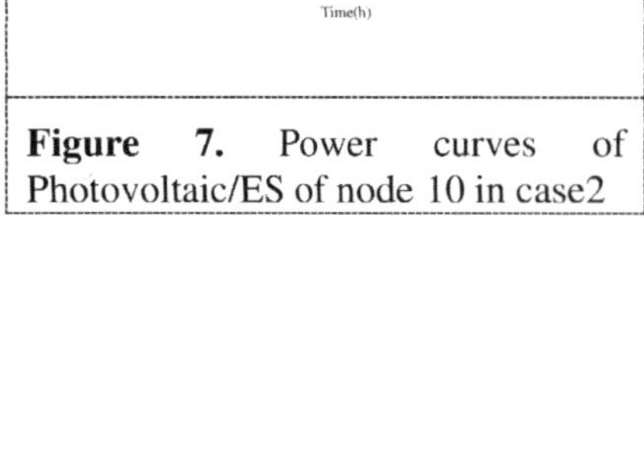

Figure 7. Power curves of Photovoltaic/ES of node 10 in case2

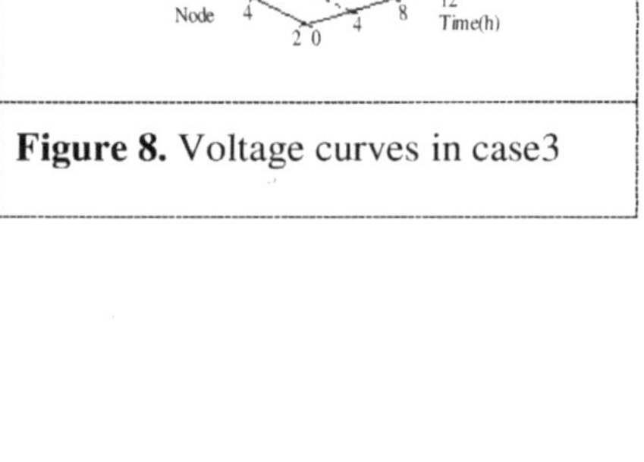

Figure 8. Voltage curves in case3

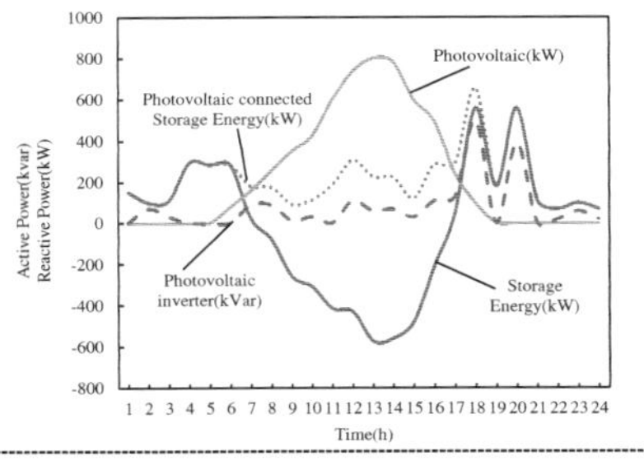

Figure 9. Power curves of Photovoltaic/ES of node 6 in case3

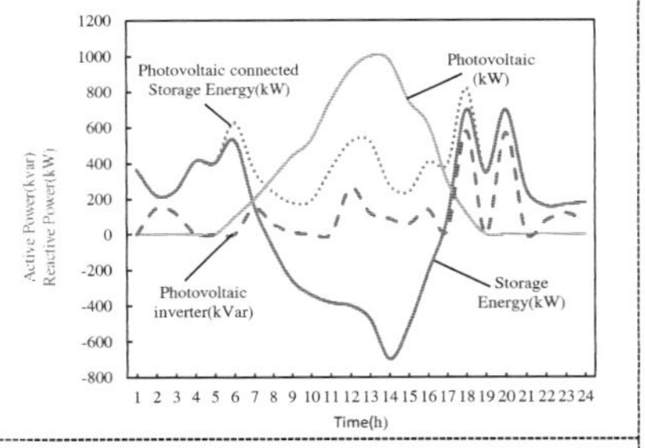

Figure 10. Power curves of Photovoltaic/ES of node 10 in case3

Table 1. Performance Comparison of Cases.

	Energy Storage Capacity of Node 6 (MW.h)	Energy Storage Capacity of Node 10 (MW.h)	Active Power Loss (kW)
Case2	1.143	0.286	0.286
Case3	1.143	0.067	0.067
C4	2.571	0.600	1.667

5. Conclusions

A) The active control method of output power for photovoltaic-storage systems based on PQ-QV-PV node is able to more directly reflect the fluctuation of photovoltaic power output and load of distribution network, owning better performance in reducing voltage fluctuation and improving the reasonability of allocating the output of ES. This method can reduce the real power loss of network.
B) The future research work will consider the rejusting cost of the ES and inverter together. In the same time, the future research will deduce the accurate mathematical model of this method.

6. References

[1] Wang Y. , Li C. B. , Qu F. Distributed wind-solar hybrid energy systems[J] . *Power System and Clean Energy*, 2010, 12(26) : 91-95 , 98(in Chinese) .

[2] Li Q. R., Zhang J. C. Solutions of voltage beyond limits in distribution network with distributed photovoltaic generators[J] . *Automation of Electric Power Systems*,2015,39(22) : 117-123 .

[3] Li H J, Xu Y, Adhikari S , et al . Real and reactive power control of a three-phase single-stage PV system and PV voltage stability[C]//*Power and Energy Society General Meeting* . San Diego , CA : IEEE , 2012 : 1-8.

[4] Tao W Q , Lu S Q , Yang G , He Q . A voltage regulation strategy of PV system in low-voltage neyworks[J] . *Renewable Energy Resources*, 2015, 33(9) : 1323-1328 .

[5] Zhou L, Chao Y, Liao B, Chao P F,Yang M . A voltage regulation strategy by grid-connected PV inverters in low-voltage networks[J] . *Power System Technology*,2013, 37(9) : 2427-2432 .

[6] Wang Y, Wen F S, Zhao B, Zhang X S . Analysis and countermeasures of voltage violation problems caused by high-density distributed photovoltaics[J] . *Proceedings of the CSEE*, 2016, 36(5) : 1200-1206 .

[7] Liu Y B, Wu W C, Zhang B M, Ju Y T, Li Z S . Overvoltage preventive control method based on active and reactive power coordinated optimization in active distribution network[J] . *Automation of Electric Power Systems*, 2014, 09 : 184-191.

[8] Lan G J, Li W Y, Yin K, Wen B, Song P C. Bi-objective optimization of capacity configuration based on grid-connected wind/photovoltaic/storage microgrid[J] . *Advanced Technology of Electrical Engineering and Energy*,2015,34(3) : 18-23.

[9] Zhu G L, Liu Z H, Liu W Z, Zhang Y J. Energy storage configuration to suppress voltage fluctuation in distribution networks with photovoltaic power[J] . *Journal of South China University of Technology(Natural Science Edition)*, 2015, 43(8) : 49-54+61.

[10] Liu Z H, Zhai S T, Zhang Y J, Yang J H, Liu W Z. A joint day-ahead scheduling for photovoltaic-storage systems based on extended QV bus-type power flow[J]. P*ower System Technology*,2015, 39(12) : 3435-3441.

Acknowledge
This work was supported in part by 2015 Science and Technology Project of China Southern Power Grid(CSGTRC-K153010-002-XT).

IOP Conf. Series: Earth and Environmental Science **67** (2017) 012017 doi:10.1088/1755-1315/67/1/012017

Performance evaluation of solar box cooker assisted with latent heat energy storage system for cooking application

Vigneswaran V S, Kumaresan G*, Sudhakar P, Santosh R,

Institute for Energy Studies, CEG, Anna University, Chennai 600025, India.
*Correspondence author: gkumaresan75@gmail.com

Abstract. Solar cooking is one of the most promising techniques to meet the cooking needs in remote areas where electricity and fuel supplies are meager. Solar box cooker is an efficient device used in solar cooking as it is simple to fabricate, easy to operate and hazard-free. In this context, the performance evaluation of a solar box cooker with varied number of reflectors has been undertaken. It was found that the time consumed for cooking in a box type solar cooker with four reflectors is lesser compared to that of a single reflector and its overall utilization efficiency increases with increase in the cooking mass. Further, a latent heat energy storage system was designed and fabricated to cook the food at off-peak hours of solar radiation. This latent heat energy storage system was combined with the solar box cooker. Oxalic acid dihydrate was used as the phase change material due to its high specific enthalpy and its melting point lying close to the cooking temperature. It was found that the solar box cooker with phase change material could be effectively utilized to cook food during off-peak hours of solar radiation.

1. Introduction

Major energy needs of domestic consumers include cooking, heating and lighting. Of this, cooking predominantly account for 36% of India's primary energy demand; hence it is necessary to meet this particular demand by using clean and efficient fuel like Liquefied petroleum gas (LPG) for cooking applications. In India, even today about 90% of rural people do not have access to modern cooking fuels and it drops to 33% in urban areas [1].

About 90% of the energy consumed by low-income people is used to satisfy their cooking demands. Hence, the need of the hour is to find out a cheap energy source which is locally available and ensure its availability to meet the energy demands of the poor and needy. As of now, biomass cook stoves and solar cookers are the best available alternatives because both biomass and solar energy are locally available renewable fuels. Among the two, solar energy is preferred as most parts of India is blessed with more than 275 sunny days with a mean daily solar radiation ranging from 5–7 kWh/m^2 [2]. According to estimates released on March 25, 2012, globally 4.3 million deaths were attributed to household air pollution (HAP), almost all in low and middle income (LMI) countries [3]. Hence it is necessary to design an improved solar cooker which can satisfy these demands either totally or partly. Though different type solar cookers are available, the solar box cooker (henceforth SBC) is easy to operate, simple and hazard-free [4]. But its slower cooking rate, lower absorber plate temperature and its inability to cook food during cloudy hours have resulted in lower adaptation. Moreover, it cannot be used in many cooking processes that involve deep frying. In spite of all these disadvantages, about 1,00,000 SBC's were disseminated in India with the help of government promotion by providing 50% subsidy [5]. Many researchers continue to toil hard to overcome the shortcomings of SBC's. The solar irradiation reaching the cooker was enhanced by using booster mirrors and it was found to be more efficient than box cookers without booster mirrors [6-7]. The primary benefit of using SBC's with booster mirrors is that it has the capacity to cook two meals per day. In addition to that, it also keeps the foodstuffs warm in the evening [8].

The temperature of the absorber plate must be as high as possible for effective cooking and this available heat energy must be efficiently transferred to the cooking vessel. The temperature of the absorber plate can be increased by using selective coating instead of normal black coating [9] and the heat transferred to the cooking vessel can be enhanced by 7%, by using fins [10]. Moreover, usage of trapezoidal shaped absorber plate helps in capturing the solar irradiation of larger incident angle [11]. Similarly, absorber plate of thickness 0.5 mm to 1 mm can be used to reduce radiative and convective heat losses [12, 13]. It is estimated that only 55% of the utensil heat is utilized to cook the food and the remaining is lost in the form of heat losses. Therefore, the utensils must be properly insulated for efficient transfer of heat [14]. Sometimes, the utensils are also covered with air tight plastic bags instead of normal metallic lids to enhance the heat transferred to it [15]. Further, use of concave lid instead of flat lid in SBC's decreases the cooking time span by about 10-13% [16]. The efficiency of the SBC increases by 5.9% when vessels with central cavity are placed in the SBC using lugs [17, 18]. As far as glazing is concerned, double glazing is found to be more effective than single glazing but, when transparent insulating material (TIM) of thickness 40 mm was introduced between this double glazing; the temperature of absorber plate increased by 40°C thereby making cooking possible even during off-peak hours [19,20]. This is because the convective loss occurring at the glazing is minimized with the introduction of TIM. Hence, the efficiency of the SBC using TIM increased to 30.5% resulting in energy savings of up to 1485 MJ of fuel equivalent per year [21].

Even after following all the above performance enhancement techniques in SBC, it is still not possible to cook food during cloudy days or off-peak hours unless energy storage is available in the SBC. Therefore, a 160 W heater must be incorporated in the SBC to make it operate throughout the year, in Indian weather conditions [22]. However, SBC's are generally used in rural areas where electricity is either not available or highly unreliable. Hence, SBC's must have an energy storage system to make it operate during cloudy days or late evening period. Though both sensible heat energy storage materials and latent heat energy storage materials can be used as energy storage medium latent heat energy storage material is preferred because of its higher energy density. The use of phase Change Material (PCM) which has its melting point around 100°C to 110°C is recommended because cooking takes place in the SBC between 95°C to 97°C [23]. Recently, a comprehensive performance assessment of a solar domestic cooking unit using high melting phase change material temperature (~167°C) of D-Mannitol PCM was carried out [24-26]. Further, many studies have been carried out by using PCM's like stearic acid, acetamide and acetanilide but none of them have melting point in the range of 100°C to 110°C [27-28]. This range of temperature is suitable for cooking by boiling which is a common need in food preparation by most Indian homes. Hence, in this paper, the performance evaluation of a SBC that uses Oxalic acid dihydrate as PCM, which has a melting point of 103°C, is carried out to study the feasibility of the designed SBC. should be indented by 5 mm.

2. System description

The cooking experiments were conducted with a double glazed (glass covers) SBC whose area is 0.49 X 0.49 m with 0.07 m depth. Fig. 1, shows the photographic view of the SBC with single reflector, multiple reflectors and fabricated PCM storage container with thermocouples. It consists of absorber plate made of aluminium coated with black paint. The experiments were carried out for both types, with and without additional reflectors; which was supported by frame and wooden blogs. A cooking vessel made of stainless steel was used for cooking inside the SBC. It was also painted with dull black paint in order to absorb the incident solar radiation. A PT100 thermocouple with an accuracy of ±0.15°C was used to measure the temperature of the food item kept in the cooking pot. The absorber plate and the upper glass plate temperatures were also measured by using thermocouples. The solar radiation was measured by using a pyranometer (Hukseflux LPO2) with an accuracy of <±1%; the ambient temperature and wind speed were measured using a weather station (WatchDog 2000). All the sensors were connected to a data acquisition system (Agilent make, Model no. 34970A) and observed values were recorded at regular time intervals during the experimentation. Oxalic acid dihydrate is used as PCM for storage of solar heat by its change of phase during sunshine hours thereby making cooking possible using a SBC, even during late evening hours. The PCM was selected by considering various aspects like phase change temperature, high latent heat and ready availability in the market.

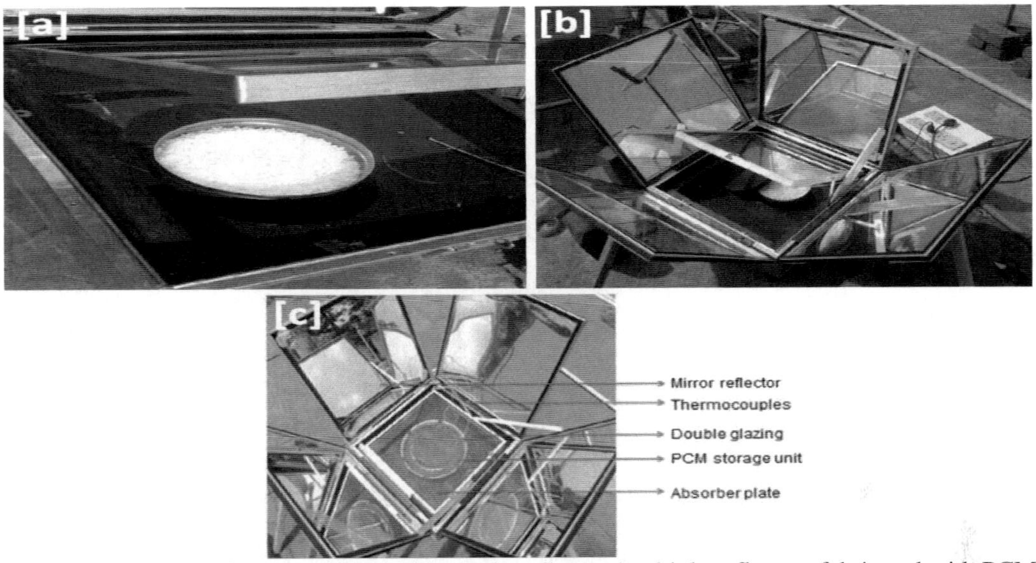

Figure 1. Photographic view of SBC with single reflector /multiple reflectors fabricated with PCM storage container with thermocouples.

The thermo-physical properties of oxalic acid dihydrate PCM are shown in Table 1.

Table 1. Thermo-Physical Properties of Oxalic Acid Dihydrate[a].

Property	Value (Unit)
Density	1.65 (kg/m^3)
Latent heat of fusion	370 kJ/Kg
Phase change temperature	101°C
Specific heat capacity	1.62 kJ/Kg°C

[a] Supplier: Excel Bio-Sciences, Chennai.

IOP Conf. Series: Earth and Environmental Science **67** (2017) 012017 doi:10.1088/1755-1315/67/1/012017

The required amount of oxalic acid PCM was evaluated as 2.9 kg by considering the energy requirement to cook half a kg of rice. For effective heat transfer, the energy storage unit must be as close as possible to the utensils. The fabrication of the PCM storage unit consists of two hollow cylinders made from aluminium with a diameter of 25.5 cm and 17.5 cm, placed one inside the other, as shown in Fig. 2. The PCM was heated to its melting point and poured into the outer container in liquid form. The outer surface of the PCM storage unit was painted with dull black paint in order to absorb the maximum solar radiation falling on it. The top cover of the PCM storage unit was also covered with aluminium sheet and was also coated with dull black paint.

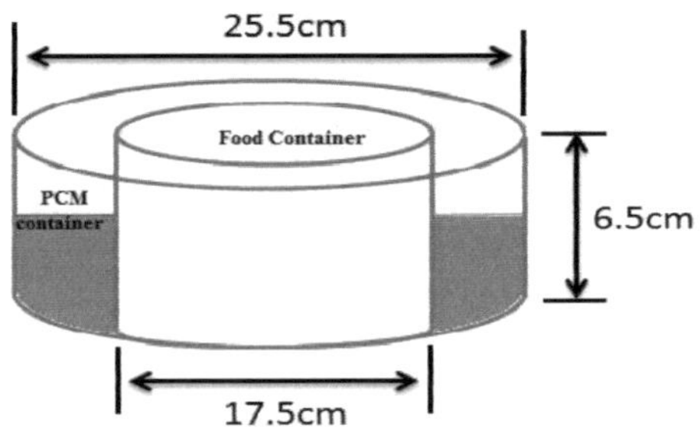

Figure 2. Schematic diagram of PCM storage unit with dimensions.

3. Experimental procedure

The performance of the SBC was studied in Anna University, Chennai, India. It is located at 13° latitude and 80° longitude. The test was initially conducted with water loaded in the SBC having single reflector and four reflectors. The experimental setup was developed to store the heat energy in the SBC during daytime hours and to utilize the stored energy for late evening time cooking requirements. In order to evaluate the maximum temperature achievable by the SBC, a known quantity of palm oil was loaded in it and tested. The temperature results confirmed the capability of the SBC to melt the selected PCM during daytime hours. The actual cooking trials and the thermal performance of the SBC integrated with energy storage system were conducted. During the experiment, a known quantity of rice kept in the SBC was cooked by using the stored energy in the PCM.

4. Results and discussion

4.1. Performance evaluation of the SBC with single and multiple reflectors

Initially, the experiment was carried out for a load of 400g of water in the SBC consisting of a single reflector and the temperature values obtained during the experimentation are shown in Fig. 4. The experiment was repeated in the same way for a higher load of 800g water hold-up in the SBC with four reflectors and the temperature values obtained during the experimentation are given in Fig. 4. Both the experiments were conducted in the afternoon, between 13.25 hrs. and 15.15 hrs. There was only a small ambient temperature variation of 2°C (31°C-33°C) during entire period of the conducted experiment. During the first half of the experiment, absorber plate temperature (T_p) increased suddenly. Then the variation in temperature was seen at slower rate. The fluctuation in glass cover temperature (T_{gu}) is due to the wind velocity effect. From Fig. 4, it is seen that due to more reflectors in the SBC, water temperature (T_f) was able to reach a higher value than the absorber plate temperature. The solar radiation intensity was highest at the start of the experiments and lowest at the end of the experiments. Fig.3 shows that the mean temperature of water is 77.5°C whereas in Fig. 4 it is 81.6°C. The overall utilization efficiency of the SBC was calculated using equation (1) [29].

$$\eta_u = \frac{m_f \; C_f \; \Delta T_f}{I_{av} \; A_c \; \Delta t} \tag{1}$$

Where η_u is the overall utilization efficiency, m_f is the mass of the fluid , C_f is specific heat capacity of the fluid, I_{av} is average solar radiation, A_c is aperture area , ΔT is temperature difference and Δt is the time taken. The overall utilization efficiency of the SBC with single reflector containing 400g of water was found to be 15.74%, whereas for 800g of water but with four reflectors, it was found to be 25.47%. It is also inferred from Fig. 3 and 4 that the time taken to reach maximum temperature for single reflector SBC is two hours whereas for four reflector SBC it is one and half hours, in spite of increase in load.

From the above results, it was decided to carry out further experiments with the SBC having four reflectors (booster mirrors).

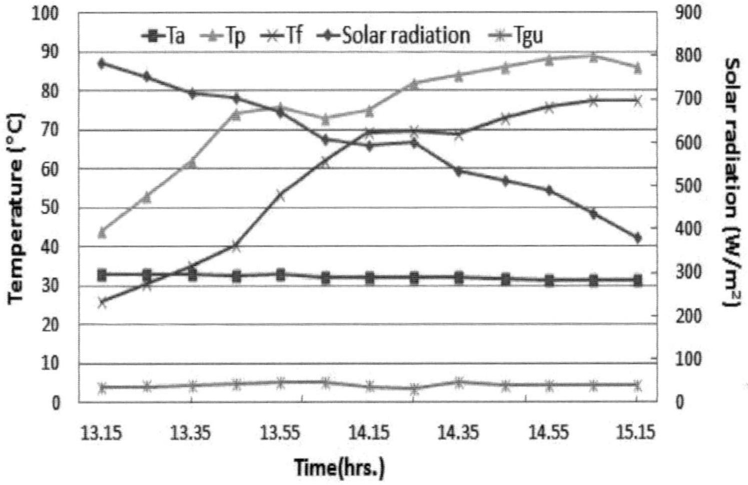

Figure 3. Various Temperatures during boiling of 400 g of water.

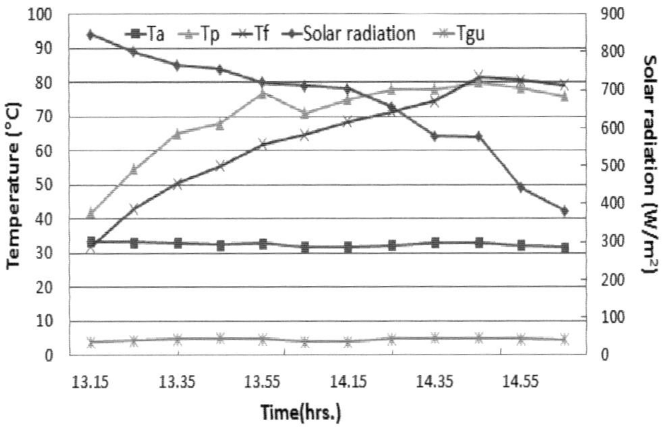

Figure 4. Various Temperatures during boiling of 800 g of water.

4.2. Performance testing of loaded sbc with palm oil

In order to check whether the SBC is capable of heating the PCM beyond its melting point of 103°C, tests were carried out by heating 250g of palm oil in the SBC with booster mirrors on four sides. Fig. 5 shows the temperature responses of the cooker with palm oil. The experiments were conducted at 1.00 p.m. and the palm oil attained maximum temperature (T_f) of 136.8°C at 14.45 hrs; thereby ensuring that Oxalic acid dihydrate PCM can be melted using the SBC with four reflectors. Quick decrease in palm oil temperature was observed at the end of the experiment due to less intensity of solar radiation.

4.3. Performance evaluation of the SBC assisted with PCM storage unit

Generally, the noon meal is cooked by using the solar radiation available up to 12.00 noon. Hence, the radiation available in the second half of the day was used for charging the PCM. The experiment was started at 12.30 hrs. Without load, i.e. nothing was placed inside the food chambers of the SBC, and temperatures were measured as shown in Fig. 6. An average ambient temperature (T_a) of 32°C and average solar radiation of 717.5 W/m^2 were recorded during experimentation.

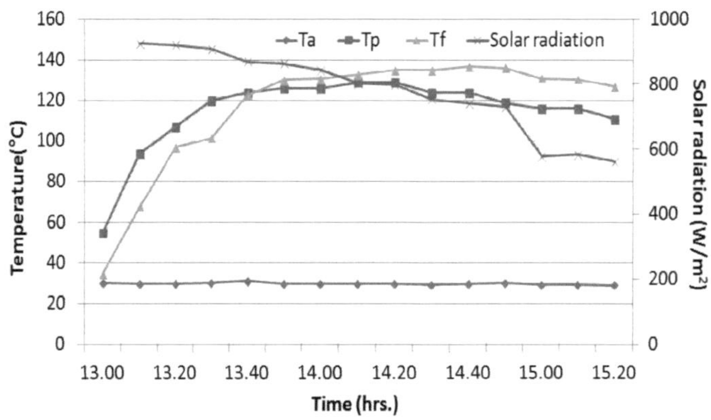

Figure 5. Various Temperatures during boiling of 250g of Palm Oil

The figure shows that the temperature of PCM (T_{pcm}) attained a maximum of 121°C, well beyond its melting temperature. It is evident from the figure that during phase change process, the temperature of PCM maintained a constant value of 103°C between the time period of 13.30 hrs. to 14.45 hrs. After charging the PCM up to 16.15 hrs in the evening, the cooker was kept closed to minimize the heat loss. The total energy stored in the PCM at the end of the charging process was calculated as 759.12 kJ at a temperature of 121°C by using equation (2).

Energy stored in the PCM = $[(m \times c_p \times \Delta T)_{solid,\ PCM} + m(\Delta H) + (m \times c_p \times \Delta T)_{liquid,\ PCM}$ (2)

Where ΔH represents the latent heat of fusion. But the cooking process was initiated from 19.00 hrs and the temperature of PCM dropped to a value of 104°C at this time. The energy contained by the PCM at this time was calculated as 720.24 kJ. The difference between these energy quantities are considered as losses in the storage unit and it was 38.88 kJ. The 140g of rice with 360g of water was loaded and cooked in the SBC by using energy stored in the PCM. The amount of energy spent to cook the food item was calculated as 410.63 kJ by using equation (3). Therefore, there was enough energy stored in the PCM to cook the food completely.

Total heat energy required to heat the whole mass to the boiling point = $(m \times c_p)_{system} \times$ (Water boiling temperature - ambient temperature) (3)

The time taken for cooking in the SBC and storage unit was 45 minutes. The food was found to be well-cooked using PCM as latent heat storage unit. The discharging efficiency (η_{dis}) of the latent heat energy storage system was calculated, using the above-stated energy quantities as 57% by using equation (4)

$$\eta_{dis} = \frac{\text{Energy required to prepare the food item}}{\text{Energy stored in the PCM}} \quad (4)$$

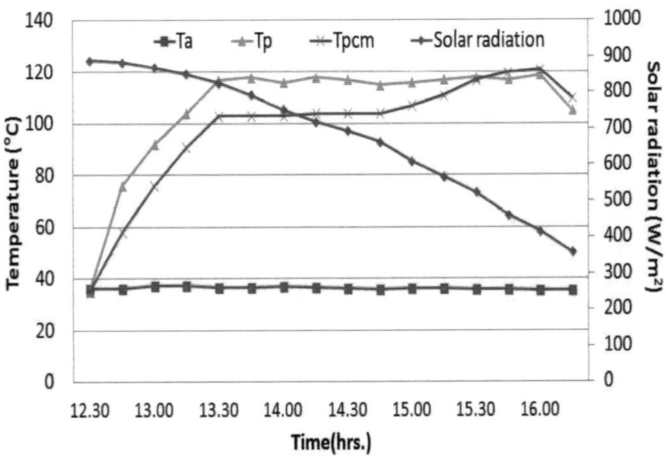

Figure 6. Temperature responses of SBC with PCM

5. Conclusion

In this work, an improved latent heat energy storage system was designed and fabricated to carry out experiments with conventional box type solar cooker for late evening cooking purposes. It was found that the efficiency of the solar cooker increased as the mass of the cooking substance increased when boosters were used. Moreover, the time taken to reach high temperature of fluid in the SBC also reduced when boosters were used. Oxalic acid dihydrate of 2.9 kg was used as phase change material. The results of the cooking experiment conducted at late evening hours showed that the selected PCM was able to supply heat energy effectively, with a discharge efficiency of 57%.

6. Refrences

[1] Reddy B S and Nathan H S K, Energy in the development strategy of Indian households—the missing half 2013 *Renewable and Sustainable Energy Reviews* **18** 203-210

[2] Sharma A, Chen C R, Murty V V S and Anant Shukla, Solar cooker with latent heat storage systems: A review 2009 *Renewable and Sustainable Energy Reviews* **13** 1599–1605

[3] Venkataraman C, Sagar A D, Habib G, Lam N and Smith K R, The Indian National Initiative for Advanced Biomass Cook stoves: The benefits of clean combustion 2010 *Energy for Sustainable Development* **14** 63–72

[4] Muthusivagami R M, Velraj R and Sethumadhavan R, Solar cookers with and without thermal storage: A review 2010 *Renewable and Sustainable Energy Reviews* **14** 691–701.

[5] Otte P P, Solar cookers in developing countries— what is their key to success? *Energy Policy* 2013 **63** 375–381

[6] Rao A V N, Rao T S and Subramanyam S,Mirror boosters for solar cookers – I 1988 *Energy Conversion and Management* **28 (3)** 265–269

[7] Saxena A, Agarwal N, Gupta P and Srivastav G, The Role of Mirror Boosters in Solar Box Type Cookers 2010 *Proc. of International conference on Jawaharlal Nehru Solar Mission,India*

[8] Mirdha U S and Dhariwal S R, Design optimization of solar cooker 2008 *Renewable Energy*. **33(3)** 530–44
[9] Shrestha V.M., Modification of the solar cooker 1987 *Solar and Wind Technology* **4(3)** 305–311

[10] Harmim A, Belhamel M, Boukar M and Amar M, Experimental investigation of a box type solar cooker with a finned absorber plate 2010 *Energy* **35(9)** 3799–3802
[11] Kumar S, Natural convective heats transfer in trapezoidal enclosure of box type solar cooker 2004 *Renewable Energy* **29(2)** 211–222.
[12] Thulasi Das T S, Karmarkar S and Rao D P, Solar box cooker: part I. Modeling 1994 *Solar Energy* **52 (3)** 265-272.
[13] Thulasi Das T S, Karmarkar S and Rao D P, Solar box cooker: part II. Analysis and simulation 1994 *Solar Energy* **52 (3)** 273-282.
[14] Lof G O, Recent investigation in the use of solar energy for cooking 1963 *Solar Energy* **7(3)** 125 133
[15] Ibrahim S M A and Elreidy M K., The performance of a solar cooker in Egypt 1995 *Renewable Energy* **6(8)** 1041–1050
[16] Gaur A, Singh O P, Singh S K and Pandey G N, Performance study of solar cooker with modified utensil 1999 *Renewable Energy* **18(1)** 121–129
[17] Rao A V N and Subramanyam S, Solar cookers – part I: cooking vessel on lugs 2003 *Solar Energy* **75(3)** 181–185
[18] Rao A V N and Subramanyam S, Solar cookers – part II: cooking vessel with central annular cavity 2005 *Solar Energy* **78(1)** 19-22
[19] Mullick S C, Kandpal T C and Kumar S, Top heat-loss factor of double-glazed box type solar cooker from indoor experiments 1997 *Energy* **22(6)** 559-565
[20] Nahar N M, Marshall R H and Brinkworth B J, Studies on a hot box solar cooker with transparent insulation materials 1994 *Energy Conversion and Management* **35(9)** 787-791
[21] Nahar N M, Design, development and testing of a double reflector hot box solar cooker with a transparent insulation material 2001 *Renewable Energy* **23(2)** 167-179
[22] Chaudhuri T K, Estimation of electrical backup for solar box cooker 1999 *Renewable Energy* **17(4)** 569–572
[23] Sharma S D, Buddhi D, Sawhney R L and Sharma A, Design, development and performance evaluation of a latent heat storage unit for evening cooking in a solar cooker 2000 *Energy Conversion and Management* **41(14)** 1497–1508
[24] Kumaresan G, Raju G, Iniyan S and Velraj R, CFD analysis of flow and geometric parameter for a double walled solar cooking unit 2015 *Applied Mathematical Modelling* **39(1)** 137-146
[25] Kumaresan G, Vigneswaran V S, Esakkimuthu S and Velraj R, "Performance assessment of a solar domestic cooking unit integrated with thermal energy storage system," Journal of Energy Storage, vol. 6, May 2016, pp. 70-79
[26] Kumaresan G, Iniyan S and Velraj R, Experimental and CFD analysis of a solar based cooking unit 2012 *World Renewable Energy Forum, Including World Renewable Energy Congress XII and Colorado Renewable Energy Society (CRES)Annual Conference* 312–317
[27] Buddhi D and Sahoo L.K., Solar cooker with latent heat storage: design and experimental testing 1997 *Energy Conversion and Management*. **38(5)** 493–498
[28] Sharma S D, Buddhi D, Sawhney R L and Sharma A, Design development and performance evaluation of a latent heat storage unit for solar cooker 2000 *Energy Conversion and Management* **41(14)** 1497–1508
[29] El-Sebaii A A and Ibrahim A, Experimental testing of a box-type solar cooker using the standard procedure of cooking power 2005 *Renewable energy* **30(12)** 1861-1871

IOP Conf. Series: Earth and Environmental Science **67** (2017) 012018 doi:10.1088/1755-1315/67/1/012018

Low-Cost energy contraption design using playground seesaw

I A P Banlawe [1,2] **and N J E L Acosta** [1]

[1] College of Engineering and Technology, Western Philippines University, Aborlan, Palawan, Philippines
[2] School of Graduate Studies, Mapùa Institute of Technology, Manila, Philippines
Email: ivaneannbanlawe@gmail.com

Abstract. The study was conducted at Western Philippines University, San Juan, Aborlan, Palawan. The study used the mechanical motion of playground seesaw as a means to produce electrical energy. The study aimed to design a low-cost prototype energy contraption using playground seesaw using locally available and recycled materials, to measure the voltage, current and power outputs produced at different situations and estimate the cost of the prototype. Using principle of pneumatics, two hand air pumps were employed on the two end sides of the playground seesaw and the mechanical motion of the seesaw up and down produces air that is used to rotate a DC motor to produce electrical energy. This electricity can be utilized for powering basic or low-power appliances. There were two trials of testing, each trial tests the different pressure level of the air tank and tests the opening of on-off valve (Full open and half open) when the compressed air was released. Results showed that all pressure level at full open produced significantly higher voltage, than the half open. However, the mean values of the current and power produced in all pressure level at full and half open have negligible variation. These results signify that the energy contraption using playground seesaw is an alternative viable source of electrical energy in the playgrounds, parks and other places and can be used as an auxiliary or back-up source for electricity.

1. Introduction

Electricity is one of the rudiments of mankind. It deals with the empowerment of machines and equipment that make the human work easier and simpler. Electricity comes from different types of source. It is either from a renewable or non-renewable energy.

Nowadays, people are in search of renewable sources of energy. With increasing demand and decreasing resources, many seek new ways of converting different forms of energy to electrical energy. Energy contraption is widely used to produce electricity by transforming different kinds of energy into electrical. One of the sources of energy contraption is from everyday human activities.

A study done by [1] about the human power conversion system based on children's play provided the basic theory behind the method wherein prototype compressed air-human power conversion system using a teeter totter (seesaw). The playground is used every day by children and produces motion which can be converted to electrical energy. Based on the results of [2] and [3], the up and down movement of the seesaw generates usable power that can be harnessed. This is the one of the alternative generation of electrical energy.

Air pressure engines, as described in [3] and [4], are prime movers that use air pressure to transmit motions to certain machines, specifically in this study, a DC motor from an old CD drive connected by the gear. The air pressure engine will be used to rotate the gear of the DC motor.

IOP Conf. Series: Earth and Environmental Science **67** (2017) 012018 doi:10.1088/1755-1315/67/1/012018

Electrical machine is the more general term for electric motor and generator since this device could operate as either a motor or generator depending on its configuration. As described in [3], the advantage offered by DC machines is their versatility. The ability to develop high starting and breaking torques, to make quick reversals of rotations, to maintain constant mechanical power output or to maintain constant torque and to permit continuous speed variation over a range as large as 4:1, make DC motors better suited to many industrial applications.
This study is about conversion of motion to electrical energy by children playing seesaw and measure the output voltage, current and power. Using pneumatic principle (hand pumps were employed in this study) the mechanical energy from the motion of the seesaw is converted to electrical energy. The objective of this study was to design a low-cost, prototype energy contraption using playground seesaw, design the said system using locally available and recycled materials, measure the voltage, current and power outputs produced by the energy contraption design at different situations and estimate the cost of the prototype.

2. Methodology
The study started with the selection of the available materials to be used for localized assembly of a small seesaw to be used as a prototype. The study proceeded with the use of an improvised check valve and acquiring materials from junk shops and old electronics equipment. Then parameters to be considered and the mechanism used to produce the energy were examined. After the system was assembled, it was evaluated. Three trials were done, with two replications each. Different pressure level of the air tank and the opening of on-off valve when the compressed air released was tested. Analysis of the data was done using a descriptive method and Coefficient of Variation.

2.1. Design considerations
The study proposed a low-cost energy contraption design using the seesaw and thus the following were considered: 1) Materials should be locally available; 2) Recycled materials are to be used; 3) No leakage of the air going into the tank; 4) Check valves could allow lower pressure of air entering the tank; 5) Air engine turbine must be suited with the air pressure to ensure rotation. Based on these criteria, the prototype was conceived.

2.2. Machine design
Fig. 1 shows the block diagram of operation of the prototype. The compressed air output from the hand pumps is stored in an air tank. When the compressed air inside the air tank reaches a set pressure level measured by the attached pressure gauge, the on-off valve is opened by the control of evaluator; it is either full-open or half-open. The released compressed air from the air tank is used to drive an air engine or air motor. A DC motor is coupled to the shaft of the air engine, resulting in conversion of the compressed air energy to electrical energy by the mechanical motion of rotation of shaft. The generated electricity was measured using a microcontroller.

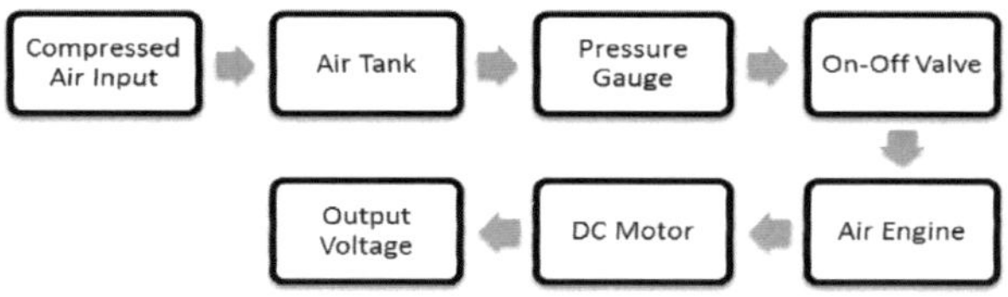

Figure 1. Block diagram of operation

2.3. Evaluation and data collection
The prototype was evaluated at Western Philippines University for data collection. The data were collected on January 20-21, 2016. Three trials were done, with two replications each. Different pressure level of the air tank (30psi, 20psi and 10psi) and the opening of on-off valve (full open and half open) when the compressed air released was tested. Output voltage and current was measured using ATMEGA 328p Microcontroller Unit (MCU), and then logged into a laptop.

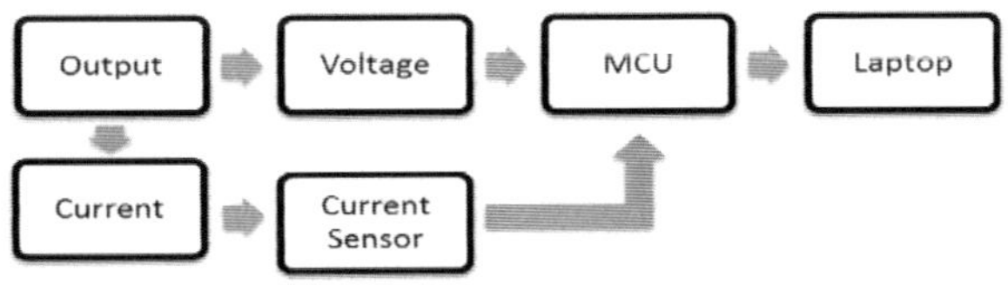

Figure 2. Block diagram of data logging

Fig. 2 shows the block diagram of the data logging process. The output voltage was directly measured using the analog input of the MCU while the current passes through the ACS712 current sensor before going into the analog input of the MCU. Data were recorded every second as long as the DC motor is rotating. The data measured from the MCU was directly recorded into a laptop through the serial cable. The average values of voltage and current for every rotation was computed and analyzed. Analysis of the data was done using descriptive method and confidence interval.

3. Results

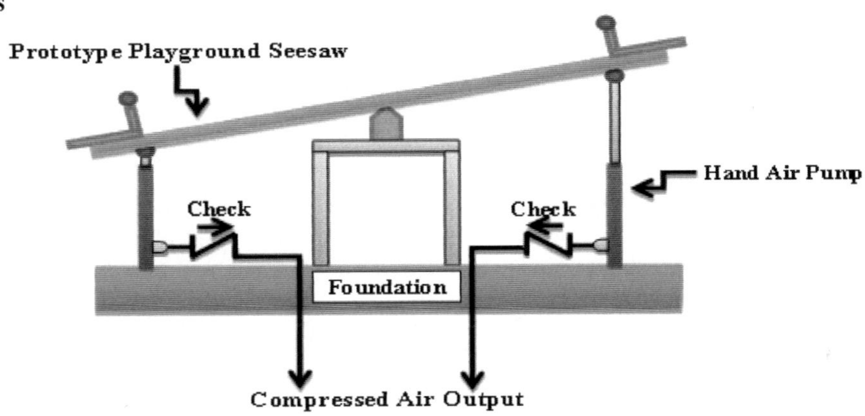

Figure 3. Compressed air generation from hand air pumps

3.1. Assembled prototype design

The prototype playground seesaw was made from ¾ inch diameter Galvanized Iron Pipe and constructed in a welding shop. Additional fittings were added for compatibility of the materials connected. Some of the parts of the system were from recycling shops. The air tank (or gas tank of mini gas stove as alternative) in Fig. 4 was brought from a junkshop. Based on the measurement, the tank has a maximum pressure of 30 psi.

The assembled air engine is shown in Fig. 5. DC generator and plastic gears from condemned CD drive were used. The DC generator produces 13volts. The gears have two different sizes, the high speed gear has ⅜ inch diameter and the low speed gear has 2 inches diameter. Bearings were brought from a shop of glass and aluminum; it was from junked sliding glass windows. The bearings have a size of ¾ inch diameter and ¼ inch diameter at the hole. Other materials were brought from a warehouse of used/old materials and equipment.

The other parts that were modified are the air engine/motor and air tank. The check valves, in Fig. 6, were devised using jackstone rubber ball inserted to ¾ inch diameter PVC straight connector.

Figure 4. Modified Air Tank.

Figure 5. Air Engine

Figure 6. Devised check valves

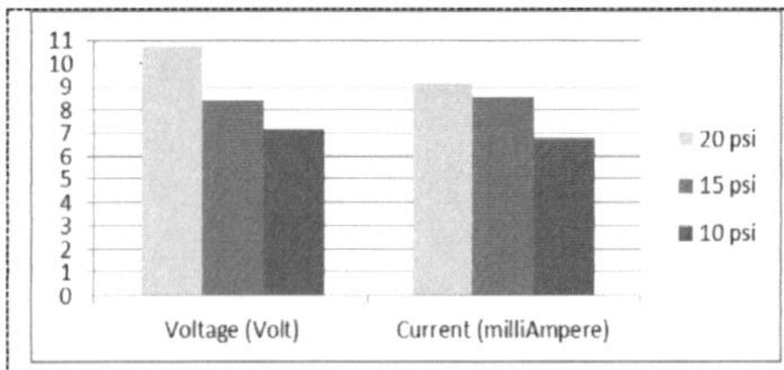

Figure 7. Voltage and current output at different pressure levels

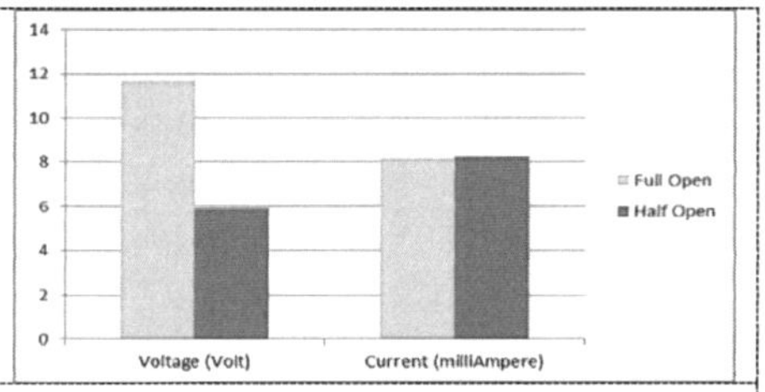

Figure 8. Voltage and current output at different valve opening

3.2. Output values

The relationship between the different pressure level of the air tank (30psi, 20psi and 10psi), the opening of on-off valve (full open and half open) and the output values of voltage, current and power were tested. Gathered data were analyzed using confidence interval analysis and statistical interaction of variables.

Results revealed that the pressure level at 20 psi produced highest voltage of 10.74V, than at 15psi with the voltage value of 8.41V and 10 psi with 7.61V as seen in Fig. 7. Current values also decrease gradually as the pressure decreases, average current values recorded are 9.18mA, 8.55mA, and 6.81mA for the corresponding 20psi, 15psi and 10psi respectively. From Fig. 8, it can also be seen that opening of the on-off valve at full open produced higher voltage (11.65V) and power (0.1W) as compared with the half open with an average voltage of 5.89V and a power of 0.05W, while current values do not vary extensively.

The interaction between the pressure levels and the opening of the valve is presented in Table I. Statistical analysis showed that all pressure level at full open produced significantly higher voltage than the half open. But, the 20 psi at half open also produced significantly higher voltage. However, the mean values of the current and power produced in all pressure level at full and half open have negligible variation. These results imply that higher output values are produced at higher pressure and at full open valve.

Table 1. Output values based on the interaction between pressure levels and opening of the valve

Pressure	On-off Valve	Voltage (Volt)	Current (Ampere)	Power (Watt)
20 psi	Full Open	$12.13^{a} \pm 1.04$	0.010061 ± 0.0037	0.123711 ± 0.0442
	Half Open	$9.36^{a} \pm 2.67$	0.008308 ± 0.0016	0.081280 ± 0.0312
15 psi	Full Open	$12.03^{a} \pm 1.28$	0.008971 ± 0.0026	0.118256 ± 0.0320
	Half Open	$4.78^{b} \pm 1.47$	0.008126 ± 0.0029	0.047627 ± 0.0281
10 psi	Full Open	$10.80^{a} \pm 1.66$	0.005433 ± 0.0024	0.060657 ± 0.0389
	Half Open	$3.53^{b} \pm 1.89$	0.008189 ± 0.0022	0.029319 ± 0.0190

Note: Means with the same letter in the same column are not significantly different ($p > 0.05$).

4. Conclusion

A prototype system of playground seesaw for energy contraption was developed in this study. Results showed that higher pressure at full open valve can be associated to higher output values (see Fig. 7). These results signify that the energy contraption using playground seesaw is an alternative viable source of electrical energy in the playgrounds, parks and other places and can be used as an auxiliary or back-up source for electricity, especially in developing countries.

The study could be further improved by providing bigger air tank to prolong the release of the compressed air, investigating the different diameters and length of the hose that can be used, use of a double-acting pneumatic cylinders to be more efficient in air production than the hand air pumps, exploring different types of the compressed air engine/motor and turbine design, and utilization of an automatic on-off valve or pressure relief valve to release air if the certain pressure level of the air tank is reached.

5. References

[1] Pandian S R 2004 *A Human Power Conversion System Based on Children's Play* (Tulane University, New Orleans, LA)

[2] Goff H 2008 *See-saw to power African schools* (BBC News – UK: education)

[3] Abad J R B, Capucao M D and Legaspi L D C 2012 Energy Contraption Design Using Playground Seesaw *Proc. 22nd ASEMEP National Technical Symposium* (Pasay City, Philippines) pp1-3

[4] Herbert L 1836 *Air pressure engines* (*The engineer's and mechanic's encyclopædia* Vol. 1) p31.

Acknowledgments

The authors gratefully recognize Prof. Sharon Rose P. Anunciado, statistician, for the analysis. Engr. Mirriam Banlawe, the dean, together with the whole faculty and staff of the College of Engineering and Technology, Western Philippines University, Aborlan, Palawan for their kind suggestion and invaluable support in this study.

IOP Conf. Series: Earth and Environmental Science **67** (2017) 012019 doi:10.1088/1755-1315/67/1/012019

Investigation of indirect benefits of PV rooftop in Thailand

T Khumkrong[1], S Chuangchote[1,*], D Chenvidhya[2] and K Kirtikara[2]

[1] The Joint Graduate School of Energy and Environment, Centre of Excellence on Energy Technology and Environment, King Mongkut's University of Technology Thonburi (KMUTT), 126 Pracha Uthit Rd., Bangmod, Thung Khru, Bangkok 10140, Thailand
[2] CES Solar Cells Testing Center (CSSC), King Mongkut's University of Technology Thonburi (KMUTT), 49 Soi Tientalay 25, Bangkhuntien, Bangkok 10150, Thailand

*E-mail: surawut.chu@kmutt.ac.th

Abstract. Other than electricity generation, which is the direct benefit of PV rooftop, cooling load reduction due to PV shading is a benefit impact in the uses of PV rooftop. This report is a study of those indirect benefits of PV rooftop. Relation of shading of PV modules and reduction of cooling load was studied in a real testing cite at the office building of CES Solar Cell Testing Center (CSSC). Several data, i.e. solar radiation, rooftop temperatures before/after PV-panel installation, and electricity consumed by equipment, were monitored and collected. This data could be further estimated for cooling load via transient heat conduction approach.

1. Introduction

Electricity is one of major important things to all human living and it is a primary factor for country development. Since, it plays significant roles in both economic and community sectors. Moreover, it helps to encourage agricultural products, industrial products, and public service parts.

In the last decades, population growth has been significantly increased but production of electricity would be insufficient to the energy consumption in Thailand. Therefore, Thailand has to find sufficient primary energy sources for electricity production. However, due to inefficient energy production technologies and energy management plans, Thai government requires huge amount of natural resources, i.e. crude oil, diesel, natural gas, and coal [1]. In addition, Thai government is also buying more electricity from Malaysia and Laos to support domestic energy demands [2-4].

In fact, if we have no a good management plan, our reserved energy will be terminated in the near future. Energy-statistic data of Thailand in 2014 showed the continuous increase of overall energy consumption. The highest electricity has spent by the household, which is mainly from the air-conditioning system. The way to reduce this electricity cause is reduction of the cooling load the building. One strategy to reduce the cooling load is decrease of heat gain from solar radiation absorbed on the rooftop. Since the rooftop directly receives heat during daytime. If there is no proper prevention, cooling load of air-conditioner will increase and cause more expenses to the resident in the future.

From previous studies, there are various methods to reduce heat gain passing through the rooftop to the building. One of interesting method is the uses of solar PV panel as shading devices. Moreover, solar energy is a suitable energy resource and shows potential to support other renewable technologies in Thailand, such as wind or hydroelectric powers, because Thailand is geographically located at the position with a high level of solar insolation. Moreover, the cost of PV-system installation is relatively low and panels are relatively visually unobtrusive compared with other renewable energy solutions such as wind power turbines [4]. Therefore, Thailand Ministry of Energy established a new Feed in

Tariff Program for solar PV rooftops [5-7] to provide incentives and draw more intention of people on PV system.

This new solar policy aimed to support new capacity of 1,000 MW (200 MW for rooftop projects and 800 MW for community projects). New (fixed) feed in tariffs was specified. For the rooftop projects FiT was set at 6.16-6.96 THB (depending on size). For community projects (>1 MW), FiT was set for (1) an initial value of 9.75 THB, (2) 6.5 THB after the 4th year, and (3) 4.5 THB during the 11th year and the 25th year [8-9].

Not only decrease of the heat gain flows into the building using PV panels as a shading device as mentioned above. Solar rooftop also has other benefits (or indirect benefits). Firstly, the expense of resident can be reduced by use of electricity from solar PV instead of that from electricity grids. Secondly, it helps to promote the energy and environment preservation and reduction of the climate change problem. Lastly, it educates the good energy and environmental attitudes to all family members and community that use solar rooftops.

As mention above, the residential solar rooftop market in Thailand is new and needs further marketing and support for future growth. Therefore, this research aims to investigate an indirect benefit, i.e. reduction of heat gain to building, of PV rooftop in Thailand in details.

2. Experimental setup

2.1. Testing site building and location

The building used in this study was a model room of CSSC building at the King Mongkut's University of Technology Thonburi (KMUTT). The simulation images of the building are shown in figure 1. The building was made from lightweight concrete (thick = 0.1 m) with a roof area of 126 m^2 made from roof tiles (CPAC Monier). Solar PV panels used in this work were amorphous silicon cells (54W) purchased from Kaneka (Japan) with the size of 996 mm x 999 mm. The panels (14 panels) were mouthed horizontally with the roof surface in 4 groups (4, 4, 4, and 2 panels) (Fig. 1).

There were 9 windows placed around the building. This building contains 2 sections, office and toilet zones. The office zone was located at the front (facing to the North), while the toilet zone was located behind the building (facing to South). The office section was divided into air conditioning and un-conditioning zones. On workdays, the building was cooled by the air-conditioning system. For the toilet, it was made of hollow-lightweight concrete cube without any air conditioning system (it was cooled by natural ventilation through the gate and windows around the building).

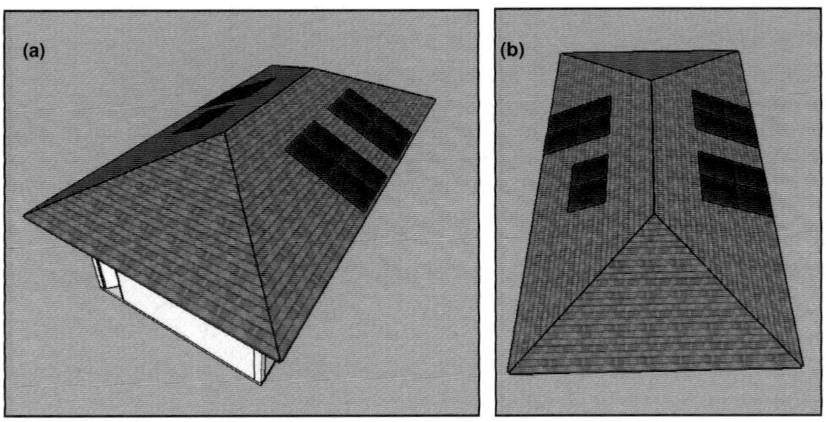

Figure 1. Model of CSSC building (facing to North) with 14 PV panels (4, 4, 4, and 2 panels in 4 groups) created by SkectUp Program: (a) eagle-eye and (b) side views.

2.2. Deta monitoring and collections

Data monitoring was carried out in 3 sections; on the rooftop, the loft under PV panels (which directly received shading from PV panels), and the conditioning zone (office). The details are in Fig. 2.

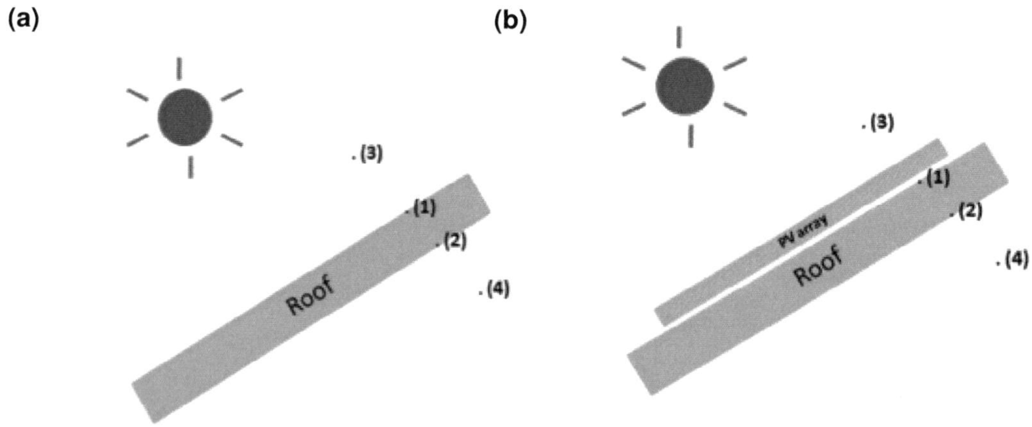

(1) Outer surface temperature of the roof
(2) Inner surface temperature of the roof
(3) Outdoor dry-bulb temperature
(4) Air temperature above ceiling

Figure 2. Temperature measurement points on: (a) a conventional room (directly exposed roof) and (b) a model room equipped with PV panels.

Interior monitoring. Data were collected at the air conditioning (office) zone and the loft zone every day from 0000 PST, Tuesday, September 27, 2016 to 1930 PST, Saturday, October 2, 2016. For the air conditioning zone, the temperature was collected by micro loggers (the results are not reported here), while for the loft zone, the temperatures (inner surface temperature of the roof and air temperature above ceiling) were collected by thermocouples.
Exterior monitoring. Solar irradiation, humidity, outside temperature (outer surface temperature of the roof), and wind speed were collected every 1 min throughout the study period using a pyranometer, a temperature probe, and a wind monitor, respectively. Outdoor dry-bulb temperature was collected by a weather station of CSSC.

3. Results and discussion

3.1. Benefits of PV rooftops (reduction of heat gain)

Figure 3 shows inner surface temperature of the roof, outer surface temperature of the roof, outoor dry-bulb temperature, and air temperature above ceiling measured in both a room without any PV panel (exposed roof) and (b) a room with PV panels. It was found that when the model room exposed to sunlight, the outer surface temperature of the west-facing roof (measurement points 1 in figures 2(a) and 2(b) showed a peak at noon and always higher than the outdoor dry-bulb temperature (measurement point 3 in figures 2(a) and 2(b) and the air above ceiling temperature (measurement point 4 in figures 2(a) and 2(b)) during daylight hours. However, the outer surface temperature was lower than the outdoor dry-bulb and air above the ceiling temperature during the night.

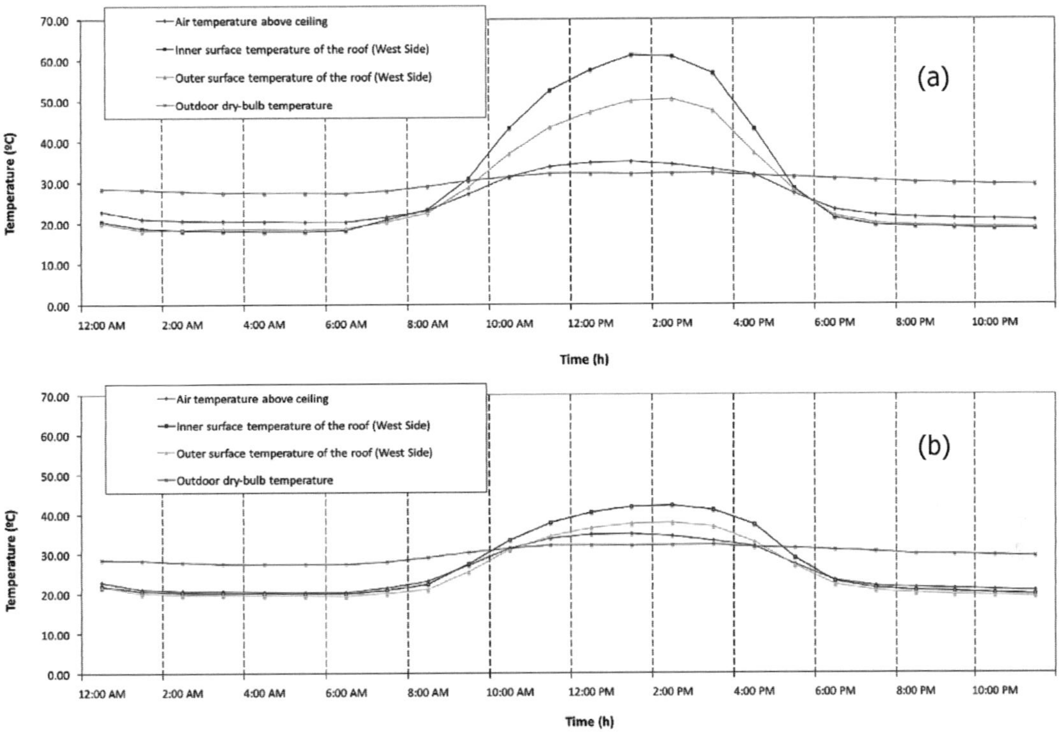

Figure 3. Inner surface temperature of the roof (West side), outer surface temperature of the roof (West side), outoor dry-bulb temperature, and air temperature above ceiling measured in both (a) a room without any PV panel (exposed roof) and (b) a room with PV panels.

For the exposed roof (no PV panel), the outer surfaces of roof (point 1 in figure 2(b)) were heated by solar radiation, then the heat could transfer to the inner surface (under the roof, point 2 in figure 2(b)). Since the air can move on the outer surface, so the outer surface was cooler than the unventilated under surface (in the loft, the air was trapped, so it heated the inner surface up). This situation also occurred in the case of the room with PV panels.

The temperatures under the PV arrays (roof with PV shading) (measurement points 1, 2, and 4 in figure 2(b)) was significantly lower than that of the exposed roof (measurement point 1, 2 and 4 in figure 2(a)). This is because surface temperature of the exposed roof was heated directly by the strong solar irradiance [9, 10]. However, under the PV arrays was shaded by the panels. So, it was not heated by the solar radiation, but only long wavelength radiation (lower energy) under the panels and weak diffuse radiation from the sky. The sum of these radiations are less than the solar irradiance from the exposed roof [9].

3.2. Thermal infrared camera

Sections A thermal infrared (TIR) image was use the confirm the temperature difference in the case of use of PV rooftop. Figure 4 demonstrated that roof surface temperatures under the PV arrays were lower than the directly exposed roof (up to 9 °C) at 1200 PST. As illustrated in figure 4, a cool area under the PV panels was obviously observed. This observed result was in line with the actual situation explained above.

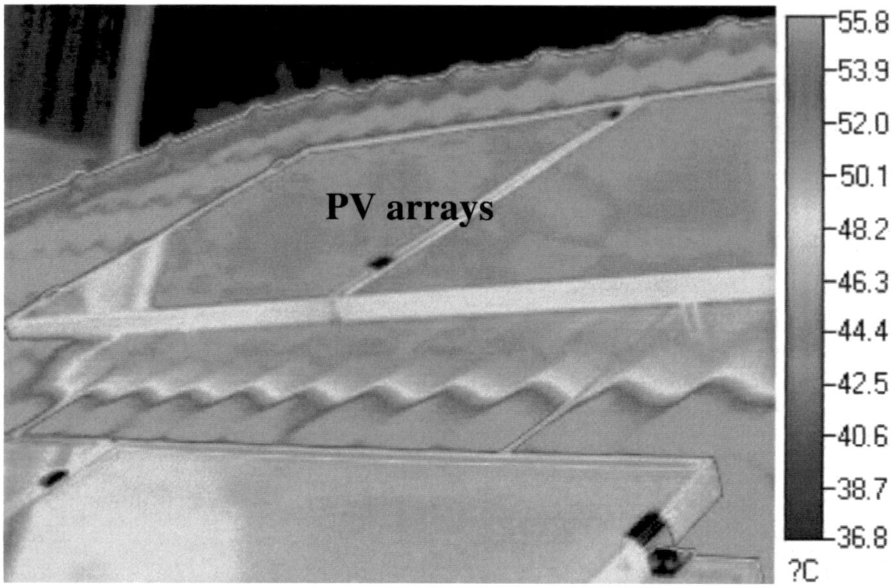

Figure 4. A thermal IR camera image of outer roof surface (measurement point 1 in figure 2) taken at 1200 PST on September 30, 2016. The color bar denotes temperatures in °C.

4. Conclusion

This study demonstrates that PV panels as building shading serve as a potential solution for the reduction of heat pass through the roof to the building in the hot climate days. All results could be considered as an estimation of PV benefits for the PV-panel equipped building. The model in this works can be use further to calculate for energy saving improvements. This data can be further estimated for cooling load via transient heat conduction approach. The electricity generation and reduction of building cooling load by shading of PV system in a bigger image (covered the recently announced solar energy policy of Thailand) should be further investigated.

5. References

[1] Department of Alternative Energy Development and Efficiency (DEDE) 2013 *Thailand PV status report 2012-2013* (Bangkok)
[2] Department of Alternative Energy Development and Energy Efficiency (DEDE) 2012 *Thailand Energy Situation 2011, Annual Report* (Bangkok)
[3] Department of Alternative Energy Development and Efficiency (DEDE) 2014 *Thailand final energy consumption 2009-2014* (Bangkok)
[4] Energy Policy and Planning Office (EPPO) 2012 *Thailand Energy Statistics* (Bangkok)
[5] Department of Alternative Energy Development and Energy Efficiency (DEDE) 2012 *Alternative Energy Development Plan 2012-2021* (Bangkok)
[6] MacDonald S 2012 Solar Photovoltaic Energy in Thailand: An assessment of government support mechanisms (Bangkok)
[7] Tongsopit S 2014 Thailand's Feed-in-Tariffs for Solar Power: Calculation, Impacts and Future Directions *ESI.* vol 7 p 11-12
[8] Energy Policy and Planning Office (EPPO) 2012 *Energy Efficiency Development Plan 2011-2030* (Bangkok)
[9] Al-Obaidi K M, Ismail M, and Rahman A M A 2014 Passive cooling techniques through reflective and radiative roofs in tropical houses in Southeast Asia: A literature review *Front. Arch. Res.* vol 3 p 283-297

[10] Dominguez A, Kleissl J, and Luvall J C 2011 Effects of solar photovoltaic panels on roof heat transfer *Sol. Energy*. vol 85 p 2255-2255

Acknowledgement
The authors would like to gratefully acknowledge The Joint Graduate School of Energy and Environment (JGSEE) and CES Solar Cell Testing Center (CSSC) for facilities and supports.

IOP Conf. Series: Earth and Environmental Science **67** (2017) 012020 doi:10.1088/1755-1315/67/1/012020

Solar PV leasing in Singapore: enhancing return on investments with options

Shuang Song, K.L. Poh

Department of Industrial & Systems Engineering, National University of Singapore, 1 Engineering Drive, Singapore 117576

Abstract. Renewable energy is getting more important nowadays as an alternative to traditional energies. Solar energy, according to Energy Market Authority, is the most viable in the context of Singapore compared to other renewable energy sources due to land constraints. In light of the increasing adoption of solar power in Singapore, this paper focuses on solar PV leasing using a case study. This paper assesses the prospect for solar PV leasing companies in Singapore through the lens of embedded real options. The recent news that solar power is becoming the cheapest form of new electricity presents the leasing company an option to expand the scale of solar PV system. Taking into account this option, the Net Present Value (NPV) of the investment increased significantly compared to the case without real options. Technological developments result in a continuously changing environment with uncertainties. Thus, decision makers need to be aware of the inherent risk associated and identify options to maximize NPV. This upside potential is realized by exercising the managerial flexibility and exploiting the uncertainty. The paper enables solar energy planners to consider possible managerial flexibilities under uncertainties, showing how option thinking can be incorporated in the valuation of solar energy.

1. Introduction

One important pressing issue we are facing now is sustainable energy. Renewable energies such as wind, solar and nuclear have gained momentum in different countries and would replace traditional energies which are not sustainable in future. For Singapore particularly, land constraints and other considerations make solar energy the most viable renewable energy. Thus, the motivation behind this paper is to utilize real option analysis and decision analysis to study the economic value of solar PV leasing for manufacturers in Singapore.

2. Valuation methods

2.1. Risk-adjusted net present value

This method is based on discounted cash flow analysis; however, the discount rate is adjusted to reflect different types of risks. Discounted cash flow analysis discounts expected free cash flows using a discount rate and then sum them up.

$NPV = \sum \frac{E(free\ cash\ flow\ in\ year\ t)}{(1+discount\ rate)^t} - I_0$, where I_0 is the present value of investment outlay for the project.

Total NPV = NPV (Revenue) – NPV (costs).

With option, the expanded NPV = Passive NPV + Value of Option to Expand.

Under this method, it is assumed that all future expected free cash flows are discounted using the same discount rate, which is not accurate especially for risky projects. It may underestimate the value of projects and thus, lead to myopic decisions [1]. Furthermore, the result is subject to sensitive factors such as thc discount factor and cash flows.

KP [2] pointed out there are generally two broad categories of risks, namely private risks and market risks. Private risks are company-specific or projects-related whereas market risks affect all similar companies. Thus, under the risk-adjusted NPV method, cash flows associated with private risks should be adjusted using risk-free rate since investors are unwilling to pay for a risk premium; in contrast, investors are willing to pay a risk premium for market risks. Using different discount rates for cash flows associated with different type of risks, the value of an investment or project can be more accurate.

2.2. Decision analysis with utility functions

Under the decision analysis approach, risk-preference is taken into account using a specific utility function. Table 1 below shows the utility function for risk averse, risk neutral and risk seeing cases.

Table 1. Risk appetite and utility function.

Risk Attitude	Utility Function
Risk averse	$u(w) = a - b\,e^{-w/\rho}$, a and b > 0
Risk neutral	$u(w) = a + bw$, a and b > 0
Risk seeking	$u(w) = a + b\,e^{-w/\rho}$, a and b > 0

In the equation, ρ represents risk tolerance. On a decision tree, each end-point represents an outcome depending on the preceding decisions. Compute NPV of each outcome and then the utilities of the NPVs. The decision tree is then rolled back using the utility of NPVs. Finally, the certainty present equivalent can be calculated from the utility function.

2.3. Risk-dependent utility

Compared to the decision analysis approach, the discounted cash flow method does not account for the risk preference. Instead of using a risk-free discount rate, a more comprehensive way would be using utility of NPVs for cash flows associated with private risks. The cash flows associated with market risks are still discounted using a risk-adjusted discount rate by adding a risk premium to the risk-free rate. This integrated method and detailed calculations will be illustrated in the next chapter.

3. Case study of solar PV leasing

Renewable energy is getting more important nowadays as an alternative to traditional energies. Solar energy, according to Energy Market Authority, is the most viable in the context of Singapore compared to other renewable energy sources such as wind and nuclear energies due to land constraints. It is clean, infinite, and free of charge.

3.1. Solar PV market landscape

In the U.S. and some European countries, roof-top solar PV has been a popular choice among residents. In a roof-top solar PV leasing arrangement, a solar PV leasing company negotiates with the owner of a residential or commercial building to lease the rooftop space at a predetermined cost over a period of time. The energy generated by solar PV is sold to the owner at a discount rate compared to the electricity tariff. If there is excess energy generated, it can be sold to the grid and profit will be given to the developer. Figure 1 below illustrates the model.

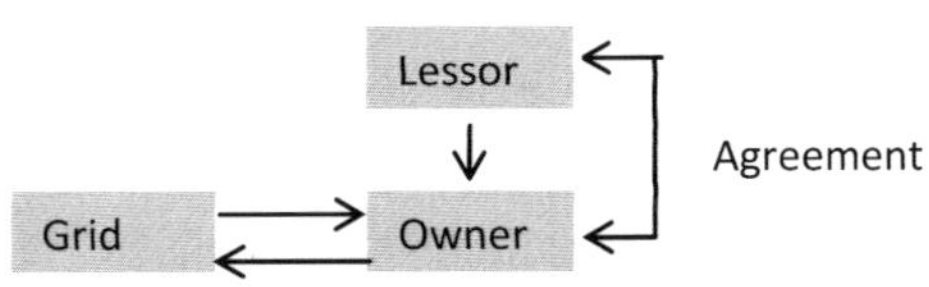

Figure 1. Solar PV leasing model.

PV stands for Photovoltaics and PV system can be made of arrays of polycrystalline silicon PV panels. The system connects to the grid and there is no battery storage. When exposed to sunlight, DC electricity is generated and the inverter, which is connected with PV arrays, converts DC to AC. In Singapore, there are a few solar PV leasing players in the market. Phoneix Solar, for example, has successfully implemented solar PV leasing projects in collaboration with SportsHub, Sheng Siong, Changi Airport Budget Terminal, INSEAD Business School, P&G, NUS-CREATE, JTC, City Square Mall, NUS High School of Math & Science, Singapore Polytechnic etc. Sunseap Leasing, another big player, has done solar PV leasing projects with Jurong Port, Housing Development Board (HDB), Marine Port Authority, Singapore American School etc. These successful projects illustrate the feasibility and future potential of solar PV leasing.
In October 2016, Mr. S Iswaran, Second Minister for Trade and Industry, announced the SolarNova Singapore programme. Being initiated by the Singapore Economic Development Board and in partnership with HDB, this plan aims to increase the adoption of solar power 350 MWp by 2020 according to the Ministry of Trade and Industry. This would lead to rising demand for solar deployment through solar leasing services. According to HDB, it has called a solar leasing tender to install solar panels across nine government organizations.

3.2. Case study: a solar PV leasing investment decision

Changi Airport (the "Host") is considering entering an agreement with Solar Leasing, a solar PV leasing company, to install the PV system for its latest Terminal in order to save electricity. In this agreement, the duration is 20 years and the Host is granted electricity at a discount rate while Solar Leasing is paid for excess energy exported to the grid. Assume both parties are risk neutral. A preliminary cash flow model using traditional discounted cash flow analysis is shown in tables 2 and 3. Table 4 below illustrates the parameters used and NPV for both counterparties.

The recent news that solar power is becoming the cheapest form of new electricity shows that the price of solar is bound to fall [3]. One opportunity in this case is the decreasing cost of solar PV system due to technological development. In the next period, there is a high chance of lower costs, which leads to potential gain in profits. At present, only the probabilities are known. However, a year from now, it will be known if the costs are lower or remain constant. The decision makers can decide if they want to expand the system. In contrast to the assumptions of the discounted cash flow (DCF) approach, expected future cash flows are uncertain due to fluctuations in the costs. This will affect future decisions.

Table 2. Cash flows for the solar PV leasing company.

Year	0	1	2	3	4	5	6	20
Revenue		52,403	51,355	50,328	49,321	48,335	47,368	35,699
Costs	- 375,000							
NPV	**18,358**							

Table 3. Cash flows for the host.

Year	0	1	2	3	4	5	6	20
Revenue		13,101	12,839	12,582	12,330	12,084	11,842	8,925
NPV	**112,243**							

Table 4. Parameters used for cash flow models.

Solar Leasing Company	
Number of kilowatt peak (kWp)	250
Cost / kilowatt peak (S$)	1.5
Electricity generated (kilowatt hours / year)	320,000
Electricity tariff (S$)	0.2047
Tariff annual growth	-2%
Price discount factor	0.8
Number of years	20
Discount rate	10%

One contingent decision we can identify in this case is the option to expand if cost is lower in the next period. The figure below illustrates the decision tree for this example.

Having learnt that the project is economically viable, the Host is offering Solar Leasing company the option to expand the project after one year if the cost of PV system is lower. Currently, the 250 kWp PV system can yield 320 MkWh out of 450 MkWh of energy required. Under the expansion plan, with additional information, it is assumed that there is a 70% probability of lower cost of PV system at S$1 per watt and 30% probability of S$1.5 per watt as before. If invests in another 250 kWp PV system, a total amount of 640 MkWh electricity can be generated; thus, 190 MkWh of electricity can be exported and sold to the grid. Using DPL software, we can represent the expansion scenario with a decision tree using a discount rate of 10% as shown in Figure 2 below.

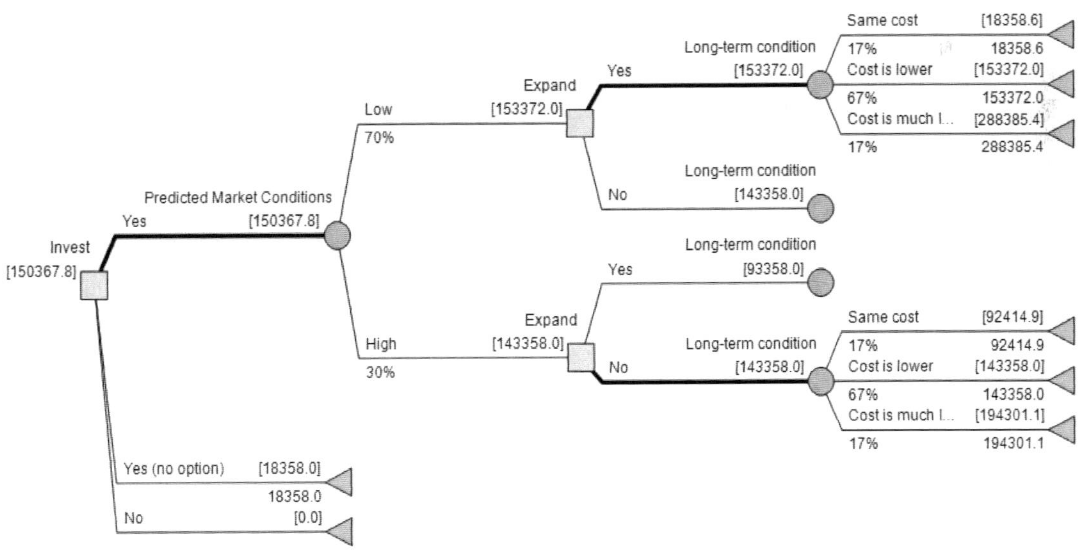

Figure 2. Decision tree.

If cost is low next year, certainty equivalent = max (0, 18,358, 150,368) = 150,368. Thus, the decision should be expanding the solar PV leasing system. A suitable valuation approach for the solar PV investment should account for uncertainty and the managerial flexibility (the option to expand in this case). To better understand the value of real options, a few scenarios are analyzed as illustrated in the scenarios shown in Table 5 below.

Table 5. Parameters used for cash flow models.

Period	**Year 0**	**Year 1**	**Year 1**
Scenario	**Decision**	**Cost**	**Decision**
1	Invest	Low	Expand
2	Invest	Low	Do not Expand
3	Invest	High	Expand
4	Invest	High	Do not Expand
5	Do not Invest	N.A	N.A
6	Invest (no option)	N.A	N.A

3.3. Risk-adjusted NPV

For the manufacturer, revenue is discounted using risk-free interest rate of 10% since it represents private risk. Costs are discounted using a risk-adjusted discount rate of 12% as it is subject to market conditions, and thus should be adjusted higher. The Certainty Equivalent obtained is higher than that using a universal discount rate.

After adjusting the model using risk-adjusted discount rate, the value of option to expand the project = Value of project with the option – Value of the project without real options = $134,581 - $18,358 = $116,223. This is a significant difference compared to the case without option, showing that options embedded in the capital investments can add value to them. Figure 3 shows the decision tree.

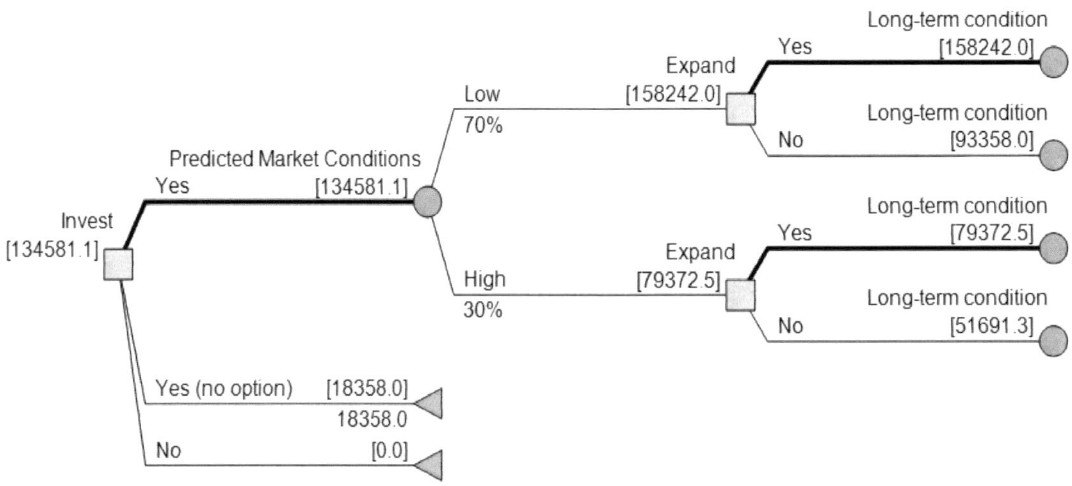

Figure 3. Decision tree for risk-adjusted NPV.

3.4. Risk-dependent utility function

Assume the risk attitude here is risk averse. Using the utility function u(x) = $1.1 - 1.1e^{\frac{-x}{57,925}}$, we can calculate Certainty Equivalent of revenues (private risks) by rolling back the decision tree. Since the costs reflect market risks, they are discounted using a higher risk premium.

If the risk attitude is risk seeking, a utility function $u(x) = x^{0.997}$ can be found. Integrating utility function and risk-adjusted NPV methods, the same option value can be derived.

4. Conclusion

From the author's perspectives, carrying out research in this field has allowed the author to gain more knowledge about solar PV leasing especially with respect to the Singapore market.

The paper illustrates several important valuation methods through the case study of solar PV leasing. The decision analysis modelling involves the use of decision tree and risk-adjusted NPV as the primary modelling tool. By integrating theory and application on solar PV leasing, new insights are obtained. Such insights are critical for decision makers to assess the value of projects and options embedded accurately.

The results of the paper have several managerial implications. Firstly, it illustrates the importance of real option thinking especially when there is uncertainty. Real option thinking encompasses a valuation model as well as strategic management decisions. In certain sector such as energy sector which involves uncertainty and is susceptible to the market conditions, having the option thinking is critical. From a real option thinking perspective, this paper argues that a decision analysis approach could better capture the real options embedded in capital investments than other methods and shows how option thinking can be incorporated in the valuation of solar PV leasing. The results showed that with real options included in the analysis, the value of the project can increase significantly compared to the case without real options. Moreover, real options are valuable in electricity markets because of high uncertainties [4]. In future, decision makers will face more uncertainties when new technologies arise. For example, biomass and tidal energy are examples of renewable energy with embedded options. It is the author's great hope that option thinking can be gradually adopted by more practitioners and this research can increase interest among the solar energy industry.

5. References

[1] Trigerorigis, Lenos 1993 Real Options and Interactions With Financial Flexibility, Financial Management, volume 22, issue 3, page 202-224

[2] Kodukula, P., & Papudesu, C.2006 Project Valuation Using Real Options: A Practitioner's Guide.J.Ross Publishing.minutes

[3] Randall, T. 2016 World Energy Hits a Turning Point: Solar That's Cheaper Than Wind. Retrieved from https://www.bloomberg.com/news/articles/2016-12-15/world-energy-hits-a-turning-point-solar-that-s-cheaper-than-wind

[4] Nouicer, A. 2015 Diffusion of new renewable power in Brazil: A Real Options Approach (Unpublished thesis). Comillas Pontifical University.

Utilization of eggshell waste as low-cost solid base catalyst for biodiesel production from used cooking oil

N P Asri[1*], B Podjojono[1], R Fujiani[1] and Nuraini[1]

[1]Chemical Engineering Department, Faculty of Engineering,, WR. Supratman University, Surabaya, Indonesia, 601191. *corresponding author: nyoman_puspaasri@yahoo.com

Abstract. A solid CaO-based catalyst of waste eggshell was developed for biodiesel production from used cooking oil. The waste eggshell powder was calcined in air at 900°C for 4 h to convert calcium species in the eggshells into active CaO catalysts. The characterization of CaO catalyst was done by XRD and BET analysis. The CaO catalyst was then introduced for transesterification of used cooking oil (UCO) for testing of its catalytic activity. The experiment was conducted in batch type reactor that consists of three-neck glass equipped by reflux condenser and magnetic stirrer. Before tranesterification process, the UCO was treated by coconut coir powder in order to reduce the free fatty acid content. The result showed that the catalyst was potentially use for transesterification of used cooking oil into biodiesel with relatively high yield of 75.92% was achieved at reaction temperature, reaction time, molar ratio UCO to methanol and catalyst amount of 65° C, 7 h, 1:15 and 6%, respectively.

1. Introduction

The energy consumption has increased every year in line with the increase in human population. This increase occurred in various sectors including transportation, power generation, industry and households. Until now, the world's energy needs are supplied by fossil fuels. By 2030, it is estimated that the increase in energy consumption reaches 53%, so that the oil reserves expected to be exhausted within the next 41-63 years [1]. In addition, the use of fossil fuels continuously, causing environmental impacts such as the increase of CO2 emissions that result in accelerated of global warming, air pollution and other environmental issues. The rapid depletion of fossil fuels and the increasing energy needs has encouraged each country to make savings and conducting various studies to obtain renewable fuels. Biodiesel is a renewable fuel that can be used as a substitute for diesel fuel.

Biodiesel is mono alkyl ester derived from many sources such as vegetable oils, animal fats or used cooking oil. The main factors affecting the cost of biodiesel production is the raw material and the type of catalyst used. In 2012, cooking oil demand in Indonesia is estimated to reach more than 4.7 million tons. After use, the cooking oil will undergo changes in chemical composition and contain carcinogenic compounds. Therefore, the used cooking oil is no longer worthy to be used for food processing. The abundant and availability of used cooking oil can be used as an alternative raw material for the manufacture of biodiesel [2]. Typically, biodiesel was produced through the transesterification process using homogeneous acid / alkaline catalyst [3], [4], [5] that usually called conventional process. This process is fast and efficient, for example more than 95% conversion is achieved at 60° C in 1 h of reaction time. However, it has many disadvantages including the formation of side products such as soaps [6], [7], [8], the complicated separation of the product with the catalyst [3]. In addition, the alkaline waste that formed in large enough quantities also requires a complicated separation process. This separation process requires considerable energy and the increasing of costs production [7], [9]. Those problems can be mitigated by replacing homogeneous catalysis with solid catalyst. The use of solid acid catalysts in biodiesel production has been studied primarily from transition metal oxides such as zirconia and titanium oxide [9], [10]. The main obstacle of use of acid catalysts requires long reaction times and high reaction temperatures [9], [11]. Solid base catalyst such as calcium oxide [12], magnesium oxide [13], alkali alumina [14], hydrotalcite [15], zeolite [16] seems to be a promising strategy to substitute acid catalyst, but most of them use of synthetic chemicals that are relatively expensive.

This study developed a solid CaO-based catalyst of eggs shell for the production biodiesel from used cooking oil. The purpose of this study is to utilize CaO contained in the eggshell into a solid catalyst instead of homogeneous catalysts that will influence decreasing of biodiesel production costs, so the price of biodiesel can compete with diesel oil. Eggshell has a small porosity and high of $CaCO_3$ content that is indispensable in the industry of biodiesel [17]. Components of the eggshell include calcium carbonate (94%), magnesium carbonate (1%), calcium phosphate (1%), and other organic materials (4%). Based on the pore structure and content of $CaCO_3$, eggshell provides opportunities for development as heterogeneous catalysts [18].

Utilization of this catalyst is expected to provide a solution to obtain a feasibility study of biodiesel production process that is effective and efficient and reduce environmental pollution. The study was conducted in the scale laboratory includes several stages, namely preparation of catalysts, catalyst characterization, and testing catalyst activity to determine the best operating conditions include amount of catalyst and reaction time.

2. Materials and Methods

2.1. Materials

Waste cooking oil was found from local street food vendors in Surabaya, East Java. The physical and chemical characteristics of UCO were analyzed using several methods based on the procedure described elsewhere [2]. The free fatty acid (FFA) content of 4.4946% was analyzed based on American Oil Chemist (AOCS) official Method (Ca 5a-71, 1993). Meanwhile, its density and water content of 0.9219 g/cm^3 and 0.78%, respectively were analyzed based on ISO 662 – 1089. FAME standards and internal standards of chromatographic grade were obtained from Sigma–Aldrich (Switzerland). Technical grade of methanol was purchased from local supplier Bratacho-Chem. Meanwhile eggshell was collected from the local bakeries in Surabaya, East Java, Indonesia.

2.2. Catalyst preparation and characterization

At first, the eggshells being crushed into small chips, then soaked and rinsed thoroughly several times with hot water to remove the impurities and other interference materials. The clean eggshell chips dried at 110°C for 24 h in the dry oven. The dried chips eggshells then powdered and sieved to the size of qualifying on 100 mesh and retained on 200 mesh. Calcination process was performed in the muffle furnace at 900°C for 4 h under static air. Surface area analysis of the catalyst was examined using Brunauer-Emmett-Teller (BET) methods. Sample was degassed at 105 °C prior to analysis and the adsorption of N_2 was measured at -196 °C. Meanwhile, X-ray diffraction (XRD) was measured by XPert MPD using Cu Ka radiation. The prepared catalyst from eggshell was assigned as ES-CaO, and then kept in a desiccator in the presence of silica gel in order to avoid water and CO_2 contacting with the catalyst.

2.3. Transesterification process

In this research, the collected UCO was pre-treated based on the procedure described elsewhere [2], [8]. First, the collected UCO was filtered to remove the food residues or the suspended particulate matters, and then it was heated at 100-105°C for 30 minutes to remove the excess of water content. Further, UCO was treated by adsorption process using coconut coir powdered (CCP) (wt.% to the UCO) to reduce the FFA content of the UCO. It has been stated in the last work [2] that the optimum amount of CCP is 7% (wt.% to the UCO), therefore it was used for treating UCO in this work. The treated UCO was analyzed to determine its final FFA content, and then it was ready subjected for transesterification process using ES-CaO catalyst.

Transesterification process was conducted in batch type reactor consists of three neck glass, equipped with reflux condenser and magnetic hot plate stirrer. The reaction was carried out at 65ºC and molar ratio of UCO to methanol 1:15. Meanwhile, the amount of catalyst and reaction time were varied of 3-7% (wt.% o oil) with 1% of range and 2-7 h with range of 1 h, respectively. The amount of ES-CaO catalyst, methanol, and treated UCO were measured and introduced into the reactor, and allowed to keep a while until the reaction temperature of 65°C was reached. The reaction mixtures were stirred thoroughly at constant speed using magnetic stirrer. After desired duration time was reached, the reactor was cold down and the reaction mixtures were withdrawn at room temperature. The mixtures were filtered to separate the catalyst from the liquid phase. The liquid phase was introduced into a separation funnel where the glycerine was decanted, and then the excess methanol was recovered using a rotary evaporator. Biodiesel concentration was analysed by a gas chromatograph GC-14B (Shimadzu) equipped with a flame ionization detector and a capillary column HP-Innowax (30 m × id 0.25 mm, 0.25 μ m). Helium was used as the carrier gas throughout the whole experiment. Biodiesel analyses can be used to calculate the yield of biodiesel (%) defined as Equation 1 [19]

$$\text{Yield of biodiesel (\%)} = \frac{w_{\text{of ab}}}{w_{\text{of oil}}} \text{ x } 100\% \qquad (1)$$

Where, w of ab means weight of actual biodiesel was calculated from weight of biodiesel produced (mg) in the experiment multiple by weight % of FAME in the sample. Meanwhile, W of oil is weight of oil (mg) used in the experiment.

3. Result and Discussion

3.1. Characterization of ES-catalyst

The specific surface area prepared ES-CaO catalyst of 125.927 m^2g^{-1} was obtained using BET methods. This result is much higher than that crab shells catalyst of 13 m^2 g^{-1} that was stated by Boey et al. (2009) [20]. Meanwhile, mud clam shell and commercial CaO have the surface area of 68.57 and 62.91 m^2 g^{-1}, respectively [21].

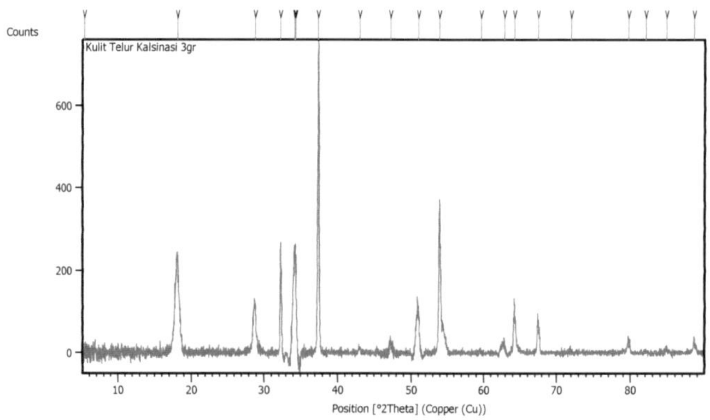

Figure 1. XRD patterns of eggshell (ES-CaO) catalyst calcined at 900°C

X-ray diffraction (X-RD) patterns of ES-CaO was depicted by figure 1. XRD results revealed that the composition of uncalcined eggshell mainly consisted of calcium–magnesium carbonate ($CaCO_3$–$MgCO_3$) with the absence of CaO peak (the figure was not shown). This result confirm with the composition of eggshell that mostly consisted by $CaCO_3$ (94%). However, for the calcined catalyst at 900° C (figure 1), $CaCO_3$ completely converted into CaO and release the CO_2 gas. The narrow and high intense peaks of the calcined catalyst showed the well crystalline structure of the catalyst that mainly consisted of the active ingredient CaO.

3.2. Effect of reaction variable on yield of biodiesel

The yield of biodiesel was affected by many factors, such as the type triglyceride, type of catalyst used, reaction temperature, molar ratio of oil to methanol, catalyst amount or reaction time. Therefore, the

effect of reaction variables was studied in the presence of ES-CaO solid base catalyst. In this work we focused on the influence of the catalyst amount used and the reaction time on yield of biodiesel.
There are 2 keys parameter including free fatty acid and moisture content that are determining the viability of vegetable oils to be used in transesterification process [2], [22]. Water content will accelerate the hydrolysis reaction and simultaneously reduce the amount of ester formation [23]. To achieve 90% yield of biodiesel, then water content of triglyceride that used in the transesterification process not more than 0.5% [24]. Moreover, it was known that higher free fatty acid contents will lead to formation of soap and water. If free fatty acid content more than 3%, the homogeneous base catalyst is not suitable for the transesterification reaction [2], [24].
The UCO that was used in this work contained 4.4946% (wt. %) free fatty acid (FFA). In order to reduce the FFA content of the UCO, it was treated by adsorption process using 7% of coconut coir powder [2]. Coconut coir is one of the agricultural waste products often used as adsorbent in wastewater treatment [25]. Due to its availability and its abundant in the environment, make it to be a good source used in the adsorption process [2], [25]. The nature of coconut coir dust such as good structural stability, high water absorptivity and high porosity enhance its effectiveness as adsorptive or ion exchange capacities. Its composition includes lignin, cellulose, pentosan and ash were 36.15, 33.61, 29.27 and ash 0.61% (wt. % of dry basis), respectively [26]. After adsorption process, the FFA content of UCO was determined using AOCS method. The result showed its FFA content decrease into 0.2079%. Furthermore, the treated used cooking oil is ready for the transesterification process.

3.2.1. Effect of catalyst amount on yield of biodiesel

Investigating the influence of catalyst amount on yield of biodiesel the experiments were performed in varied catalyst amount of 3-7% with range of 1% (wt.% to UCO) under constant stirring speed using reaction temperature, reaction time and molar ratio of UCO to oil at 65 °C, 1 hr and 1:15, respectively. Figure 2 shows the influence of catalyst amount on yield of biodiesel. As depicted by figure 1, the amount of catalyst was significantly influence on yield of biodiesel. Yield of biodiesel increases with increasing the amount of catalyst. Without the presence of catalyst, no biodiesel was formed (data yield of biodiesel using 0 % catalyst amount was not shown in the figure).

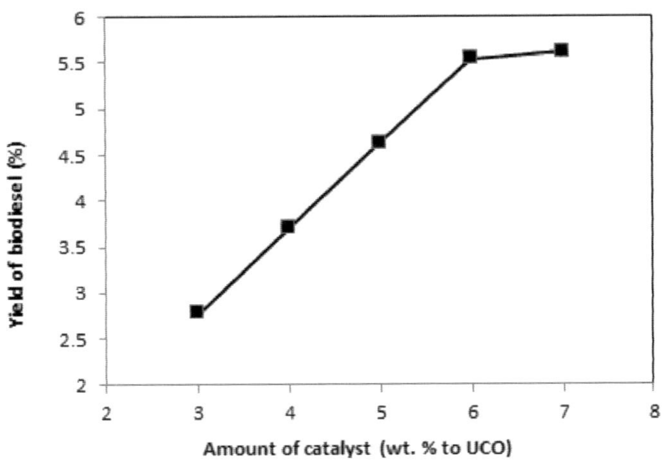

Figure 2. Effect of amount of ES-CaO catalyst on yield of biodiesel

In contrary, yields of biodiesel linearly increase with the increasing the amount of catalyst at 3 up to 6%. It is strongly evidence that the presence of catalyst affected the reaction rate [2]. In this work, the maximum yield of biodiesel of 5.62% was found at 7% catalyst amount of ES-CaO. However, at catalyst loading amount of 6-7% there is no increased significantly on yield of biodiesel. Therefore, for the economic point of view, 6% amount of catalyst was decided as the optimum condition, and it was selected for further studies.

3.2.2. Effect of reaction temperature on yield of biodiesel

Evaluating the influence of reaction time on yield of biodiesel on transesterification of UCO using ES-CaO catalyst, the experiments were conducted by varying time of reaction from 1 to 7 h with range of 1 h. Meanwhile, the others conditions were kept constant, at 65°C, molar ratio treated UCO to methanol of 1:15, and amount of catalyst of 6% (wt. % to UCO).

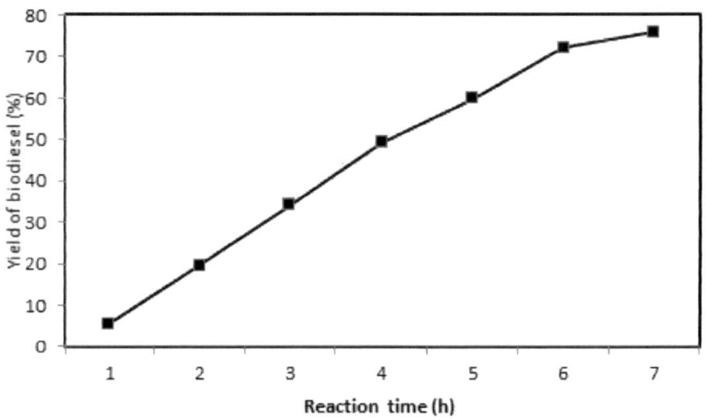

Figure 3. Effect of reaction time (h) on yield of biodiesel (%)

Figure 3 depicted that yield of biodiesel significantly increased with increasing of the time of reaction. In 1 h of reaction, yield of biodiesel was low, because it is a heterogeneous reaction and the fact that mass transfer is slow in the earlier of reaction. Huaping et al. (2006) reported the similar result [27]. After one up to 6 h, yield of biodiesel linearly increase from 5.54 up to 72.05%. At reaction time of 6 up to 7 h, biodiesel yields were gradually increased from 72.05 to 75.92 %. It seems that the equilibrium condition was achieved at 7 h with highest yield of 75.92%. It proved that the yield of biodiesel increased up to certain of reaction time. Refaat et al. (2008) and Asri et al. (2013) stated the similar statement [19], [28]. Therefore, the reaction time of 7 h was decided as the optimum reaction time on transesterification of used cooking oil into biodiesel using ES-CaO catalyst.

4. Conclusion

Calcium oxide (CaO) solid base Catalyst has been successfully prepared from eggshell waste by calcination process at high temperature (900° C). The eggshell waste (ES-CaO) catalyst has been introduced to transesterification process of used cooking oil with refluxed methanol. The high content of free fatty acid in used cooking oil reduced using 7% (wt.% to UCO) of coconut coir powder. The transesterification reaction was carried out under constant stirring speed at 65 °C and molar ratio of UCO to methanol 1:15. The result shown, that the amounts of catalyst and reaction time have significant effect on yield of biodiesel. The high enough biodiesel yield of 75.92% was obtained using 6% of catalyst amount within 7 h of reaction time. It proved that the eggshell waste catalyst seems to be very promising for producing biodiesel from used cooking oil. However, the activity of the eggshell catalyst should be improved to achieve the higher biodiesel yield (>95%) in a shorter time of reaction. Therefore, the study of developing and engineering of CaO solid base catalyst derived from shell waste is still need to be encouraged to enhance the efficiency of biodiesel production from used cooking oil.

5. References

[1] Kiakalaieh A T, Amin N A S, and Mazaheri H 2013 A review on novel processes of biodiesel production from waste cooking oil *Applied Energy* **104** 683–710

[2] Asri N P, Puspitasari D A, Poedjojono B, and Suprapto 2015a Pre-treatment of waste frying oils for biodiesel production *Modern Applied Science* **9** **7** 99-106

[3] Freedman, B, Butterfield RO, and Pryde EH 1986 Transesterification kinetics of soybean oil *J American Oil Chemists Society* **63** 1375-80

[4] Canakci M 2007 The potential of restaurant waste lipids as biodiesel feedstocks *Bioresource Technology* **98 1** 183–190.

[5] Asri N P, Machmudah S, Wahyudiono W, Suprapto S, Budikarjono K, Roesyadi A, and Goto M 2013a Non catalytic transesterification of vegetables oil to biodiesel in sub-and supercritical methanol: a kinetic's study *Bulletin of Chemical Reaction Engineering & Catalysis* **7 3** 215-223

[6] Ilgen O, Akin A N 2008 Development of alumina supported alkaline catalyst used for biodiesel production *Jourrnal of Turkey* **33** 281–287

[7] Zabeti M, Daud W H A M, and Aroua M K 2009 Activity of solid catalysts for biodiesel production: A review *Fuel Processing Technology* **90** 770-777

[8] Asri N P, Poejojono B, Puspitasari D A, Suprapto S, and Roesyadi A 2015b Biodiesel production using double promoted catalyst CaO/KI/ γ-Al_2O_3 in batch reactor with refluxed methanol *Springer Procedding in Energy, Energy System And Management* Germany Springer Verlag, ch 15

[9] Furuta S, Matsuhashi H and Rata K 2006 Green diesel fuel production with solid amorphous-zirconia catalyst in fixed bed reactor *Biomass And Bioenerg* **30** 870-873

[10] Jitputti J, Kitiyanan B, Rangsunvigit P, Bunyakia K, Attnatho L and Jenvanipanjakul P 2006 Transesterification of kernel oil and crude coconut oil by different solid catalysts *Chemical Engineering Journal* ***116*** 61-66

[11] Marchetti J M, Miguel V U and Errasu A F 2007 Heterogeneous esterification of oil with high amount of free fatty acids *Fuel* **86** 906–910

[12] Granados M L, Poves M D Z, Alonso D M , Mariscal R, Galisteo F C, Moreno-Tost R, Santamaría J and Fierro J L G 2007 Biodiesel from sunflower oil by using activated calcium oxide *Applied Catalysis B: Environmental* **73** 317–26

[13] Wang Y, Zhu S, He H and Liu X 2007 Transesterification of soybean oil to biodiesel using SrO as a solid base catalyst *Catalayst Communication* **8** 1107-11

[14] Xie W and Li H 2006 Alumina supported potassium iodide as a heterogeneous catalyst for biodiesel production from soybean oil *J Molecular Catalysis A: Chemical* **255** 1-9

[15] Kim M J, Park S M, Chang D R and Seo G 2010 Transesterification of triacetin, tributryin, and soy bean oil with methanol over hydrotalcite with different water content *Fuel Process Technology* **91** 618-24

[16] Noiroj K, Intarapong P, Luengnaruemitchai A and Jai-In S 2009 A comparative study of KOH/Al_2O_3 and KOH/NaY catalyst for biodiesel production via transesterification from palm oil *Renewable Energy* **34** 1145–50

[17] Empikul, V N, Krasae P, Puttasawat B, Yoosuk B, Chollacoop N and Faungnawakij K 2010 Waste shell of mollusk and egg as biodiesel production catalyst *Bioresource Technology* **101** 3765-67

[18] Stadelman, W J 2000 Eggs and egg products. In: Francis, F.J. (Ed.) *Encyclopedia of Food Science and Technology 2nd Ed.* New York John Wiley and Sons 593–599

[19] Asri N P, Machmudah S, Wahyudiono W, Suprapto S, Budikarjono K, Roesyadi A and Goto M. 2013b Palm oil transesterification in sub and supercritical methanol with heterogeneous base catalyst *Chemical Engineering and Processing: Process Intensification* **72** 63-67

[20] Boey P L, Maniam G P, and Hamid S A 2009 Biodiesel production via transesterification of palm olein using waste mud crab (*Scylla serrata*) shell as a heterogeneous catalyst *Bioresource Technology* **100** 6362–68

[21] Ismail S, Ahmed A S, Anr R, and Hamdan S 2016 Biodiesel production from castor oil by using calcium oxide derived from mud clam shell *Journal of Renewable Energy* Article ID 5274917, 8 pages

[22] Gashaw A and Teshita A 2014 Production of biodiesel from waste cooking oil and factors affecting its formation: A review *International Journal of Renewable and Sustainable Energy* **3 5** 92-98

[23] Leung D Y C and Guo Y 2006. Transesterification of neat and used frying oil: Optimization for biodiesel production. *Fuel Processing Technology* **87 10** 883–90

[24] Gnanaprakasam A, Sivakumar V M, Surendhar A, Thirumarimurugan M, and Kannadasan T 2013 Recent strategy of biodiesel production from waste cooking oil and process influencing parameters: A review. *Hindawi Publishing Corporation Journal of Energy,* 10 pages

[25] Okafor P C, Okon P U, Daniel E F, Ebenso E E 2012. Adsorption capacity of coconut *(Cocos nucifera L.*) shell for lead, copper, cadmium and arsenic from aqueous solutions *International Journal Electrochemical Science* **7** 12354 –12369

[26] Etim U J, Umoren S A and Ednok U M 2012 Coconut coir dust as a low adsorbent for the removal of cationic dye from aqueous solution *Journal of Saudi Chemical Society*

[27] Huaping Z, ZongbinW, Duan Shjie CY, Yuangxiong C, Ping Z, Xiaohua L and Zongqiang M. 2006 Preparation of biodiesel catalyzed by solid super base of calcium oxide and its refining process *Chinese Journal Of Catalyses* **27 5** 391-96

[28] Refaat A A, Attia N K, Sibak H A, El Sheltawy S T and El Diwani I 2008 Production optimization and quality assessment of biodiesel from waste vegetable oil *International Journal of Environmental Science and Technology* **5 1** 75-82

6. Acknowledgments

The authors would like gratitude to Directorate of Research and Community Services, Directorate General of higher education, the Ministry of Research Technology and Higher Education of Republic Indonesia, for the financial support through the research grant in the fiscal year of 2017.

Formation of hydrocarbon compounds during the hydrocracking of non-edible vegetable oils with cobalt-nickel supported on hierarchical HZSM-5 catalyst

L Marlinda[1,2], M Al-Muttaqii[2], A Roesyadi[2] and D H Prajitno[2*]

[1]Department of Chemistry, Faculty of Science and Technology, University of Jambi, Jambi, 36361, Indonesia
[2]Chemical Reaction Engineering Laboratory, Department of Chemical Engineering, Faculty of Industrial Technology, Sepuluh Nopember Institute of Technology, Sukolilo, Surabaya 60111, Indonesia

*Corresponding Author: dana@chem-eng.its.ac.id (Danawati Hari Prajitno)

Abstract. The hierarchical Co-Ni/HZSM-5 catalyst with hierarchical pore structure was prepared by desilication and incipient wetness impregnation. Hydrocracking of non-edible vegetable oils at temperature of 400 °C, 20±5 bar for 2 h was performed in the presence of this type of catalyst under hydrogen initial pressure in pressured batch reactor. Non-edible vegetable oils, such as Reutealis trisperma (Blanco) airy shaw (sunan candlenut) and Hevea brasiliensis (rubber seed) were chosen to study the effect of the degree of saturation and lateral chain length on hydrocarbon compounds obtained through hydrocracking. Cerbera manghas oil was also tested for comparison because the composition of fatty acid was different with the other oils The hydrocracking test indicated that liquid product produced has a similar hydrocarbon compounds with petroleum diesel. The most abundant hydrocarbon is pentadecane (n-C15) and heptadecane (n-C17). The high aromatic compounds were found in liquid product produced in hydrocracking of Sunan candlenut oil.

1. Introduction

The non-edible vegetable oils used as feedstock in biofuel production is considerable because of their chemical structure due to their energy capacity ability. The carbon chains of triglycerides are similar to the carbon chains of petroleum diesel. However, as reported by previous study [1], the vegetable oil has properties which are not beneficial if used directly as fuel. The high vicosity, high flash point and low heating value compared to the fossil diesel were related with the presence of oxygen atoms in the form of carboxyl or carbonyl groups. Several technology processes have been applied to reduce a high viscosity of vegetable oil such as transesterification in methanol. However, fatty acid methyl esters (FAMEs) produced through this technology have a poor energy density and oxidation stabillity compared to fossil diesel because of an excess of oxygen. In addition, according to Kim et al. [2] the impurities produced such as saturated free fatty acid, unreacted triglycerides, diglycerides and catalyst can cause problems on a machine seperti clogging fuel filters and swelling. Hydrocracking of non-edible vegetable oils with HZSM-5 catalyst is considerable to obtain free-oxygen biofuel containing paraffin-rich hydrocarbons as an alternative fuel. The content of olefins, iso-paraffins, cycloparaffins and aromatics is found in a small amount during hydrocracking process through dehydrogenation, isomerization, cyclization and aromatization reactions.

The previous result stated that a certain level of poly-unsaturation of the fatty acids is required to produce relatively high aromatics content. When Pt/Al_2O_3/SAPO-11 catalyst is used, about 15 wt.% of aromatics are produced via hydrotreating of the poly-unsaturated soybean, sunflower and camelina oils while hydrotreating of palm or castor oils produces aromatics of 2 wt.%. In addition, the reaction condition and type of catalyst influence type and amount of hydrocarbon compounds [3]. The presence of hydrocarbon compound will determine biofuel properties so that it is important to understand dan control these reactions during hydrocracking.

Da Rocha Filho et al. [4] stated that cyclization reaction was promoted at high temperature of hydrogenation favoured cracking and deydrogenation and finally aromatics ware formed. Sotelo-Boyas et al. [5] reported that Pt/zeolite catalyst used in hydrocracking of rapeseed oil produced more iso- than n-paraffin in the range of C5-C22. The presence of isoparaffin can improve cold flow properties. Chen et al. [6] produced C5-C18 of liquid paraffins from catalytic hydroprocessing of FAMEs using Ni/HZSM-5 catalyst. According to Marlinda et al. [7], when hydrocracking of Cerbera manghas oil with Co-Ni/HZSM-5 was applied at temperatur of 350 °C, a small alkylbenzenes were found such as n-pentylbenzene and hexylbenzene and cycloparaffins were detected in form of n-nonylcyclohexane and isobutylcyclopentane. A more high n-paraffins were produced at this reaction condition. Al-Muttaqii et al. [8] reported that when Co-Ni/HZSM-5 was used in hydrocracking of Sunan candlenut, a slightly n-paraffins were produced at temperature of 350 °C and aromatic hydrocarbons had been found in liquid product. Because the decarboxylation and/or decarbonylation reactions were endothermic reaction (Table 1), the reaction temperature above 350 °C was needed to produce more n-paraffins [6]. The presence of hierarchical pore structure on the Co-Ni/HZSM-5 catalyst increased n-paraffins (C11-C18) from 11.95 to 34.18 area% through hydrocracking of Sunan candlenut oil at 375 °C and produced a more high aromatics of 46.32 area% in liquid product, compared with HZSM-5 catalyst. It was different with a previous result [8], a various aromatic compounds were found and (1-methyldecyl) benzene was predominant in liquid product. A small cycloparaffin was also found [9]. Hierarchical pore structure was expected to solve problems about mass transport of tryglyceride molecules. The hierarchical HZSM-5 has a mesoporous material properties (fast diffusion and accessibility of large molecules) and micropores (strong acidity and high hydrothermal stability). These hierarchical pore structure of zeolite allows large molecules that can not diffuse into the micropores will continue to experience a catalytic reaction in the mesoporous wall. According to Sandstede and Lehmann [10], the hydrophobic properties owned zeolite can crack vegetable oils containing triglycerides.

Table 1. Plausible reactions during hydrotreatment [6].

Liquid phase reactions					ΔG_{573} (kJ/mol)	ΔH_{573} (kJ/mol)
I.	Decarboxylation :	R-COOH	→	R-H+CO_2(g)	-83.5	9.2
II.	Decarbonylation :	R-COOH	→	R'-H+CO(g)+H_2O(l)	-17.0	179.1
		R-COOH+H_2(g)	→	R-H+CO(g)+H_2O(l)	-67.6	48.1
III.	Hydrodeoxygenation :	R-COOH	→	R-CH_3+2H_2O(l)	-86.1	-115.0
	(HDO)	R=saturated alkyl group, R'=unsaturated alkyl group				

The hierarchical HZSM-5 catalyst was well prepared in order to enhance transport of reactan and product molecules from to zeolite pore. Desilication or alkaline treatment process was chosen to create intracrystalline in ZSM-5 framework. Although, the properties of the catalyst have been used in hydrocracking of vegetable oil to biofuel which had hydrocarbon compunds similar to petroleum diesel, the effects of different degrees of saturation and chain lengths on hydrocarbon compounds distribution obtained through hydrocracking of non-edible vegetable oil are still unclear for the hierarchical HZSM-5 catalyst. In this work, the reutealis trisperma (Blanco) airy shaw (sunan candlenut) and hevea brasiliensis (rubber seed) were chosen to investigate the effects. Cerbera manghas oil was also tested for comparison because the composition of fatty acid was different with the oils used in this work. In addition, according to the result in previous work [8], [9], the effect of the high temperature (at 400 °C) in forming the hydrocarbon compounds, such as paraffin-rich hydrocarbon, cycloparaffins, aromatics, and olefins in liquid product were also studied.

2. Material and Methods

2.1. Material
Sunan candlenut oil was extracted from seeds, as shown in Figure 1. The fruits was collected from Bogor, West Java, Indonesia and afterwards the seeds was pressed with screw press mechine to obtain oil. The oil content obtained was about 50 wt.% of dried seeds, as reported by Syafaruddin and Wahyudi [11]. While, rubber seeds was collected from Jember, East Java, Indonesia and pressed with screw press mechine to obtain rubeer seed oil. Non-edible vegetable oils were analyzed using gas chromatography-mass spectrometry (GC-MS) to determine composition of fatty acids.

Figure 1. Sunan candlenut (a) fruit, (b) seed before skin removed, (c) seed after skin removed, (d) oil.

2.2. Catalyst preparation and characterization
The Ammonium-ZSM-5 zeolite was purchased from Zeolyst International (CBV 8014, a surface area of 400 m^2/g, Na_2O content less than 0.05 wt.%) and calcined at 550 °C for 5 h to HZSM-5 zeolite. Another chemicals were $Ni(NO_3)_2.6H_2O$ and $Co(NO_3)_2.6H_2O$ purchased from Merck with ≥98 % purity. The hierarchical Co-Ni/HZSM-5 catalyst used in this work had been prepared with procedure as reported by Marlinda et al. [9].The first step, the hierarchical HZSM-5 catalyst was prepared in stainless steel vessel through desilication on HZSM-5 zeolite within 0.2 M NaOH solution and kept under stirring at 65 °C for 2 h. The second step, the sample was modified by designed concentration of $Ni(NO_3)_2.6H_2O$ and $Co(NO_3)_2.6H_2O$ aqueous solution with incipient wetness impregnation, then calcined at 400 °C for 2 h with air and reduced at 450 °C for 3 h to obtain metallic phase. Finally the hierarchical Co-Ni/HZSM-5 catalyst was obtained. Furthermore, the detailed characterizations of the hierarchical Co-Ni/HZSM-5 catalyst have been described in previous report [9]. The creating mesoporosity in this catalyst was also effected by cobalt-nickel supported on hierarchical HZSM-5.

2.3. Hydrocracking process
As reported by previous study [12], the batch reactor equipped with a mechanical stirrer was used to make biofuel from non-edible vegetable oil by hydrocracking process. The hydrocracking process was conducted under hydrogen initial pressure in pressured batch reactor equipped with a mechanical stirrer with procedure as reported by Marlinda et al. [12]. The non-edible vegetable oil of 200 ml was placed into the reactor, which was containing the catalysts of 1 g. Then, a reactor was purged with nitrogen to remove air dissolved in oil for 30 min, as reported in previous study [13]. The catalyst was activated under flowing H_2 for 1 h. In previous study, Martinez-Grimaldo et al. [14] activated hydrocracking catalyst at 350 °C for 1 h. The reaction temperature was carried out at a temperature of 400 °C for 2 h. In practically, the reaction pressure changed between 10 bar and 15 bar.
Liquid products were analyzed by gas cromatography-mass spectrometry with standard GC : Agilent HP 6890 models 19091S-433, HP–5MS capillary column 30 m x 250 µm x 0.25 µm. This analyze was held at nominal initial pressure 13.3 psi and the temperature for the oven was at 150 °C for 2 min, then it increased to 240 °C at a rate of 10 °C/min for 11 min. The Wiley275 and NIST02 mass spectral library of data were used to determine hydrocarbon compounds. The hydrocarbon components with a probility match equal to or higher than 80% were considered. N-paraffin, aromatic, cyloparaffin, and

olefin contained in liquid product were grouped as gasoline-like hydrocarbon (C5-C9), kerosene-like hydrocarbon (C10-C13), and gasoil-like hydrocarbon (C14-C22), as reported by Barron et al. [15].

3. Result and Discussion

3.1. Characterization of non-edible vegetable oils

According to previous work [9], [17], a high content of poly-unsaturated fatty acids found in both the sunan candlenut and rubber seed oils were around 50 % and 95.2 %, respectively. Linoleic acid was the main component. In addition, palmitic acid as saturated fatty acids was found in around 3.67 - 20.3 area%. While, cerbera manghas oil contained mono-unsaturated fatty acid [7].

3.2. Analysis of Biofuel

GC-MS chromatogram of Sunan candlenut oil and liquid product produced at temperature of 400 ^{o}C are shown in Figure 2. The compounds with retention time of 10-20 min in the Sunan candlenut oil chromatogram disappear after hydrocracking reaction [9]. As shown in Figure 2 (b), the chromatograms of liquid product showed that peaks at retention time of 0-12 min appeared and indicated the increasing a large amount of hydrocarbon compounds. It implied that triglycerides were converted into short chain molecules by several reaction sequence [16], including interconnected reactions of triglyceride molecules and intermediates such as hydrogenolysis forming fatty acids and propane, deoxygenation of carboxylic groups in fatty acids, cyclization, dehydrogenation of cyclic ring to aromatic ring, hydrocracking of alkyl chains, and isomerization of alkyl chains [3].

The similar phenomenon can be also observed in GC-MS chromatogram of rubber seed oil and liquid product produced at temperature of 400 ^{o}C, as shown in Figure 3. A small peaks intensity of fatty acid such as palmitic acid and stearic acid were detected in GC-MS chromatogram of rubber seed oil liquid product so that it can be implied that trilinoleic was successfully hydrogenated and cracked into saturated fatty acids at 400 ^{o}C with hierarchical Co-Ni/HZSM-5 catalyst. It was interesting to be studied later, a peaks intensity of aromatic (e.q. 1-methyldecylbenzene) were detected in GC-MS chromatogram of sunan candlenut oil liquid product. In addition, pentadecane and heptadecane, the most abundant n-paraffin were detected at retention time of 4.73-4.77 and 6.67-6.86 respectively, as presented in Figure 2 and Figure 3.

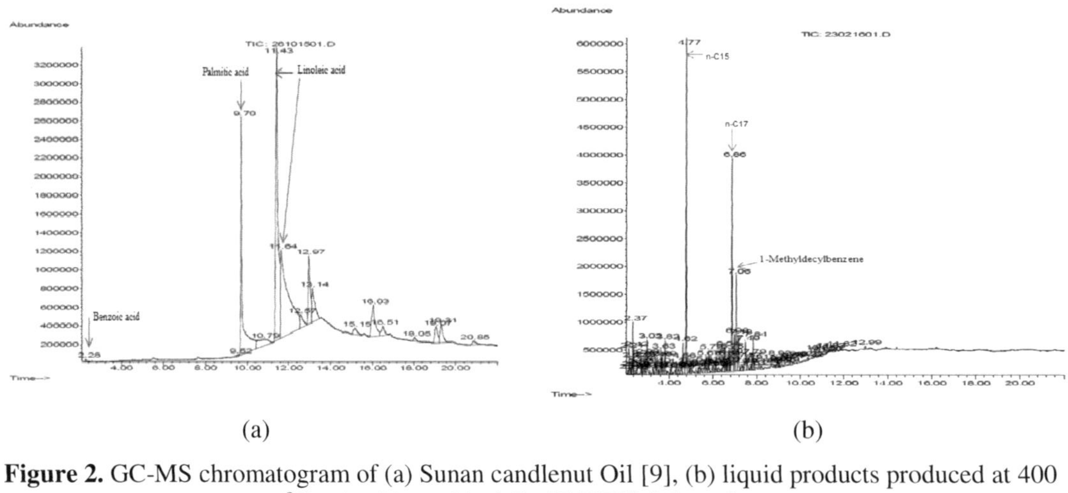

(a) (b)

Figure 2. GC-MS chromatogram of (a) Sunan candlenut Oil [9], (b) liquid products produced at 400 ^{o}C using hierarchical Co-Ni/HZSM-5 catalyst.

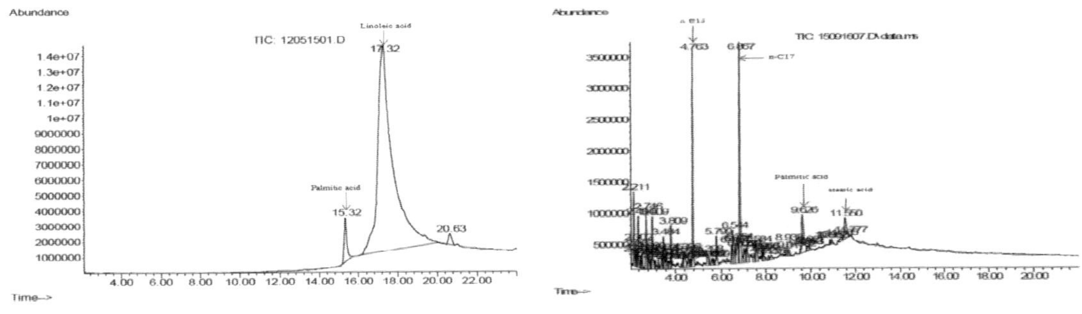

(a) (b)

Figure 3. GC-MS chromatogram of (a) Rubber seed oil [17], (b) liquid products produced at 400 °C using hierarchical Co-Ni/HZSM-5 catalyst.

3.3. *Effect of non-edible vegetable oil composition on formation of paraffins, cycloparaffins and alkylbenzenes*

The presence of n-paraffins was identified on mass spectra in Figure 4 (a). It was found that pentadecane (molecular weights of 212) and heptadecane (molecular weights of 240) were produced from hydrocracking of all non-edible vegetable oils with the hierarchical Co-Ni/HZSM-5 catalyst. While hydrocracking of cerbera manghas oil with the Co-Ni supported on HZSM-5 catalyst also indicated the presence of n-C15 and n-C-17 in liquid product, as listed in Table 2. The abundant carboxylic acids in liquid product from hydrocracking of cerbera manghas oil showed that the HZSM-5 catalyst without hierarchical pore structure less increased the conversion level of carboxylic acid to hydrocarbon compounds. In the previous report [7], [8], a high n-paraffin content in liquid product was reached for hydrocracking of cerbera manghas oil. According to Da Rocha Filho et al. [4], fatty acid composition of oil was significant factor on the formation of n-paraffin. It can be also observed in Table 2, large amount of alkylbenzenes were only produced via hydrocracking of sunan candlenut oil. While, long chain saturated fatty acids and olefins were not found. Small amount of olefins (less than 4 area%) and aromatic were found in cerbera manghas oil liquid product. It can be said that degree of saturation of vegetable oil and reaction temperature also effected the hydrocarbon product distribution. As reported by previous study [7], [12], olefin content decreased with the increasing reaction temperature. While, the normal paraffins were produced via decarboxylation and/or decarbonylation with elimination of oxygen atoms as CO_2, CO and H_2O. This phenomenon coresponds to equation I and II which are attributed to endothermic reaction, as presented in Table 1.

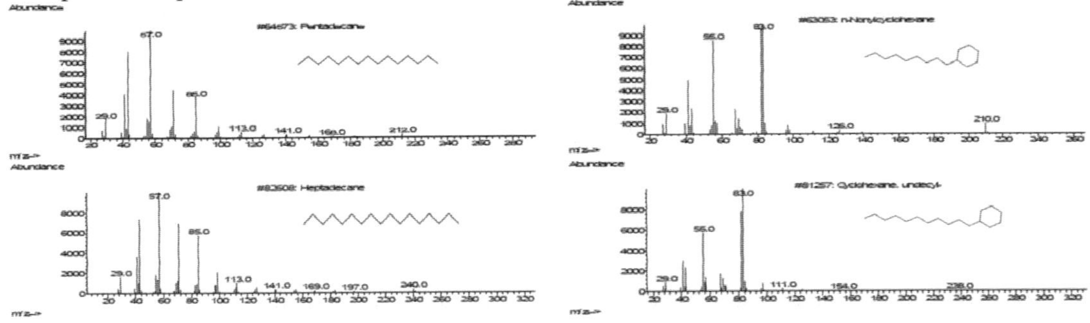

(a) (b)

Figure 4. Mass spectra of (a) n-paraffins, (b) cycloparaffins in liquid product produced from hydrocracking of non-edible vegetable.

IOP Conf. Series: Earth and Environmental Science **67** (2017) 012022 doi:10.1088/1755-1315/67/1/012022

Table 2. Hydrocarbon composition of liquid product (area% of liquid product) at temperature of 400°C, 20±5 bar for 2 h.

Compound/Formula	Structure	Abundance (area%)		
		I[a]	II[b]	III[c]
n-paraffin				
Dodecane/$C_{12}H_{26}$		1.53	1.05	2.03
Tridecane/$C_{13}H_{28}$		2.06	2.02	2.96
Tetradecane/$C_{14}H_{30}$		1.60	1.75	1.86
Pentadecane/$\mathbf{C_{15}H_{32}}$		11.49	17.48	11.04
Hexadecane/$C_{16}H_{34}$		1.43	1.58	2.25
Heptadecane/$C_{17}H_{34}$		7.24	13.42	14.67
Octadecane/$C_{18}H_{38}$		0.50	0.52	0.65
Nonadecane/$C_{19}H_{40}$		1.12	0.65	0.82
Eicosane/ $C_{20}H_{42}$		0.6	0	0.55
Heneicosane/$C_{21}H_{44}$		0.91	0	0
Tricosane/$C_{23}H_{48}$		0.69	0	0
Pentacosan /$C_{25}H_{52}$		0	0	0.31
Aromatic				
Pentylbenzene/$C_{11}H_{16}$	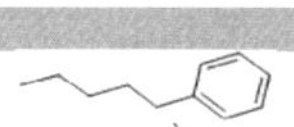	0.38	0.54	0
(3-Methylbut-2-en-2-yl)benzene/$C_{11}H_{14}$	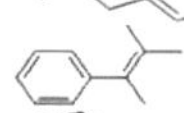	0	0.26	0
Cyclopentacycloheptene/Azulene/$C_{10}H_8$		0	0.57	0
Hexylbenzene/$C_{12}H_{18}$	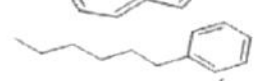	0	0.82	0
1,3-Dimethylbutylbenzene/$C_{12}H_{18}$	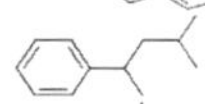	0	0.86	0
1-Methylnaphthalene/$C_{11}H_{10}$	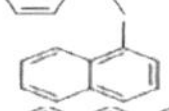	0	0.39	0
2-Methylnaphthalene/$C_{11}H_{10}$	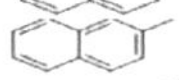	0.47	0.66	1.42
1-Heptylbenzene/$C_{13}H_{20}$	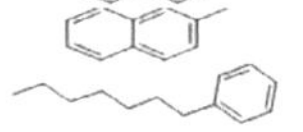	0	1.39	0
1-Methyl-2-n-hexylbenzene/ $C_{13}H_{20}$		0	0.89	0
1-Ethylnaphthalene/ $C_{12}H_{12}$	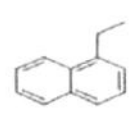	0	0.31	0
2-Ethylnaphthalene		0	0.55	0
1,3-Dimethylnaphthalene/ $C_{12}H_{12}$	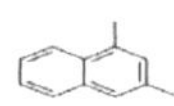	0	0	0.35
Octylbenzene/ $C_{14}H_{22}$	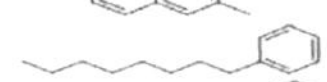	0	1.05	0
n-Nonylbenzene/ $C_{15}H_{24}$		0	0.89	0
1-Methyldecylbenzene/$C_{17}H_{28}$	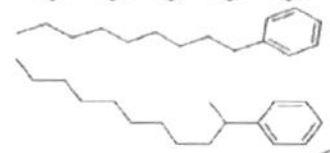	0	9.58	0
n-Undecylbenzene/$C_{17}H_{28}$		0	0.88	0
Cycloparaffin				
(1,3-Dimethylbutyl)cyclohexane	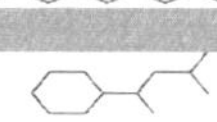	0	1.07	0
Nonylcyclohexane/ $C_{15}H_{30}$		1.46	0.71	1.46
n-Undecylcyclohexane/ $C_{17}H_{34}$	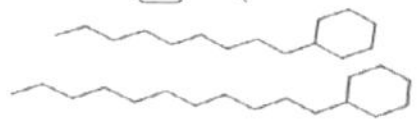	1.80	1.02	2.05
1-(Cyclohexylmethyl)-2-ethylcyclohexane /$C_{15}H_{28}$		0	0	1.42
Cyclopentadecane		0.69	0	0
Cyclopentane, nonyl-		0.65	0	0
Cyclotetracosane		0.31	0	0
Olefin				
1-Heptadecene/$C_{17}H_{34}$	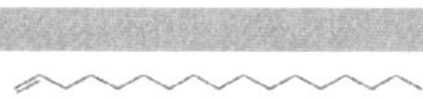	0.51	0	0.68
1-Nonadecene		0	0	0.47
1-Docosene		0	0	0.11
4-Dodecene		0.51	0	0
2-Tridecene		1.05	0	0
3-Heptadecene		1.41	0	0
Carboxylic acid				
Octanoic Acid/ $C_8H_{16}O_2$		1.79	1.22	3.38
Nonanoic acid/ $C_9H_{18}O_2$		3.03	1.05	3.64
n-Decanoic acid/$C_{10}H_{20}O_2$		3.29	1.19	1.94
Undecanoic acid/$C_{11}H_{22}O_2$		1.33	0	1.78
n-Hexadecanoic acid		19.03	0	6.8
Octadecanoic acid		11.21	0	4.6
Oleic Acid		4.71		

IOP Conf. Series: Earth and Environmental Science **67** (2017) 012022 doi:10.1088/1755-1315/67/1/012022

Continued . .

Oxygenated compound				
6,10,14-Trimethyl-2-pentadecanone/ $C_{18}H_{36}O$	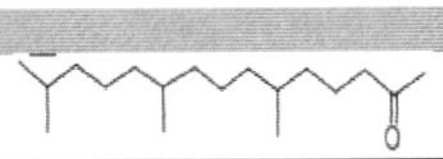	0	0.36	0

[a]Liquid product produced from hydrocracking of cerbera manghas oil over Co-Ni/HZSM-5 catalyst,
[b]Liquid product produced from hydrocracking of sunan candlenut oil,
[c]Liquid product produced from hydrocracking of rubber seed oil.

The cyclization and C-C bond cleavages were enhanced by the presence of a large amount of unsaturated component in vegetable oil, as reported Da Rocha Filho et al. [4]. Rabaev et al. [3] also stated that the formation of alkylbenzenes were determined by poly-unsaturation fatty acids in vegetable oil. As presented in Figure 4 (b), cycloparaffins with molecular weights of 210 and 238 found in liquid product for all vegetable oils showed that sites migration of unsaturation have occured. Only in liquid product produced from hydrocracking of sunan candlenut oil, some of alkylbenzenes with molecular weights of 176, 190, and 232 were found, as shown ini Figure 5. Small amount of naphthalene compounds found in liquid product for rubber seed oil were 2-methylnaphthalene and 1,3-dimethylnaphthalene.
It is interesting to study that the presence of alkylcycloparaffin was proportional to the presence of corresponding n-paraffin, such as if a large amount of n-$C_{17}H_{36}$ with molecular weight of 240 was present, a large amount of the corresponding n-undecylcyclohexane with molecular weight of 238 could be observed. This phenomenon can be explained with multistep sequence for the hydrocracking of the vegetable oils over a NiMo/Al_2O_3 catalyst [4]. In practically, triglyceride hydrogenation is uncomplete reaction at certain reaction temperature so that unsaturated fatty acids will produce and the decarboxylation and/or reduction on them can produce unsaturated hydrocarbons due to proceed cyclization into alkylcycloparaffins. In addition, alkylcycloparaffin was obtained via cyclization of alkenes produced by the thermally and/or catalytically dehydrogenation on n-paraffins as the main product hydrocarbons.

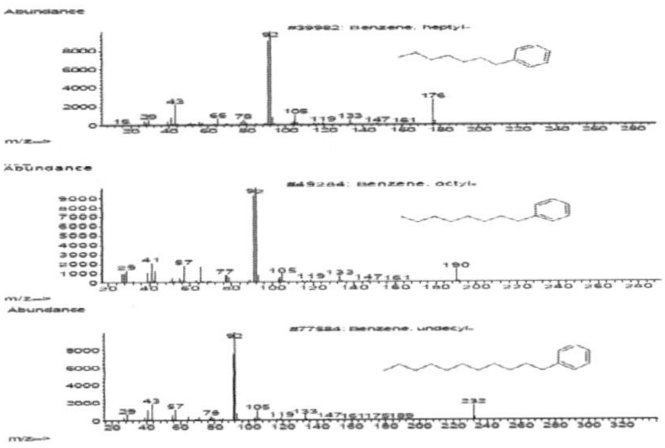

Figure 5. Mass spectra of alkylbenzenes in liquid product produced from hydrocracking of sunan candlenut oil.

4. Conclusion

According to the result, the hierarchical Co-Ni/HZSM-5 catalyst with hierarchical pore structure can contribute to control various reactions occurred in hydrocracking process at 400 °C, 20±5 bar for 2 h, including hydrogenation, cracking, cyclization, and aromatization. Large amount of alkylbenzenes were produced via hydrocracking of sunan candlenut oil. It can be said that degree of saturation of vegetable oil and reaction temperature effected the hydrocarbon product distribution. The most abundant hydrocarbon is pentadecane (n-C15) and heptadecane (n-C17). The future work including biofuel purification into gas oil and development catalyst modified was expected to be applied in low reaction temperature with remain keep the quality of biofuel.

5. References

[1] Romero M, Pizzi A, Toscano G, Casazza A A, Busca G, Bosio B and Arato E 2015 Preliminary experimental study on biofuel production by deoxygenation of Jatropha oil. *Fuel Processing Technology* **137** 31–7

[2] Kim S K, Brand S, Lee H, Kim Y and Kim J 2013 Production of renewable diesel by hydrotreatment of soybean oil: Effect of reaction parameters *Chemical Engineering Journal* **228** 114–123

[3] Rabaev M, Landau M V, Vidruk-Nehemya R, Koukouliev V, Zarchin R and Herskowitz M 2015 Conversion of vegetable oils on Pt/Al_2O_3/SAPO-11 to diesel and jet fuels containing aromatics *Fuel* **161** 287–294

[4] Da Rocha Filho G N, Brodzki D and Djega-Mariadasso G 1993 Formation of alkanes, alkylcycloalkanes, and alkylbenzenes during the catalytic hydocracking of vegetable oils. *Fuel* **72** 543-9

[5] Sotelo-Boyás R, Liu Y and Minowa T 2011 Renewable Diesel Production from the Hydrotreating of Rapeseed Oil with Pt/Zeolite and NiMo/Al_2O_3 Catalysts *Ind. Eng. Chem. Res.* **50** (5) 2791–9

[6] Chen L, Li H, Fu J, Miao C, Pengmei Lv and Yuan Z 2016 Catalytic hydroprocessing of fatty acid methyl esters to renewable alkane fuels over Ni/HZSM-5 catalyst *Catalyst Today* **259** 266-276

[7] Marlinda L, Al-Muttaqii M, Roesyadi A and Danawati H P 2016 Production of Biofuel by Hydrocracking of *Cerbera Manghas* Oil Using Co-Ni/HZSM-5 Catalyst : Effect of Reaction Temperature *J. Pure App. Chem. Res.* **5** (3) 189-195

[8] Al-Muttaqii M, Marlinda L, Roesyadi A and Danawati H P 2017 Co-Ni/HZSM-5 Catalyst for Hydrocracking of Sunan Candlenut Oil (Reutealis trisperma (Blanco) Airy Shaw) for Production of Biofuel. *J. Pure App. Chem. Res.* **6** (2) pp - *and published on May 3, 2017.*

[9] Marlinda L, Al-Muttaqii M, Roesyadi A, Danawati H P 2016 Preparation of hierarchical mesoporous Co-Ni/HZSM-5 catalyst and its application in hydrocracking of Sunan candlenut oil (Reutealis trisperma (Blanco) Airy Shaw). *The International Conference on Engineering, Science and Nanotechnology 2016* (ICESNANO 2016), Sebelas Maret University, August 3-4, 2016.

[10] Sandstede G and Lehmann T 2014 Catalyst With Supplement Component for Hydroprocessing of Bio-Feedstock. *Pub. No. US 20140058182 A1*, Pub date : Feb. 27, 2014.

[11] Syafaruddin and Wahyudi A 2012. Potency of Kemiri Sunan Varieties for Biofuel as a Source of Energy. *Perspektif* **11** (1) 59 – 67 ISSN 1412-8004

[12] Marlinda L, Al-Muttaqii M, Gunardi I, Roesyadi A, and Danawati H P 2017 Hydrocracking of Cerbera manghas Oil with Co-Ni/HZSM-5 as Double Promoted Catalyst. *Bulletin of Chemical Reaction Engineering & Catalysis (BCREC)*, University of Diponegoro : *Article in Press*

[13] Veriansyah B, Han J Y, Kim S K, Hong S, Kim Y J, Lim J S, Shu Y W, Oh S and Kim J 2012 Production of renewable diesel by hydroprocessing of soybean oil: Effect of catalysts *Fuel* **94** 578–585

[14] Martinez-Grimaldo H, Ortiz-Moreno H, Sanchez-Minero F, Ramirez J, Cuevas-Garcia R, and Ancheyta-Juarez J 2014 Hydrocracking of Maya crude oil in slurry-phase reactor. I. Effect of reaction temperature *Catalysis Today* **220-222** 295–300

[15] Barrón C A E, Melo-Bandaa J A, Dominguez E J M, Hernández M E, Silva R R, Reyes T AI and Meraz M M A 2011 Catalytic hydrocracking of vegetable oil for agrofuels production using Ni–Mo, Ni–W, Pt and TFA catalysts supported on SBA-15 *Catalysis Today* **166** 102–110

[16] Zheng X, Chang J and Fu Y 2015 One-pot catalytic hydrocracking of diesel distillate and residual oil fractions obtained from bio-oil to gasoline-range hydrocarbon fuel *Fuel* 157 107–114

[17] Al-Muttaqii M, Marlinda L, Roesyadi A and Danawati H P 2016 Hydrocracking of non-edible vegetable oil with Co-Ni/HZSM-5 catalyst to gasoil containing aromatics. *The 2nd International Seminar on Chemistry* (ISoC 2016). Sepuluh Nopember Institute of Technology, Surabaya, July 26-27, 2016

Catalytic hydrocracking of Kapuk seed oil (*Ceiba pentandra*) to produce biofuel using Zn-Mo supported HZSM-5 catalyst

Y W Mirzayanti[1], D H Prajitno[1] and A Roesyadi[1*]

[1]Chemical Engineering, Industrial Technology Faculty, Sepuluh Nopember Instutite of Technology, Surabaya, Indonesia, 60111. *corresponding author: aroesyadi@yahoo.com

Abstract. In the present paper Kapuk seed oil (KSO) was considered as a potential biofuel for alternative fuel from inedible oil. Catalytic hydrocracking of Kapuk seed oil using Zn-Mo supported on the HZSM-5 catalyst in a slurry pressure batch reactor at various temperature with reactor pressure in range 10-15 bar. The Zn-Mo/HZSM-5 catalyst was prepared by incipient wetness impregnation. The physicochemical properties of the catalyst were investigated by X-ray diffraction (XRD), scanning electron microscopy (SEM) and Brunauer-Emmett-Teller (BET) method. The best catalyst performance on catalytic hydrocracking of KSO using Zn-Mo/HZSM-5 (Si/Al = 25) with loading 2.92%wt for Zn and 7.55%wt for Mo. It displayed the highest hydrocarbon content decarboxylation and/or decarbonylation were 35.51 area% of n-paraffins and the highest content for gasoil-range alkanes was 17.24 area% at 400^0C. The liquid product predominant is n-C_{15}, and the second component is n-C_{17}. The catalyst Zn-Mo/HZSM-5 was helpful for the catalytic hydrocracking of KSO for hydrocarbon biofuel production.

1. Introduction

As a results of the rapid economic development, the energy consumption is also growing rapidly. Increasing of energy demands causes declining of fossil oil production for the world and negative environmental issues, such as emission of nitrogen oxides, sulphur oxides, unburned hydrocarbons, and particulate matters [1]. Moreover, global demands for renewable fuel and regulation to reduce consumption of fossil fuel are also increasing. Biofuel, which is composed of various triglycerides, is a type of renewable fuels and an alternative carbon reduction strategy.

Kapuk seed oil (*Ceiba pentandra*) is inedible vegetable oil. This oil is extracted from seeds of *Ceiba pentandra* plant. It grows naturally humid and sub-humid tropical regions. Kapuk seeds make up about 25-28 wt% of each pod with an average potential oil yield of 1280 kg/ha annually. Air dried seeds have typically 58.0% kernel, 3.4% moisture, 28.7% oil and 23.0% protein. *Ceiba pentandra* is one of the most effective CO_2 absorbers [2]. This plant is potential to development of renewable raw materials of biofuel. Nevertheless, it could not be used directly as biofuel because it still contains high oxygen atoms, elevated viscosity, high freezing point, low heating value, and thermal instability [1], [2]. There are several common methods to convert vegetable oil to biofuel, such as transesterification, catalytic cracking and hydroprocessing (catalytic hydrocracking). Catalytic hydrocracking is an effective method to convert vegetable oil to biofuel even though the introduction of hydrogen contributes a portion of the processing cost [3]. The selection of catalyst in the catalytic hydrocracking process is very important. The catalyst type is a bimetal catalyst with supported by HZSM-5. Zeolite HZSM-5 of MFI (Mobile five) structure has hydrophobic site to cracking triglyceride [4]. HZSM-5

has strong acid site, high surface area, appropriate pore diameter and shape selectivity. Metals used as a promote are zinc and molybdenum. The zinc (Zn) metal might promote the hydrogen transfer during the catalytic upgrading of biofuel process and molybdenum (Mo) metal was selectively a deoxygenation process through the path hydrodeoxygenation process [5]. Further, the use of bimetallic catalyst for catalytic hydrocracking of vegetables oils were investigated.
The main aim of this research is to study catalytic hydrocracking of KSO as potential biofuel for alternative fuel from inedible oil in a slurry pressure batch reactor at various temperatures (300, 325, 350, 375, 400 ^{0}C). The effect of Zn-Mo/HZSM-5 catalyst on the product of biofuel, including n-paraffin, the main abundant with hydrocarbon was discussed in this paper. The Zn-Mo/HZSM-5 catalyst was prepared by incipient wetness impregnation. That catalyst was characterized using a variety of techniques, such as Brunauer-Emmett-Teller (BET) method, X-ray diffraction (XRD) and scanning electron microscopy-energy dispersive X-ray (SEM-EDX) for morphology catalyst.

2. Material and Methods

2.1. Raw Material
Kapuk seed oil (KSO) used as raw material in this research was obtained from Pandaan district, East Java Province, Indonesia. Linoleic acid (C18:2) and palmitic acid (C16:0) found in KSO were 78.85% and 21.12%, respectively. It suggested that the composition of KSO was dominated by polyunsaturated fatty acids.

2.2. Catalyst preparation and characterization
A commercial ammonium-ZSM-5 zeolite (a surface area of 425 m^2/g, Na_2O of 0.05 wt.% and mole ratio of SiO_2/Al_2O_3 = 80) purchased from Zeolyst International was used as a catalyst support and calcined at 550°C for 5 hours under an air atmosphere to obtain HZSM-5. A double promoted Zn-Mo/HZSM-5 catalyst was prepared by incipient wetness impregnation methods. A zinc nitrate salt is of $Zn(NO_3)_2.6H_2O$ and ammonium heptamolybdatetetrahydrate of $(NH_4)_6Mo_7O_{24}.4H_2O$ purchased from Merck was used as metallic precursors. To obtain Zn-Mo/HZSM-5, in this experiment 0.015 M of zinc nitrate and 1.0318 M of ammonium heptamolybdatetetrahydrate was dissolved in the volume of solution corresponding to the total pore volume of the HZSM-5 support. With the successive impregnation, the solution was poured over 4.75 g of HZSM-5 and stirred for 3 hours to ensure that the attachment to the mouth of HZSM-5 pores to occur. After that, the catalyst was dried at 120°C for 12 hours and calcined at 400°C in a flowing air for 5 hours. Finally, the metal impregnated catalyst was reduced at 450°C for 2 hours in hydrogen atmosphere to obtain an active metal phase.
The phase analysis of Zn-Mo/HZSM-5 catalyst was characterized by X-ray diffraction (XRD). Morphology catalyst, actual metal content and dispersion on zeolite were performed by scanning electron microscopy-energy dispersive X-ray (SEM-EDX). In addition, the total surface area, pore volume and pore size were determined by N_2 adsorption-desorption isotherms at 77K. Prior to measurement, the samples were out gas at 300°C for 3 hours. Brunauer-Emmett-Teller (BET) method was applied to calculate the total surface area. Pore size distribution of catalyst was determined by Horvath-Kawazoe (H-K) method.

2.3. Catalytic hydrocracking of KSO
The catalytic hydrocracking of kapuk seed oil was carried out in a slurry pressure batch reactor (Parr USA 4563) with a mechanical stirrer. The reaction conditions were applied in catalytic test, i.e. reaction temperatures of 300, 325, 350, 375 and 400°C, reaction time of 2 hours, and reactor pressure at the range of 10-15 bar after flowing hydrogen for about one hour. Approximately 0.56 wt.% of the catalyst to KSO (200 ml) were loaded into the reactor. After that, nitrogen was purged into a reactor to remove air that might have been dissolved in oil for twice. After processing for 2 hours, liquid product was collected and analyzed by gas chromatography-mass spectrometry (GC-MS). The liquid product

was considered as hydrocarbon bio fuel such as gasoline (C_5-C_{12}), kerosene (C_{10}-C_{13}), and gas oil (C_{14}-C_{22}) [6], [7].

3. Result and Discussion

3.1. Characterization of catalyst

The characterization of catalyst was done by XRD, SEM-EDX and BET methods. The X-ray diffraction patterns of the HZSM-5 catalyst used for catalytic hydrocracking of KSO. The characteristic's peaks of HZSM-5 at 2θ of 7.87°, 8.74°, 23.04°, 23.24°, 23.64°, 23.91° and 24.31° indicate that HZSM-5 have a framework type MFI. The XRD patterns of catalyst showed small peak intensity at 2θ of 36.0899° for ZnO, while for MoO_2 detected on three main diffraction peak at 2θ i.e. 26.0182°, 36.9332° and 53.3027° as reported by previous study [8]. Molybdenum catalyst can facilitate the hydrodeoxygenation reaction at catalytic hydrocracking triglyceride of KSO [9].

Figure 1a and Figure 1b shows electron microscopy images of the calcined HZSM-5 and Zn-Mo/HZSM-5. The HZSM-5 (Si/Al=25) crystals assemble to uniform small cuboid particle (Figure 1a). The SEM picture of the Zn-Mo/HZSM-5 to show evidence that crystals of Zn and Mo can be impregnation into HZSM-5 (Figure 1b). The presences of Zn and Mo are evidenced by EDX analysis. Based on EDX, Zn and Mo loading was approximately 2.92%wt for Zn and 7.55%wt for Mo.

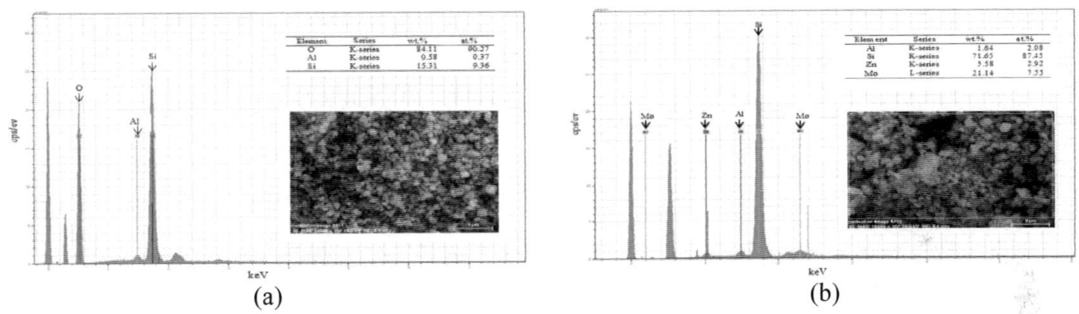

Figure 1. SEM image and the corresponding EDX patterns of (a) HZSM-5 obtained from the calcined NH_4-ZSM-5 zeolite, (b) Zn-Mo/HZSM-5 catalyst.

Table 1 shows properties of catalyst from BET analysis. The surface area and total pore volume about HZSM-5 significantly decrease after calcined from 425 m^2/g to 375.121 m^2/g. The characterization of Zn-Mo was modified in HZSM-5 through impregnation given by analyzes obtained BET surface area of 250.704 m^2/g. The addition of metals led to micropore surfaces are and total pore volumes were reduced through blocking.

Table 1. The properties of catalyst.

Catalyst	BET surface area, m^2/g	Total pore volume, cm^3/g	Average pore diameter, nm	Metal content[a], wt.%	
				Zn	Mo
HZSM-5	375.121	0.2712	2.89	-	-
Zn-Mo/HZSM-5	250.704	0.1613	2.57	2.92	7.55

[a]EDAX analysis

3.2. Analysis biofuel

Figure 2 shows the GC-MS spectra of kapuk seed oil (KSO) and liquid product produced at 400°C using Zn-Mo/HZSM-5 catalyst. Figure 2a show that the linoleic and palmitic acid from KSO was predominant, so KSO include as polyunsaturated fatty acid. Figure 2b show that the various compounds in biofuel after hydrocracking reaction using Zn-Mo/HZSM-5 catalyst. The first most abundant n-paraffin is pentadecane (n-C_{15}), and the second most abundant n-paraffin is heptadecane (n-C_{17}). This shows the reaction of decarboxylation or decarbonylation dominated KSO hydrocracking process. The existence of palmitic acid indicates that the process hydrogenation and cracking linoleic acid successfully performed at 400°C using Zn-Mo/HZSM-5. As shown in Figure 2a, some peaks of linoleic acid disappear at retention time of 11-15. After hydrocracking reaction (Figure 2b), palmitic acid and n-paraffin appeared at retention time of 9.6 and in range of 3-8, respectively. The presence of the n-C_{15}-nC_{19} showed biofuel contained gasoil-like hydrocarbon. It is interesting to study later, compound14-beta-h-pregna and 1,2-Benzisothiazole, 3-(hexahydro-1H-azepin-1-yl)-,1,1-dioxidewere detectedon biofuel grouped in gasoil-like hydrocarbon at retention time of 10.8-11.6, which corresponding to GC-MS analysis of petroleum diesel [10].

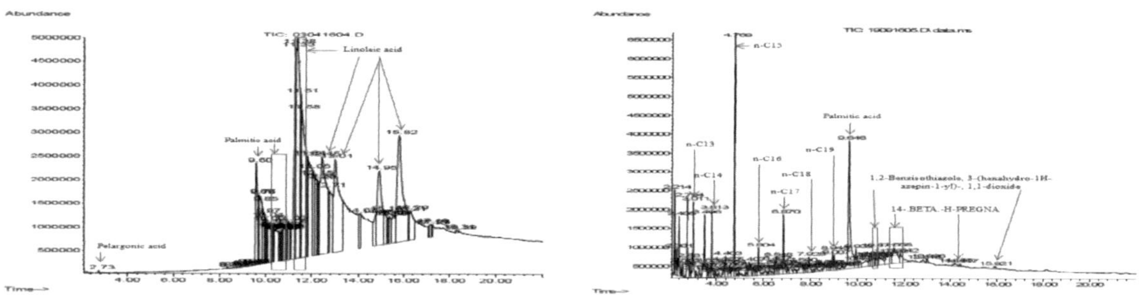

(a) (b)

Figure 2. GC-MS spectra of (a) Kapuk seed oil (KSO), (b) liquid product produced at 400°C, using Zn-Mo/HZSM-5 catalyst.Saturated chain hydrocarbons (n-paraffin): C_{12} to C_{23}.

Figure 3 shows that the effect of reaction temperature on hydrocarbon compounds of liquid product. The hydrocarbon, including n-paraffin, cycloparaffin and aromatic increased with the increasing temperature. A decrease in carboxylic acid of 72.25-36.69 area% with increasing temperature was observed. This indicated that the first of triglycerides were converted into free fatty acids, likely through β-elimination mechanism [11]. The carboxylic acids as intermediate compound according to GC-MS analysis at 400°C, the most abundant hydrocarbonin liquid product during decarboxylation and/or decarbonylation was 35.51 area% of n-paraffins. In addition, a small amount of cycloparaffins and aromatics also found were 2.92 area% and 0.80 area%, respectively. It is implied that has occurred the cyclization and dehydration reaction. It was interesting to be observed that olefin was only found in liquid product produced at 375°C. The olefin was produced via dehydrogenation and/or hydrogenolysis of n-paraffin [12]. When temperature of 400°C was applied, the olefin was not observed. It suggested that the cracking of olefins occurred to produce light hydrocarbon compounds.
Analysis of biofuel was compared against the results of GC-MS commercial diesel [10]. The biofuel was identified based on carbon number of hydrocarbon compounds obtained from the GC-MS analysis. It was defined as gasoline-range hydrocarbon (C_5-C_9), kerosene-range hydrocarbon (C_{10}-C_{13}), and gasoil-range hydrocarbon (C_{14}-C_{22}) [6]. The contents of various hydrocarbon in liquid product over Zn-Mo/HZSM-5 catalyst at the reaction temperature ranging from 300 to 400°C are shown in Table 2. Gasoil-range alkanes produced increased with increasing temperature, as reported in previous study [12, 13]. A decrease in the n-$C_{>18}$ compound with increasing gasoil-range alkanes was observed at temperature in range of 375-400°C. It implied that the cracking of long chain hydrocarbon molecules into short-chain hydrocarbon molecules occur as reported in previous study [9], [13], [14]. With Zn-Mo/HZSM-5 catalyst, the highest content of gasoil is 17.24 area% at a temperature of 400°C. According gasoil range alkanes, reaction decarboxylation and/or decarbonylation conducted at temperatures above 350°C and prefers to take place at temperatures above 400°C for improving the conversion of carboxylic acids to hydrocarbons. This corresponded with the reported that the decarboxylation and/or decarbonylation reaction was endothermic reaction [10].

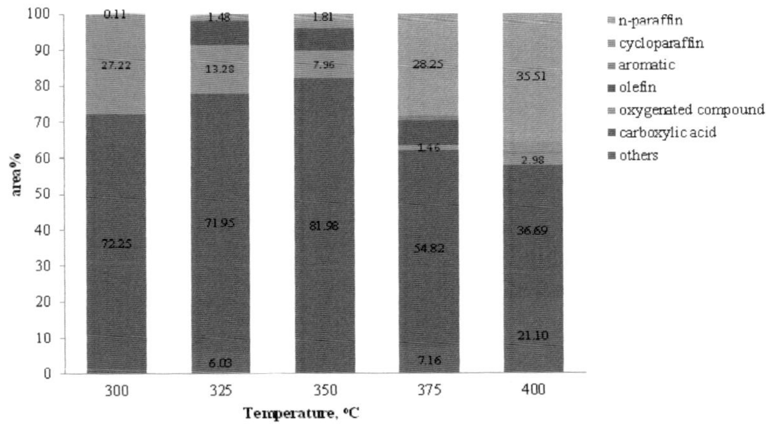

Figure. 3. Effect of reaction temperature on hydrocarbon compounds of liquid product produced using Zn-Mo/HZSM-5 catalyst. Others were *14-.BETA.-H-PREGNA and 1,2-Benzisothiazole, 3-(hexahydro-1H-azepin-1-yl)-,1,1-dioxide.*

Table 2. Contents of various hydrocarbon in liquid product over Zn-Mo/HZSM-5 catalyst

Temperature (°C)	Contents of various hydrocarbon in liquid product (area%)				
	nC_5-nC_9	nC_{10}-nC_{14}[a]	nC_{15}-nC_{18}[b]	$nC_{>18}$	Cycloparaffin, olefin, and aromatic
300	0	0.12	0.07	0	0.36
325	0	0.96	1.15	0	15.33
350	0	0.96	1.34	0	6.71
375	0	7.4	15.02	2.88	11.11
400	0	9.24	17.24	1.09	18.24

[a]*kerosene range alkanes*
[b]*gasoil range alkanes*

4. Conclusion

Inedible kapuk seed oil can be hydroprocessed to produce hydrocarbon biofuel by catalytic hydrocracking using the Zn-Mo/HZSM-5 catalyst in a slurry pressure batch reactor. With the Zn-Mo/HZSM-5 catalyst, the most abundant hydrocarbon in liquid product through decarboxylation and/or decarbonylation was 35.51 area% of n-paraffins, and the highest content for gasoil-range alkanes was 17.24 area% at a temperature of 400°C. The first most abundant compound is n-C_{15}, and the second most abundant component is n-C_{17}. The decarboxylationand or decarbonylation are the dominant reaction pathways. The Zn-Mo/HZSM-5 catalyst is an effective catalyst to produce biofuel by hydrocracking of kapuk seed oil, including n-paraffin, the main abundant with hydrocarbon.

5. References

[1] Bokhari, A., Chuah, F, L., Yusup, S., Ahmad, J., Shamsuddin, R, M., and Teng, K, M., 2015, "Microwave-assisted methyl esters synthesis of Kapuk (*Ceiba pentandra*) seed oil: parametric and optimization study", Biofuel Research Journal, 7, 281-287.
[2] Kathirvelu, S., Moorthi, N. ShenbagaVinayaga., Krishnan, S. Neela., Mayilsamy, K., Krishnaswamy, T., 2014, "Production of biodiesel from non edible ceibapentandra seed oil having high FFA content", ARPN Journal of Engineering and Applied Sciences, Vol. 9, No.12, December, pp. 2625-2634.
[3] Zhao, X., Wei, L., Cheng, S., Kadis, E., Cao, Y., and Boakye, Eric., 2016, "Hydroprocessing of carinata oil for hydrocarbon biofuel over Mo-Zn/Al_2O_3", Applied Catalysis B: Environmental, Vol. 196, pp. S 41-49.

[4] Sandstede, G and Lehmann T., 2014, "Catalyst with Supplement Component for Hydroprocessing of Bio-Feedstock". *Pub. No. US 20140058182 A1*, Pub date: Feb. 27, 2014.
[5] Anderson, J.R., andBoudart, M., 1981, "Catalysis: Science and Technology", Vol. 2, Springler-Verlag Berlin Heidelberg New York.
[6] Barron, A.E., Melo-Banda, J.A., Dominguez, J.M., Hernandez M, E., and Silva R, R., 2011, "Catalytic hydrocracking of vegetable oil for agrofuels production using Ni-Mo, Ni-W, Pt and TFA catalyst supported on SBA-15", Catalysis Today, 166, pp. 102-110.
[7] Hsu, C.S., and Robinson, P.R., 2006, "Practical advance in petroleoum processing", Springer Science Business Media, Inc, USA.
[8] Mirzayanti, Y.W., Roesyadi, A., and Prajitno, D.H., 2017, The 7th Annual Basic Science International Conference, Malang, 7-8 March 2017.
[9] Ancheyta, J., Trejo, F., Rana M.S., 2009, Asphaltenes: Chemical Transformation during Hydroprocessing of Heavy Oils, New York: CRC Press.
[10] Rasyid, R., Prihartantyo, A., Mahfud, M., Roesyadi, A., 2015, Hydrocracking of Calophylluminophyllum oil with Non-SulfideCoMo catalyst, Bulletin of Chemical Engineering and Catalyst, 10 (1), pp. 61-69, ISSN: 1978-2993.
[11] Romero, M., Pizzi, A., Toscano, G., Casazza, A.A., Busca, G., Bosio, B., Arato, E., 2015, Preliminary experimental study on biofuel production by deoxygenation of Jatropha oil. *Fuel Processing Technology* 137, 31–37.
[12] Kim S K, Brand S, Lee H, Kim Y, Kim J., 2013, Production of renewable diesel by hydrotreatment of soybean oil: Effect of reaction parameters, *Chemical Engineering Journal,* 228, 114–123
[13] Prajitno, D. Hari, Budianto, A., Iqbal, M., Purnomo, V., 2015, Modification of Ni/Zn-HZSM-5 Double Promote Catalyst for Biofuel Production from Cerbera Manghas Oil, *The 9th Joint Conference on Chemistry*, 12-13 November 2014, Semarang.
[14] Verma, D., Rana, B.S., Kumar, R., Sibi, M.G., Sinha, A.K., 2015, Diesel and aviation kerosene with desired aromatics from hydroprocessing of jatropha oil over hydrogenation catalyst supported on hierarchical mesoporous SAPO-11, *Applied Catalysis A: General*,490, 108–116
[15] Zheng, X., Chang. J, Fu, Y., 2015, One-pot catalytic hydrocracking of diesel distillate and residual oil fractions obtained from bio-oil to gasoline-range hydrocarbon fuel, *Fuel,* 157: 107–114

6. **Acknowledgments**

The authors would like greatfully acknowledge to the head of chemical reaction engineering laboratory, Department of Chemical Engineering, Faculty of Industrial Technology, Sepuluh Nopember Institute of Technology for the facilities and also thanks for the crews of laboratory, i.e. Varantika Eka Pramesti and Muhammad Faishal Razin.

IOP Conf. Series: Earth and Environmental Science **67** (2017) 012024 doi:10.1088/1755-1315/67/1/012024

The development of control and monitoring system on marine current renewable energy
Case study: strait of Toyapakeh - Nusa Penida, Bali

I S Arief[1], I H Suherman[2], A Y Wardani[1] and A Baidowi[1]
[1]Dept of Marine Engineering, Institut Teknologi Sepuluh Nopember, Surabaya, Indonesia
[2]Center of Research and Development of Marine Geology, Ministry of Energy and Mineral Resources, Bandung, Indonesia

Abstract. Control and monitoring system is a continuous process of securing the asset in the Marine Current Renewable Energy. A control and monitoring system is existed each critical components which is embedded in Failure Mode Effect Analysis (FMEA) method. As the result, the process in this paper developed through a matrix sensor. The matrix correlated to critical components and monitoring system which supported by sensors to conduct decision-making.

1. Introduction

Strait of Toyapakeh is located between Bali and Nusa Penida island (shown in Fig. 1). This strait has a good potential of Marine Current energy. Based on the measurement which has been conducted by the researchers in 2015, the Strait of Toyapakeh has the potential Marine Current energy up to 320 mw. The wave amplitude in this strait reached more than 2.6 meters. As other regions in Indonesia, tides type is twice up and down per day [7]. From the reason, Strait of Toyapakeh can be used as the alternatives area to generate the Marine Current energy.

The basic principle of marine current energy is to convert kinetic energy to electricity [7]. To support the conditions, communication infrastructure has an important role. The communication infrastructure includes control and monitoring system of Marine Current energy. As the standard of the research, communication for monitoring and control is based on international standard IEC 61400-25. The standard is can be applied on Marine Current Renewable Energy.

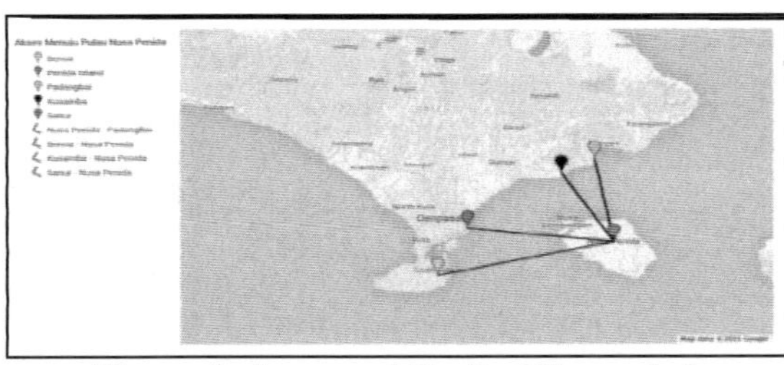

Figure 1. Access of Strait of Toyapakeh
(source: googlemap.com)

2. Methode scheme of research

Design review is crucial in order to determine failure potential and position by Failure Mode Effect Analysis (FMEA) method. In order to, FMEA describes each part of Marine Current components specifically. Then, the process is continued in control and monitoring system.
The design of control and monitoring system of Marine Current Renewable are described in picture bellow:

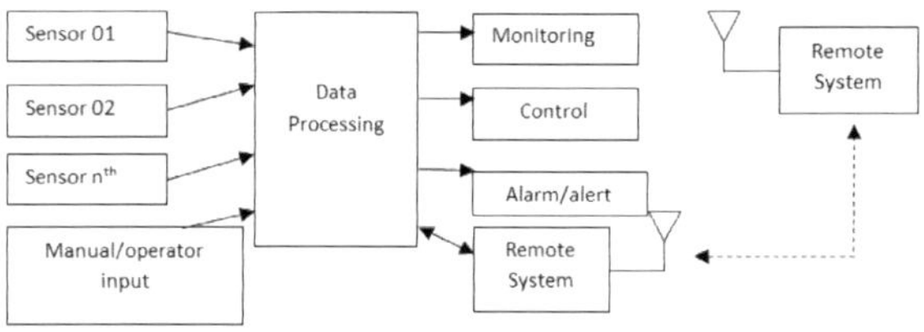

Figure 2. The Scheme of Control and Monitoring System

According to the scheme, specified function of control and monitoring is designed as early warning system. Mostly, the functions work as local and long distance area. To support long distance area, remote system is required as data communication by telemetry.
In this paper, control and monitoring system is divided into data monitoring, data analysis and processing. The physical data monitoring are split as follows:

- Direction and wind speed
- Oceanography.
- Mechanical condition and structure
- Bending Load, torsional, and vibration.
- Stress, strain, and displacement.
- Temperature operation and gearbox.
- RPM of main shaft.
- Position / turbine orientation relative to platform.
- Voltage characteristic
- The quality of power
- additional data monitoring
- relative motion platform to the reference coordinate

Based on the description, the type of sensors level position can be determined on the table follows:

Table 1. Sensor Level Position

No.	Sensor/transducer	Level Position	Explanation
1	Wind speed sensor	Highest position of platform	Wind Speed
2	Wind direction	Highest position of platform	Angina speed
3	Rainfall transducer	Highest position of platform	Humidity
4	Air temperature sensor	Highest position of platform	Wet and dry temperature
5	Relative humidity sensor	Highest position of platform	Humidity
6	Evaporation rate	Highest position of platform	Evaporation
7	Sea water current meter	As level as turbine	Speed water flow moves the turbine

For mechanical and structural conditions, the type of sensors level position can be determined on the table follows:
The characteristics voltage characteristic can be measured from the side of the power station as well as from the side of the load. The signal information characteristic of the electrical voltage is available on the system and plant distribution. Thus, the control system and monitoring lives the signals and process them. As for signals that are required are as follows:

Table 2. Sensor Level for mechanical and structural conditions.

No.	Sensor/transducer	Level Position	Explanation
1	Rotary encoder	Main shaft turbine	Turbine speed rotation
2	Rotary encoder	Generator shaft	Generator speed rotation
3	Rotary encoder	Motor shaft turbine lifting mechanism	speed rotation of motor lifting mechanism
4	Rotary encoder	Shaft turbine lifting mechanism	Angel position and speed rotation of turbine lifting mechanism
5	3-axis accelerometer	Far-end turbine main shaft	Mechanic vibration and displacement
6	3-axis accelerometer	Mid-end turbine main shaft	Mechanic vibration and displacement
7	3-axis accelerometer	Near-end turbine main shaft	Mechanic vibration and displacement
8	Dynamic torque sensor	Turbine main shaft	Torque of turbine shaft

Table 3. Sensor level for voltage characteristics

No.	Sensor/transducer	Level Position	Explanation
1	Voltmeter	in the plant and load	Power quality
2	Ampere meter	in the plant and load	Power quality
3	Current transformer	in the plant and load	Power quality

Additional monitoring data to describe the movement, position and orientation platforms are obtained on Table 4:

Table 4. Sensor level of additional monitoring data

No.	Sensor/transducer	Level Position	Explanation
1	Hybrid IMU sensor: accelero, gyro, magneto	Metacenter	Platform moving in 3D
2	GPS and hybrid IMU	afore platform	Platform position and orientation

3. Result and discussion

Data processing is change the value based on factors calibration of measurement or change data formats input to others format. In accordance is needed by applying integration, differentiation, Fourier transformation or other methods of data input.

Data analysis is conducted to obtain the condition of the system based on sensor data which has been adapted to reflect the needs. Analysis data divided into:

1.Data analysis directly, measured to:
- Temperature gearbox
- Speed of main shaft

2. Data analysis indirectly, measured to;
   a) The main shaft torsion: increase or decrease of torsional load can is predicted based on the ratio between the speeds of water flow v_w. In order side, speed blades at a certain point v_b which described as the factor ratio $f_v = \frac{v_w}{v_b}$. If the f_v up, it can be considered that the load of main torsional rose as well.
   b) The results of the analysis show unreflect on normal operations will be forwarded to control and algorithm alarm (warning or early warning).

Control system undertakes the function of control, in order to the process of the system and equipment safety. Control system works on commandment which embedded in processor to a condition reading sensors mounted on equipment on demand. The process control system started with a source of up to a produced output. Therefore, it is divided into the critical component, those are:

1. Battery.
   Control and monitoring system of battery in several conditions, such as:
   c) Short circuit
   d) Overload

e) High temperature of battery
2. Bearing
3. Gearbox
4. Inverter
5. Generator
6. Power control
7. Main shaft
8. Shaft main drive
9. Turbine

According to critical component in the research developed through a matrix sensor. The matrix correlated components critical and monitoring system which supported by sensors to do decision-making. The matrix sensor of the components can be determined on the tables below:

Table 5. Matrix Sensor of Current

Correlation of Sensor and Reaction	Current Sensor (Up)	Current Sensor (Down)
Gearbox Temperature Sensor (Up)	ALARM & ACTUATOR ACTIVE	ALARM
Gearbox Temperature Sensor (Down)	ALARM	-
Sensor of Blade Speed Rotation (Up)	ALARM & ACTUATOR ACTIVE	ALARM
Sensor of Blade Speed Rotation (Down)	ALARM	-
Sensor of Shaft Speed Rotation (Up)	ALARM & ACTUATOR ACTIVE	ALARM
Sensor of Shaft Speed Rotation (Down)	ALARM	-
Sensor of Load Torsional (Up)	ALARM & ACTUATOR ACTIVE	ALARM
Sensor of Load Torsional (Down)	ALARM	-

Table 6. Matrix Sensor of Voltage

Correlation of Sensor and Reaction	Voltage Sensor (Up)	Voltage Sensor (Down)
Gearbox Temperature Sensor (Up)	ALARM & ACTUATOR ACTIVE	ALARM
Gearbox Temperature Sensor (Down)	ALARM	-
Sensor of Blade Speed Rotation (Up)	ALARM & ACTUATOR ACTIVE	ALARM
Sensor of Blade Speed Rotation (Down)	ALARM	-
Sensor of Shaft Speed Rotation (Up)	ALARM & ACTUATOR ACTIVE	ALARM
Sensor of Shaft Speed Rotation (Down)	ALARM	-
Sensor of Load Torsional (Up)	ALARM & ACTUATOR ACTIVE	ALARM
Sensor of Load Torsional (Down)	ALARM	-

Table 7. Matrix Sensor of Flow Speed

Correlation of Sensor and Reaction	Sensor of Flow Speed (Up)	Sensor of Flow Speed (Down)
Gearbox Temperature Sensor (Up)	ALARM & ACTUATOR ACTIVE	ALARM
Gearbox Temperature Sensor (Down)	ALARM	-
Sensor of Blade Speed Rotation (Up)	ALARM & ACTUATOR ACTIVE	ALARM
Sensor of Blade Speed Rotation (Down)	ALARM	-
Sensor of Shaft Speed Rotation (Up)	ALARM & ACTUATOR ACTIVE	ALARM
Sensor of Shaft Speed Rotation (Down)	ALARM	-
Sensor of Load Torsional (Up)	ALARM & ACTUATOR ACTIVE	ALARM
Sensor of Load Torsional (Down)	ALARM	-

4. Conclusion

1. The design of control and monitoring system needs to involve failure effect analysis to determine each critical component.
2. The critical component that developed through matrix sensor.
3. The matrix correlated critical components and monitoring system which supported by sensors to do decision-making.

5. References

[1] A. G. Marbun, Failure Mode Implementation an'd Effect Analysis (FMEA) to Increase Reliability of Boiler Pipe, Jakarta: Teknik Industri, Universitas Mercubuana, 2013.
[2] D. Isdarto, Risk Analysis of Operational Failure by Risk Failure Mode and Effect Analysis, Surabaya: Magister Manajemen Teknologi, Institut Teknologi Sepuluh Nopember, 2014.
[3] K. B. Artana, Risk Assesment Subsea Gas Pipeline, Surabaya: Guna Widya, 2013.
[4] M. Mraz, FMEA-FMECA, Ljubljana: Ljubljana Spring, 2005.
[5] Tuncel and Alpan, Risk Assessment and Management for Supply Chain Networks : Case Study, Izmir: Dokuz Eylul University, 2010.
[6] A. P. Adnyasari, "Geography Condition," Pemerintahan Kabupaten Klungkung, 9 11 2015. [Online]. Available : http://www.klungkungkab.go.id/index.php/profil/14/Kondisi-Geografis. [Accessed 2 12 2015].
[7] T. P., "Basic Design of Marine Current Renewable Energy, case study: Strait of Toyapakeh," LPPM ITS, Surabaya, 2015

Acknowledgments
The research project is part of the consortium program between Institut Teknologi Sepuluh Nopember and The Center of Marine Geology, Ministry of Energy and Mineral Resources.

IOP Conf. Series: Earth and Environmental Science **67** (2017) 012025 doi:10.1088/1755-1315/67/1/012025

Daylight strategies for architectural studio facilities: the literature review

Muhammad Anas Bin Othman, Nur Azfahani Ahmad, Azizah Md Ajis

Faculty of Architecture, Surveying and Planning, UiTM Perak Branch, Seri Iskandar, Perak, Malaysia

Abstract. The implementation of daylighting strategies in buildings is a common aspect in architecture. However, due to the availability of inexpensive electricity, natural lighting strategies became insignificant, and been overlooked by designers. With the current concern over rapid increment on electricity cost, many designers now try to revitalized daylighting strategies in buildings. This includes educational buildings. In Malaysian cases, it is a norm that universities; especially during lecture and studio sessions, used artificial lighting throughout the day. Definitely, this is not parallel with the "green" aim made by the Government in the Malaysian Plan. Therefore, this paper aims to explore the impact of daylight strategies for educational studios in universities, by maximising the penetration of natural daylight into the space towards creating a more green-conducive studio. The paper review literature about the types, criteria and benefits of daylight strategies. This paper also presented a pilot study that has been performed in one university in Perak, Malaysia, by selecting architectural studios as the main subject.

1. Introduction

Nowadays, sustainability and green architecture concept has become the common interest of numerous disciplines [1]. The reason for this popularity is to perform the sustainable development and construct buildings with environmentally friendly features [2] that able to reduce carbon dioxide emission and energy wastage. Green architecture was introduced in order to minimizes harmful effects on human health and the environment [3].

Among the characteristic of green architecture is an energy-efficient lighting and indoor environmental quality (lighting comfort) [4]. The Malaysian Green Building Index (GBI), (2015) for Interiors space assessment, has stated that daylighting is one of the elements to create a good Energy Efficiency (EE) and Indoor Environmental Quality (EQ). In Malaysian Standard (MS 1525), there is also a clause that highlights the importance of daylighting towards achieving an energy efficiency building. The purpose of both act is to encourage the new and existing buildings in Malaysia to comply with energy efficiency approaches [4].

However, despite all the efforts made by the Government in producing green approach for building in Malaysia, the energy consumption is still increased. It is reported that, for the past 10 years, energy usage in Malaysia, has increased to 109,294GWh per year [5].

In Malaysian cases, artificial lighting is switch on even during the daytime and this happened in many buildings, especially government and university buildings. A report made by [6] has mentioned that the government has paid over MYR 2.7 billion for electricity bills for the government buildings, that mainly comes from government offices, particularly universities and hospitals.

For educational building, such as schools and universities, it is reported that the usage of artificial lighting during daytime has increased the electricity bill [7]. This is all due to the weakness of passive daylighting strategies [8] where most of the windows are tinted or blocked with furniture.

According to Malaysian GBI standards, the daylight factor for a classroom or educational studios need to comply within the range of 1.0- 3.5% in the working plane interior. In addition, the working plane interior also must be installed with a blind or screens that are fitted on all glazing to reduce discomfort of glare from natural light [9].

This paper aims to revitalized daylight strategies involving architectural studio in the campus. The key strategy is to create a green space area that reduces the artificial lighting usage in the space and maximize the use of natural lighting. This paper focuses on previous studies on the daylight strategies in academic space.

2. The issue

In architecture education, studio is the most important space where student develop their learning process such as drawing projection, exercises on grammar of design and visual composition. Besides, studio also a place for students sharing information, learning and interacts with their studio mates [10]. Students will spend more time in the studio.

However, there is an issue, as artificial lighting is used actively through day and night which has increased the utilities bill. In electricity data recorded in one of the university in Perak, Malaysia has shown that electricity bills increased by 8% with the increase numbers of architectural studio in that university [11]. This may concurrence with the frequent use of lighting and air-conditioning in those studios through day and night.

Therefore, there is a need to emphasized and empowered natural lighting for these studios; in order to allow energy to be utilized effectively. The proper natural lighting strategies can be made by increasing daylight usage that improve student performance and increase their comfort levels [12]. In addition, high frequency of studio usage will lead to higher consumptions of electricity energy [10]. Daylight is the most obvious contributions for energy saving in building is in exploiting the most abundant source of light available to us [13].

The use of natural daylight in buildings significantly reduces the electric energy consumption by minimizing the use of artificial lighting during the day [8]. Previous research has shown that daylight utilization (replacement of electric light by daylight) can reduce the use of electric lighting by 50%- 80% in office buildings [14]. Malaysian Minister of Energy, Communication and Multimedia, Tan Sri Datuk Leo Moggie said that one of the ways to save energy was through the efficient use of energy in building by the use of extensive day-lighting techniques that would reduce the amount electricity produced by lighting fixtures [15].

3. Research objectives and aims

The aim of this research is to propose several daylighting strategies that can be integrated into studio educational building designs, allowing the penetration of natural daylight while creating a comfortable studio environment in order to achieve a green- conducive studio.

4. Literature review

Daylighting is a passive strategy that able to help in improving energy performance and user's visual comfort (Lim *et al.*, 2012). Lighting can be divided into two types such as natural light and artificial light [16]. The good natural lighting strategies will reduce the energy consumption spent on lighting while providing adequate illuminance [8].

4.1. Daylight strategies for educational space

Daylight system can be set-up from simple static features (louvers, light- shelves, fixed overhangs, laser-cut panels, prismatic elements, anidolic systems, etc.) to adaptable dynamic elements (blinds, movable lamellae, advanced glazing, holographic optical elements, etc.) and/ or a combination of these elements. This features can, advance distribution and uniformity, increase daylight penetration indoor spaces, and control direct sunlight and reduce glare [8]. Recently, a study by [17] has suggested that it is wise to reduce the consumption of electricity through the utilization of illumination level. [18] proposed that the usage of daylight factor (DF) is significant in achieving more sustainable design of a building.

One of the most significant strategies for reducing the energy consumption effectively in a building is by the design and selection of a suitable window system[19]. For architecture studio facilities, there are certain window designs to be accomplished such as the window head should be as high as possible; at least 2 meter above the floor level to provide a clearer view and space upon designing a project. Next, the window sill must not be higher than 1 meter from level because it can prevent someone to see out when sitting. Furthermore, the window surface are should be evenly distributed over the outside wall and the window heights and widths must not be too small in relating to the window wall because this reduces the uniformity of lighting and produces undesirable shadows. Shadows in studios will affect students' eye-comfort and may attract them to use artificial lighting. Lastly, better natural lighting can be score with windows on opposite walls since the illumination produced by individual windows are superimposed. The region with the minimum daylighting factor is then displaced towards the centre of the room. The illuminations can increases when the useable depth of the room is increases. These window criteria can enhance an illumination level and reduce heat gain.

This criteria of window design has been determined in BS 8206-2, (2008) to create a good window design than can allows a suitable illuminance inside a space. If the studio is a single sided room; if the depth of room from outside wall is lower than 8 meter, the percentage of window wall as seen from inside should be a minimum of 20 percent. When it is equal or between 8 and 11 meter, the percentage of window wall should be minimum 25 percent. If it is > 14meter, the percentage of window wall should be 35 percent.

Besides of installing a window, the skylight also is one of the daylight strategies. If the studios located at the upper level, a skylight can be considered. The skylight can distribution the light uniformly to the space [20]. A solar shading device is an envelope component of daylight strategies. The application of skylight as the daylight strategies has been supported by [21]. The function of the skylight is to monitor lighting [21]. The solar shading device can control glaring problems and reduce solar gains, and thus avoid overheating. It can overcome the issues of glazing application in building facade such as overcooling, glazing and overheating [22]. Next, the uses of low-E double glazing as the glass facade can help to reduce heat gain [21].

4.2. Visual comfortness in educational space

To produce a good design for natural lighting devices in studios, the natural lighting strategies must be combined with the daylight strategies. [8] stated the natural lighting strategies is spreading out of natural daylight in such a way that it creates a good distribution of the light inside the building and facilitates its penetration in a given space. Next, the design must consider about protection against solar radiation or glaring light. It can produce a negative consequence on the spatial activities, such as reflected sunshine glaring on the desk, chalkboard and walls. Lastly, controlling of the natural daylight can manage the quantity and distribution of the light in order a given space according to the needs of its users.

The daylight strategies can be planned by determined the floor plan with an east- west axis, and south-facing windows and roof monitor and skylight for daylighting. [23] said the overhangs and interior baffles to shield direct sunlight while using daylight. There are various act (BSI, 1992; BS 8206, 2008) regulations (Hong Kong Government, 1997; Building Department, HKSAR, 2005) and design handbooks (Illuminating Engineering Society (IES) of North America, 1993) that present the precautions and conditions for good daylighting in building. USGBC, LEED-NC (2009), green building rating system from the US Green Building council, suggest a DF is 2 percent for the minimum daylight level a minimum illuminance level of 269 lux on the equinox at 9am and 3pm under the International Commission on Illumination (CIE) clear sky conditions. The environmental rating system BREEAM awards a credit where occupied spaces have an average DF of over 2 percent. The [24] stated that an illuminance level of 300lx for offices, classrooms and library type's space, occupied hours from 8am to 6pm. According to Chartered Institution of Building Services Engineers (CIBSE), in educational buildings context, the minimum illuminance level is 300 lx for classrooms and computer practice where the minimum illuminance level is 750lx for technical drawing classrooms and 500lx for conference and meeting rooms [25]. Figure 1 and 2 shows the types of daylight strategies. Figure 1 stated that the different of devices and features can be installed to the educational building window. Figure 2 showed daylight strategies by used a material application, opening (windows) devices, roof features and volume space.

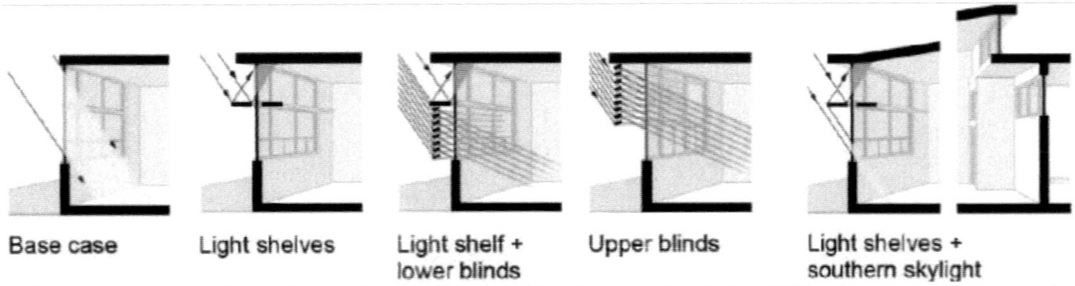

Figure 1. Daylight strategies for educational space
Source: Moreno & Labarca (2015)

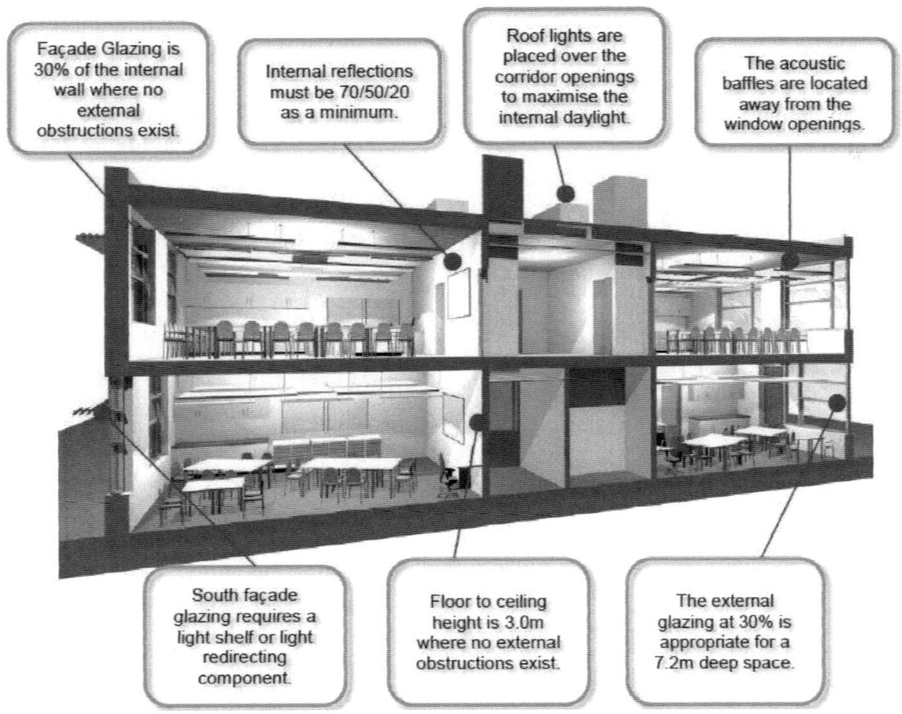

Figure 2. Daylight strategies for educational space
Source: Education Funding Agency (2014)

4.3. Measurement

In order to obtain suitable visual comfort in the studio, two variables need to measure:

- illumination (E)
- daylight factor (DF)

Illumination = illuminance

Illuminance is a total of illumination on a bright surface. Its defined as the luminous flux. Standard unit for illuminance is lux which is lumens per square meter (lm/m^2):

- $E = \frac{F}{A}$
- . E = illuminance of a surface (lm/m^2 or lx)
- F = luminous flux incident on the surface (lumen)
- A = the area of the surface (m^2).

Daylight can be measure in two calculation:

- Using luminous quantities (flux, illuminance), i.e. by a set of outdoor conditions and calculating the resulting interior illuminances.
- Using relative values (the DF) that compare indoor to outdoor regarding illuminance. For a given position, this factor is constant under widely varying outdoor lighting conditions.

The lighting design for illumination normally takes into consideration diffuse light only. The most common metric for measuring daylight- the DF- is valid under the CIE overcast sky model.

- The DF is = the relation between the indoor illuminance at a point on working plane (E_i) and the outdoor horizontal illuminance (E_o) for a CIE overcast sky conditions (Commission Internationale de l'Eclairage (CIE), 1970):
- $DF = (E_i / E_o) \times 100\%$

The DF can be effect by various parameters such as:

- Size and orientation of the glazing.
- The dimensions of the space.
- The reflectance of the surfaces.
- The external obstruction.

No electricity lighting is needed if the space has an average DF of 5 percent or more. The space needs more electric lighting if it is below 2 percent [26]. The DF of 5 percent or more will make an interior looks great daylight and DF below 2 percent will produce a dull interior (BS 8206-2, 2008).

5. Conclusions

This paper review previous studies on the effect of daylight strategies on the educational space in terms of types, criteria, and benefits of daylight strategies. Daylight strategies can reduces the building energy consumption (electrical energy), glare effect, reduce the heat gain (natural overcooling system) and increase daylight penetration indoor spaces. This study provides valuable information for academic institutions on development of green educational space by using natural light in studio. Moreover, this review can help architect and designers in designing daylight strategies in kindergarten, school and universities, which generally provides inadequate natural light.

6. References

[1] I. Sa, D. Subramaniam, H. Sulaiman, A. Latif, W. Omar, and M. Razman, *"Institutionalize waste minimization governance towards campus sustainability : A case study of Green Of fi ce initiatives in Universiti Teknologi Malaysia,"* vol. 135, 2016.
[2] E. Setyowati, A. Rochma, and Y. Nurul, *"Green Building Design Concepts of Healthcare Facilities on the Orthopedic Hospital in the Tropics,"* *Procedia - Soc. Behav. Sci.*, vol. 101, pp. 189–199, 2013.
[3] A. Ragheb, H. El-shimy, and G. Ragheb, *"GREEN ARCHITECTURE : A CONCEPT OF SUSTAINABILITY,"* vol. 216, no. October 2015, pp. 778–787, 2016.
[4] "Malaysia Standard (MS1525)," 2014.
[5] M. Maria and T. Mannhart, "*Analysis of the Power System of Malaysia*," no. 3635605, 2014.
[6] K. Jamaluddin, "*UIAM bantu usaha penjimatan tenaga elektrik*," *Utusan Malaysia*, 26-Oct-2012.
[7] F. E. Tang, "*An Energy Consumption Study for a Malaysian University*," vol. 6, no. 8, pp. 534–540, 2012.
[8] C. Marenne and C. Semidor, "*Daylighting Strategy for Sustainable Schools : Case Study of Prototype Classrooms in Libya*," vol. 3, no. 3, pp. 60–67, 2010.
[9] G. S. D. N. Bhd, J. Tangsi, and K. Lumpur, "*Gbi Assessment Criteria Contents*," no. March, pp. 0–57, 2011.
[10] F. H. Husain *et al.*, "*Environmental Psychology : An Analysis on Lighting Efficiency of the Architecture Studio in UiTM Perak*," no. October, pp. 7–8, 2015.
[11] Bahagian Pembangunan Fasiliti UiTM, "*Energy Consumption at UiTM Perak*," Seri Iskandar, 2016.
[12] S. Olson and S. Kellum, "*The impact of sustainable buildings on educational achievements in K-12 schools*," *Leonardo Acad. Clean. Greener Progr. ...*, 2003.
[13] Tugba Inan, "*An investigation on daylighting performance in educational institutions*," 2013.
[14] N. Gentile, "ScienceDirect *Lighting control systems in individual offices rooms at high latitude : Measurements of electricity savings and occupants ' satisfaction*," vol. 127, pp. 113–123, 2016.
[15] H. R. K. Razman, "*Reduce Energy Consumption, Malaysians To*ld," *Bernama News*, 13-Jan-2004.
[16] E. Jago and K. Tanner, "Influence *of the School Facility on Student Achievement - Visual (lighting and color)*," no. 1982, pp. 1–5, 1999.
[17] B. Ng and Z. A. Akasah, "*Post Occupancy Evaluation of Energy-Efficient Buildings in Tropical Climates – Malaysia*," vol. 7, no. 2, pp. 8–21, 2013.
[18] N. D. NIZARUDIN, M. R. M. HUSSAIN, and I. TUKIMAN, "*The Application of the Green Building Index (GBI) on Sustainable Site Planning and Management for Residential New Construction : Prospects and Future Benefits The Application of the Green Building Index (GBI*) on," in *Sustainable Tropical Environmental Design Conference 2010 (SUSTED 2010)*, 2010.
[19] S. Mirrahimi, N. L. N. Ibrahim, and M. Surat, "*Effect of daylighting on student health and performance*," *Comput. Methods Sci. Eng.*, pp. 127–132, 2012.
[20] J. McCreery and T. Hill, "*Illuminating the Classroom Environment*."
[21] L. Jones and W. Wong, "*More than just a green building*," 2016.
[22] T. İnan, "*An investigation on daylighting performance in educational institutions*," *Struct. Surv.*, vol. 31, no. 2, pp. 121–138, 2013.
[23] "Healthy Schools Network, Inc.," pp. 1–4, 2012.
[24] BCA:Green_Mark, "*Certification Standard for New Buildings*," *Build. Constrution Auth.*, pp. 1–211, 2012.
[25] G. K. Cook, "*Artificial lighting in classrooms-results of a lighting survey*," *Build. Environ.*, vol. 25, no. 4, pp. 329–331, 1990.

[26] E. Damaskou, "*conservation in California higher education Management strategies for sustainability education , planning , design , energy conservation in California higher education*," 2015.

Acknowledgments
Highest gratitude to my advisors, Sr Dr Nur Azfahani Ahmad and Dr Azizah Md Ajis for the guide on this paper.

IOP Conf. Series: Earth and Environmental Science **67** (2017) 012026 doi:10.1088/1755-1315/67/1/012026

Solar public engagement: the prospective study on FELDA community in Malaysia

Siti Jamiah Tun Jamil, Nur Azfahani Ahmad

Faculty of Architecture, Surveying and Planning, UiTM Perak Branch, Seri Iskandar, Perak, Malaysia

Abstract. Malaysia Energy Outlook 2016 has highlighted that Malaysia's electricity generation mix has always been highly dependent on fossil fuels. There is a concern on energy security for Malaysia recently, since the depletion of fossil fuel occurs and its effect increases the price of electricity tariff. Nevertheless, the energy demand continues to increase, which make the non-fossil renewable energy sources is back on demand. Malaysia's highest potential for renewable energy comes from solar energy and the large roofs of rural houses offer potential to contribute solar electricity for the people. Indeed, the engagement of solar energy to the public is very important in allowing this energy to be accepted by the locals. The paper will review the related literature on public engagement for solar energy project. This paper also tries to prospect the potential of implementing solar electricity for a well-known rural organization in Malaysia, known as FELDA.

1. Introduction

The demand of electricity in Malaysia increased at 4% annually [1]. This rapid intensification electricity demand is due to the high economic development rate of Malaysia today [1]. For renewable energy sector, the capacity is expected to reach 2,080MW by 2020, contributing to 7.8% of total installed capacity in Peninsular Malaysia and Sabah. In the Malaysia Eleventh Plan, focus will be on promoting new renewable energy sources. On 2014, total installed capacity for renewable energy capacity is expected to reach 2,080MW by 2020, contributing to 7.8% of total installed capacity in Peninsular Malaysia and Sabah. In the Malaysia Eleventh Plan, focus will be on promoting new renewable energy sources. The total installed capacity for solar photovoltaic is 66% out of 243MW and by the year 2020, the total installed capacity for solar photovoltaic is expected to increase by 9% [1]. The demand of solar energy is growing every year and it signify that the engagement, acceptance and the awareness level towards the development of solar energy is needed in Malaysia, particularly among the people. However, a study has shown that this engagement and awareness are greatly comes from the commercial sector and urban populations only [2].

In Malaysia, rural area has a large scale of people, and it is divided into traditional settlements and government-partnership settlements, such as FELDA community. Even though solar technology is vastly developed in Malaysia, it is still predominantly unfamiliar within the rural communities. A survey in Malaysia [3] with 214 respondents with different background areas in Malaysia. The study shows that the professionals knew about the technology, however, there are gap of knowledge within the domestic populations; especially in terms of its implementation and dissemination at the national level. Most of the rural households are still unfamiliar with the technology. Therefore, this paper tries to explore the potential of introducing solar technology into a large rural community like FELDA.

Later part of this paper will introduce on the FELDA community and the potential of engaging this technology within the community.

2. The solar public engagement: why it is important?

In realizing a solar energy project for any communities, an engagement is needed. Public engagement and community engagement is a synonymous term [4]. The public engagement involves the variation of public effort to make the institution more consistent to achieve the aspirations, values and needs of the citizens who fund it and work for it [4]. In order for public engagement to be expressive, the public should be involved, interact and have the opportunity to influence the outcome before it is determined and have an opportunity to affect future policy or decisions [5].

Public engagement within solar energy context involves public dialogue, interaction and participation, a two-way communication and collaboration partnership to develop a shared understanding, to design, and to implement the related solar program that meets the users' needs [6]. The flow of information the public engagement needed involved three (3) levels; (i) one way; begin with the information from funder to public (communication), (ii) two-way; public to funder with informal dialogue and information flow back without any discussion (consultation) and (iii) two way exchange; information flow in two direction between funder and public (participant) [7],[8] (see Fig. 1).

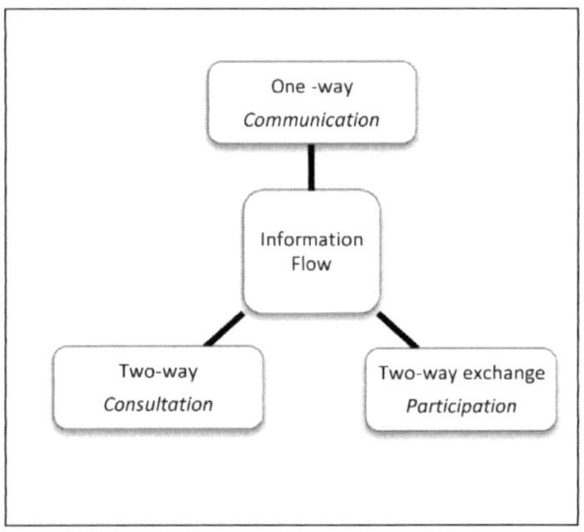

Figure 1. Public engagement Information flow [7] [8]

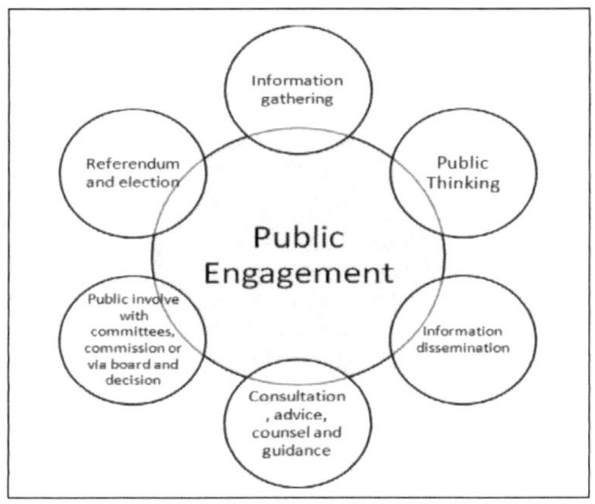

Figure 2. Public engagement approaches [4]

Solar public engagement can develop a public trust and more importantly, it establish the needs of performance feedback from customer perspectives [5]. It also authorizes an action produce from public conversations in order to make a decision or the policy in align with the future perspective. Public engagement assist the people with strategic planning, show the options on solar energy management that a community can handle and at the same time, will resolve the conflict on specific issue occurs in the future which can be used in developing a mutual strategic. However there are several of public engagement approaches (see Fig. 2) [4]. One or more approaches involve in public engagement. Merging the approaches of public engagement can improve the public engagement.

The effective public engagements depend on the ability of the right move engages with the public. The following point are the principle of public engagement [9], which start with listening, engaged to people's leading concerns, reach beyond the "usual condition", deliberation on time-frame issues, provide the information at the right time, help people in every way, expect problems and have resistancy, create various opportunities for deliberation and discussion, react thoughtfully and believe to the public's involvement and build a long-term ability so that it can gain public trust. One report highlighted that principles of public engagement are about involvement of people right, to make a community stronger, to build up their trust and establish a partnership, through accountability, sustainable services and reliable resources [5].

3. Public engagement of other countries

Many researchers engaged the solar energy project to the public on various aspects and criteria. A study done in countries like India, Bangladesh, Nepal and Sri Lanka had involved the public with selected programs linked to dissemination a solar energy program for the people [10]; it involves (i) *innovative financing schemes promotion, (ii) dissemination of solar information, (iii) sizing technical design, finance and service delivery models, (iv) cost of system, monitoring and maintenance, and (v) shares the experience and lessons of solar PV programs through public talks or seminars.* These types of engagement efforts were conducted in India, Bangladesh, Nepal and Sri Lanka which involve the technical, financial, institutional, governance and residential sectors [11].

One of the scholar highlighted that to engaged in a solar energy program, it needs to be managed under the community organizations in order to find significant strategies that suits the community lifestyles. Examples of the strategies are involving the discussions with reliable people in the community, e.g. local clinic and school staff with villagers in order to expose the community with the technology [11]. Public can be persuaded if they trust the people that disseminate the information. In this context, people always put their trust into local medical and education staff [10]. Public buildings in the community is selected to be the hub for solar PVs point so that it can be shared to the locals, namely the community schools, the clinic and the school laboratory. The social-economic data is collected from the local communities and being monitored under a special program by the Government and NGOs [12].

4. The case study – FELDA in Malaysia

Malaysia have two regions: West and East Malaysia and situated in the South East of Asia. It encompasses of 13 states. Total area of Malaysia is 329, 847 km2 [13],[14]. Malaysia which is located in tropical region receives solar radiation every day, with the average irradiance per year of 1643 kWh/m^2 [2]. Typically, roof pitches in Malaysia are angled at 30° or less, with wide area of roofs that allow PV panels to be arrayed effectively.By using the large roof area, solar irradiance can generate more solar electricity [15].This is a good potential for FELDA houses since a house in FELDA community has an average area of 100 m^2 (see Figure 3 and Figure 4) [16].

Figure 3. FELDA community; The first generation house [16].

Figure 4. FELDA community; The second generation house [16]

FELDA; known as The Federal Land Development Authority is the settlement areas that has been established by the Malaysian Government with the objective to establish the resettlement of rural poor into newly developed areas and to provide smallholder farms for growing crops, which basically is rubber trees and palm oil trees [16],[17]. Each FELDA settlement is supplied with electricity and water and community buildings (A mosque, a school, a public clinic and a public hall).
In Malaysia, 11 provinces of FELDA situated at Negeri Sembilan, Johor, Kelantan, Pahang, Perak, Terengganu, Kedah and Sabah (see Figure 5). The provinces has its own settlements. 317 FELDA settlements with almost 10,000 houses throughout Malaysia [16] (see Table 1), which indicates the community-based approach is considered significant to disseminate the solar technology and can be chosen as the key mechanism for public engagement for rural households in Malaysia.

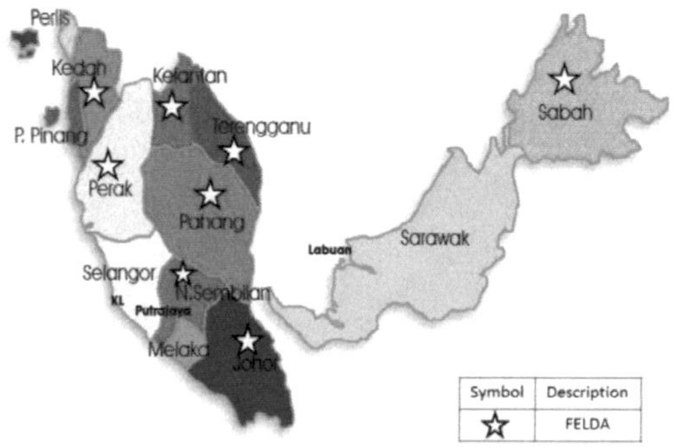

Figure 5. FELDA Malaysia

Table 1: Provinces and settlements of FELDA [16]

No	Province	No of Settlements
1	Johor Bahru	42
2	Segamat	36
3	AlorSetar	13
4	Jengka	37
5	Kuantan	42
6	Mempaga	36
7	Trolak	21
8	Raja Alias	49
9	Terengganu	21
10	GuaMusang	11
11	Sahabat	9
	Total	317

For this paper, certain generic criteria of houses have been considered based on typical common features of the roof shapes. Shadings are not an issue since typically the house yard is wide and trees are located far from the house. The electricity generated is based on the assumption that the roof is using a 2kWp PV panels. For economic purposes, only half of the roofs are covered with PV panels. Based on a simple calculation, the PV area of 50m^2 can provides about twice as much electricity as the household demand over 24 hours [17]. This is because, rural houses only used limited demand of electricity, e.g. for lighting, watching television and cooking purposes.

Considering that for each 2 kWP PV panels will produce a 10 kW solar production/daily (based on calculation from a typical solar energy calculator), the total solar production for a month is predicted at 300kW. Since the electricity consumption for FELDA houses will only be between 100-150 kW per month [17], there are 50% surpluses of solar energy production which can be used for Malaysia. This indicates that FELDA community has a large potential in implementing solar energy project.

The Calculation:
Average households electricity consumption/monthly
– 150 kW
Monthly solar production (from 2 kWp panel)
– 300 kW
Surplus of energy = 300 – 150 = 150 kW (50% surplus)
No. of FELDA Houses = 10,000 houses x 150 kW
= 1500000 kW
Therefore, 1500000 kW of electricity can be produced if all houses in FELDA are been installed with PV panels. This shows a huge potential on deploying a solar energy project for the community. This is a main reason why it is significant to have public engagement for the FELDA community in order to make the people aware and able to give full commitment.

5. Conclusion

Proactive attitude, effective promotion, good dissemination of information, informal or formal discussion and dialog from all the public and stakeholder needed to be produced in order to the benefit the solar energy industry and engaged the people with the technology. Without strategic steps, the engagement from the people is impossible and it is hard to disseminate this technology to the people.

6. References

[1] Energy Commision. (2016). Peninsular Malaysia Electricity Supply Industry Outlook 2014. *Peninsular Malaysia Electricity Supply Industry Outlook*, 73. Retrieved from http://www.st.gov.my J. Clerk Maxwell, A Treatise on Electricity and Magnetism, 3rd ed., vol. **2**. Oxford: Clarendon, 1892, pp.68–73.
[2] Chua, S. C., & Oh, T. H. (2012). Solar energy outlook in Malaysia. *Renewable and Sustainable Energy Reviews*, *16*(1), 564–574. https://doi.org/10.1016/j.rser.2011.08.022
[3] Muhammad-Sukki, F., Ramirez-Iniguez, R., Abu-Bakar, S. H., McMeekin, S. G., & Stewart, B. G. (2011). An evaluation of the installation of solar photovoltaic in residential houses in Malaysia: Past, present, and future. *Energy Policy*, *39*(12), 7975–7987. http://doi.org/10.1016/j.enpol.2011.09.052
[4] Marlowe, H. A. (2005). Theory and practice Public engagement : Working Paper, 1–40
[5] Newfoundland Labrador Office of Public Engagement. (n.d.). Public Engagement Guide. Retrieved from http://ope.gov.nl.ca/publications/pdf/OPE_PEGuide.pdf
[6] K. Elissa, "Title of paper if known," unpublished.
[7] Rowe, G., Frewer, L.J., 2005: A typology of public participation mechanisms. In: Science, Technology and Human Values, Vol. 30, 251-290
[8] Devine-Wright, P. (2011), Public engagement with large-scale renewable energy technologies: breaking the cycle of NIMBYism. WIREs Clim Change, 2: 19–26. doi:10.1002/wcc.89
[9] Agenda, P. (2008). Public engagement: A primer from Public Agenda, 1–12.
[10] Palit, D. (2013). Solar energy programs for rural electrification: Experiences and lessons from South Asia. *Energy for Sustainable Development*, *17*(3), 270–279. https://doi.org/10.1016/j.esd.2013.01.002
[11] Noll, D., Dawes, C., & Rai, V. (2014). Solar Community Organizations and active peer effects in the adoption of residential PV. *Energy Policy*, *67*, 330–343. https://doi.org/10.1016/j.enpol.2013.12.050

[12] Frame, D., Member, S., Tembo, K., Dolan, M. J., Strachan, S. M., & Ault, W. (2011). A Community Based Approach for Sustainable Off-Grid PV Systems in Developing Countries. *Environment*, 1–7.

[13] F. Muhammad-Sukki, R. Ramirez-Iniguez, S. H. Abu-Bakar, S. G. McMeekin, and B. G. Stewart, "An evaluation of the installation of solar photovoltaic in residential houses in Malaysia: Past, present, and future," *Energy Policy*, vol. 39, no. 12, pp. 7975–7987, 2011

[14] Central Intelligence Agency, 2016. The World FactBook [Homepage of Central Intelligence Agency], [Online]. Available from: /https://www.cia.gov/library/ publications/the-world-factbook/geos/my.htmlS [2016, 12/12/2016].

[15] N. A. Ahmad *et al.*, "Solar Village in Malaysia – A Route Map for Financing Mechanism," vol. 57, pp. 2–10, 2016.

[16] http://felda.net.my/

[17] Ahmad, N.A (2014), Solar 'Kampong': Community Power Supply For Low-Income Households In Rural Malaysia (Doctoral Dissertation), Retrieved from researchspace.auckland.ac.nz (Accession No 2292/25564)

IOP Conf. Series: Earth and Environmental Science **67** (2017) 012027 doi:10.1088/1755-1315/67/1/012027

The energy audit process for universities accommodation in Malaysia: a preliminary study

Muhammad Hilmi Dzulkefli

Faculty of Architecture Planning and Surveying, Universiti Teknologi MARA (Perak), Seri Iskandar, 32610, Perak, Malaysia.

Abstract.The increase of energy consumption in the Malaysian Universities has raised national concerns due to the fact that its consumption increase government fiscal budget and at the same time contributes negative impacts towards the environment. The purpose of this research is to focus on the process of energy audit conducted in the Malaysian universities and to identify the significant practice that can improve energy consumption of the selected universities. The significant criteria in energy audit may be found by comparing the energy implementation process of selected Malaysian universities through the investigation of energy consumption behavior and the number of electrical appliances, equipment, machinery and buildings activities that have an impact on energy consumption that can improve energy-efficiency in building. The Energy Efficiency Index (EEI) will be used as an indicator and combined with the suggested application of HOMER software to obtain solution and possible improvement of energy consumption during energy audit implementation. A document analysis approach will also be obtained in order to identify the best practice through the selected energy documentations. The result of this research may be used as a guideline for other universities that consume high energy in order to help improving the implementation of energy audit process in their universities.

Keywords – energy audit & universities accommodation

1. Introduction

The increase of private and government universities has a great impact on educational development for the bright national future but the problem is it also increases the energy demand. The major concern is most of the universities use high energy consumption and will impose a huge cost burden on universities. Recognizing that the development should be conducted in parallel with building sustainability, various methods and policies should be introduce in order to help the building performance in achieving energy efficiency. To motivate sustainable building practice and guidance to manage energy efficiently, a few codes related to energy policy have been introduced, for instance The Malaysian Standard MS1525:2014, Code of Practice on Energy Efficiency (EE) and Use of Renewable Energy for Non-Residential Building (Department of Standard Malaysia, 2014), Energy Efficiency and Conservation Guidelines [1], ISO 50001: Energy Management System [2] and Uniform Building By Law (UBBL). In 2008, the Efficient Management of Electrical Energy Regulation 2008 [3] also has been introduced. It stated that any installation which receives electrical energy from a supply authority with a total electricity consumption equal to or exceeding 3,000,000 kWh as measured at one metering point over any period of six consecutive months must comply with requirements in the regulations. For the development of electrical energy efficient, the organization must conduct an energy audit as a first step.

Since Malaysian utility company, Tenaga Nasional Berhad (TNB) has raised the electricity tariffs in 2015 [4], the biggest concerns for many universities is it will affect the universities energy expenditure. Therefore, energy audit is essential in order to increase saving and improve energy efficiency. A building sector should invest in new energy efficient technologies to optimize the energy use [5]. Malaysia's energy consumption increases every year. In 2008, the total energy demand was 522,199 GWh, where the industrial and transport sectors were the largest consumer around 3/4 of total demand [6]. The residential and commercial sector was the third largest consumer (14%) and only 1% consumed by the agriculture sector [7]. The energy consumption in Malaysia rises rapidly in Malaysia every year, with an average of 2,533 GWh per year [6] which involves universities as major energy users. University buildings consume high energy in the classification of commercial buildings due to its activities and populations [8].

Despite the recommendations by The Malaysian Ministry of Education (MOE) and The Malaysian Ministry of Higher Education (MOHE) to all educational institutions to undertake the responsibility to save energy; which was in line with the efforts to implement energy saving programs in government buildings [9], there are still huge energy wastage in this sector. Unnecessary energy wastage may occur due to lack of awareness on the importance of energy saving building user. This is where energy audit is significant. Energy audit can be defined as the analysis of energy flows in buildings, systems, processes or equipment to reduce the amount of energy input into the system without negatively affecting the energy output [10]. Energy audit is conducted to comprehend the buildings energy performance and facilities so the potential for improvements will be found [11]. The Malaysia Ministry of Education has insist all education centers to lessen energy consumption due to the expensive monthly commitment of electricity bill that become the major concerns of all parties [12]. The government has encouraged consumers to practices on energy saving activities such as switching off all electrical equipment when not in use [13].

Energy wastage in Malaysian university mainly cause by inefficient energy use and lack of awareness among the users, therefore it is importance to raise awareness among consumers [14]. Without awareness, there will be a difficulties to sustain an energy efficiency. Energy efficiency strategy may reduce running energy costs that is related with the life of the buildings [15]. Improving energy efficiency will produce better building with comfortable working environment, more satisfaction and improved productivity [15]. This is in line with the attempt by the Ministry of Energy, Green Technology and Water (2009) to begin energy saving programs in all government departments. There are multiple reasons why implementing energy audit and improving energy consumption in local university is important. Universities have a large number of building users and large number of facilities compared to other industries. The typical energy bill in education sector is approximately 5% of the total expenditure of the university [16]. There is a possibility of saving 5-10% of the energy expenditure by implementing improve energy-use behaviour [17].

Lack of awareness and failure to fully implement the highly energy efficient technologies during design stage has result the high energy consumption. In overall, the University had to bear a high fiscal cost in order to run the university. Without effective energy audit, the total energy consumption will increase and also effect the cost. At the moment, the energy audit implementation in Malaysian universities is still lacking. Most of the universities, fail to implement proper energy audit. For the success of energy audit implementation, energy efficient awareness is necessary and commitment of top management and involvement of entire stuff in organization is a must [18]. Thus, energy audit in the building sector should be recognize as a comprehensive approach in order to gain more energy efficient.

From statistics released by the Ministry of Higher Education (2016), there are 21 public universities, 71 private universities and six foreign universities branch campuses in Malaysia. According to the Planning and Research Unit (2011), there are 937,231 students enrolled in Malaysian higher education institutions. This indicates that the number of accommodation facilities for students also plays an important role in universities [19]. It can be considered that student are the largest consumers of energy in universities [20]. Therefore the implementation of energy audit in residential colleges is expected to be very significant in reducing overall energy consumption. Eventhough energy use in residential college is usually 10 to 20 times lower compared to office buildings, the average energy consumption of non-air-conditioned residential colleges in tropical regions is in the range between 20 to 60 kWh/m2/year [21]. This data shows that the energy usage in student accommodation is significant towards the increase of electricity use in university. Due to this, this research will focus on how energy audit can be implemented towards achieving more energy efficient usage.

2. Literature review

The development of higher education institutions is in line with the development of Malaysia to meet the needs of local and abroad students. Sustainable development is a development that meets the requirement of current situation without affect the future generations need [22]. The sustainability concept is widely known in different sectors such as government, private and also in education sector [23]. In order to achieve this sustainability, energy audit is important for all buildings. The objective of an energy audit is to regulate where, when, why and how energy is used in the buildings and to identify opportunities to improve. Energy audits is a systematic inspection of existing energy systems to reduce overall energy inputs to the systems without negatively impact on the output of the systems [10]. An energy audit is developing a compassionate of the specific energy-using patterns of a particular facility [24]. Energy audits are implemented to identify the areas with potential energy saving and understand the energy performance of the buildings [25]. Energy audits are basically classified as preliminary walk-through and comprehensive audits, each differentiated by the detail included and the depth of analysis done [18].

1. Energy audit process

The appraisal of energy consumption patterns and recognition of energy saving measures are the most vital part of energy management activities, which can be reached via energy audits [15]. Generally, there are four levels of analysis of energy audits suggested by ASHRAE (2004).

(1.1) Level 0 – Benchmarking: analysis of the existing annual utility bill and consumption, then compare the performance of the building with similar building for benchmarking.
(1.2) Level I – Walkthrough audit: preliminary audit to evaluate building energy efficiency and energy conservation. This research is based on visual verification, installed equipment and detailed recorded energy consumption collected during benchmarking process.
(1.3) Level II – Energy survey and analysis: More detailed analysis on building performance by conducting field measurements using tools and devices.
(1.4) Level III – Detailed energy audit analysis: By referring results of pre-audit, this audit is more comprehensive on analysis of energy consumption. Using software to identify the possible improvement of energy consumption.

National Energy Policy was formulated to achieve cost-effective supply of energy and efficient utilization of energy [26]. Malaysian universities should be aware on this policy and implement energy management system however there is a missing link in the energy management process since they still consume high energy annually. Thus, the question arise on how well Malaysian universities apply an effective energy audit? The argument is the management may be focusing more on energy planning and implementation phase but lack on energy checking phase. Therefore, this research aims to explore and improve this whole energy process for the universities.

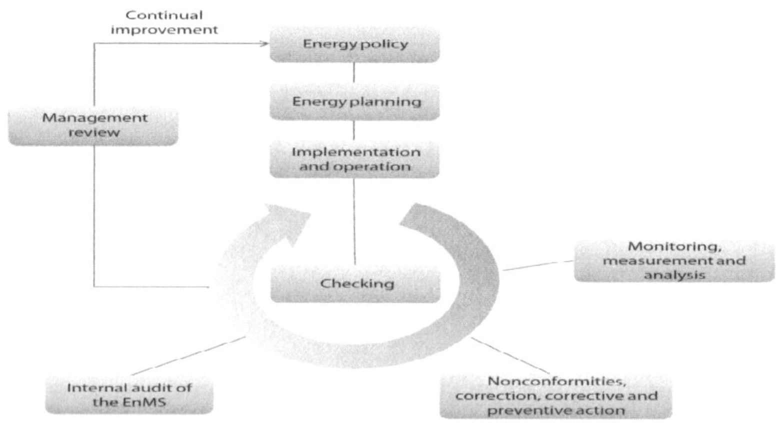

Figure 1. ISO 50001:2011(E) – energy management system

3. Research objectives

Below are the research objectives:-

1. To identify the energy consumption level in universities accommodation is being used based on the actual load profile.
2. To identify the current practice of energy audit in Malaysian universities' accommodation.
3. To determine the best practice of energy audit in achieving energy efficiency for Malaysian universities' accommodation

4. Research questions

Below are the Research Questions:-

1. What is the level of energy consumption used in Malaysian universities accommodation?
2. How is the current practice of energy audit in Malaysian universities' accommodation?
3. A Malaysian universities use best practice in energy audit in order to improve energy efficiency?

5. Problem statement

Public universities in Malaysia are being challenge with wastage of energy due to the lack of awareness between students and staff, and no particular guidelines for facilities to improve the situation. Lack of energy audits and minimal in-depth monitoring in universities also lead to higher energy consumption. Without proper energy audit, universities are unaware of the energy wastage situation [27]. There are correlation between poor energy audit and the increase of energy wastage among the people. Therefore, it is significant for this study to distinguish energy audit process that can leads towards energy efficiency in the universities. It is wise to highlight that *"with ineffective energy audit for universities accommodation, Malaysian universities will faced inefficient used of energy which hindered the energy efficiency process"*.

6. Gap

There are numerous research of energy audit on multi-residential buildings, especially in tropical regions [28]. However, there are limited numbers of research focus on energy audit for student accommodation in universities. The reason might be due to difficulty to access the level of energy use by the student. It is because that are many students in universities that use energy on a different period of time [21].

7. Research methodology

For research methodology, mixed method approaches will been used based on guideline proposed by Krarti [29] in his book "Energy Audit of Building Systems" Based on Krarti, energy audit can be divided by two types: a walkthrough audit that include a simple study by viewing on the major systems and equipment, and a detailed audit, which require a comprehensive and time consuming. Therefore, this research methodology will be divided into several stages. This method consists of 4 main phases which are:

1. **Phase I**: Data Collection
2. **Phase II:** Data Analysis
3. **Phase III:** Result and Finding
4. **Phase IV**: Reliability of Validity Data

The approach used in this research began by collecting secondary data from various sources. Secondary data includes information from books, previous research (articles and journals), websites, energy consumption data and information, the agencies involved in building conservation and other information sources.

1.1 Sequential mixed-method approaches techniques

The approach used in this research began by collecting secondary data from various sources. Secondary data includes information from books, previous research (articles and journals), websites, energy consumption data and information, the agencies involved in building conservation and other information sources.

Phase 1 : Data collection

A walkthrough survey will be carried out to survey equipment use (the type, power & consumption). Energy consumption collection of buildings will be done with data logger and ammeter. An observation on yearly electrical bills will be carried out and the use of EEI formula in order to calculate the building energy index. The last part is a research survey (Questionnaire) and the respondents will be the person in charge and the building users in order to understand method or practices on energy audit and the pattern of electricity consumption. Lastly, to analyze the energy standard use by the selected university and to study the significant energy audit in the report.

Phase 2 : Data analysis

This phase is to evaluate the characteristics of the pattern of energy systems in the buildings by using HOMER software to produce load profile and energy usage pattern. HOMER (Hybrid Optimization Model for Electric Renewables), developed by NREL (National Renewable Energy Laboratory, USA) as preferred tools due to several application on literature review. This software will assist on energy management during preliminary audit. The Energy Efficiency Index (EEI) being used in this research to demonstrate the performance of electric consumption at residential colleges in the unit of kWh/m^2/year [30]. Energy Efficiency Index (EEI) is an evaluating method from Saidur [31], who estimated energy intensity, EEI, in kWh/m^2 by using following equation:

$$\text{EEI} = \frac{\text{Energy Consumption/Input (kWh/year)}}{\text{Gross Floor Area } (m^2)}$$

Basically, the energy use per unit floor area can be described as " Normalised Performance Indicators" (NPI), also known as Energy Efficiency Index (EEI) or Building Energy Index (BEI) [32]. It basically calculates the ratio of total energy used against the total built-up area to calculate building yearly consumption. The energy consumption in buildings normally in term of Energy Efficiency Index [33]. According to MS 1525 standard, recommended energy efficiency index in Malaysia is 135kWh/m^2/year. The saving targets are based on the lowest EEI. A documentation analysis will also be obtained in checking related energy documents in the organizations. SPSS software also will be used for statistical analysis of questionnaire done. Identify best practice of energy audit implementation at universities by comparing energy audit reports.

Phase 3 : Result and finding

After completed both phase 1 and phase 2, there will be an outcome of a most significant procedure and practice in energy audit.

Phase 4 : Reliability of validity data

The reliability of validity data will be depend on energy audit manager expertise input, data and report of energy audit for the selected universities.

8. Selected case study

There are 21 public universities in Malaysia [34], whilst University Technology Mara (UiTM) comprises highest students' population. UiTM consists total of 168,000 students' enrollment with ¼ of the national population. University Technology Mara is Malaysia's largest institution of higher education in terms of size and population. Apart from the main campus in Shah Alam, the university has expanded nationwide with 13 state campuses and 24 state satellite campuses. Efficient Management of Electrical Energy Regulation 2008 (EMEER) [35] has been introduced in 2008 stated any installation which a total electricity consumption equal to or exceeding 3,000,000 kWh as measured at one metering point six consecutive months must comply with requirements in the regulations.

It requires a building owner to appoint a Registered Electrical Energy Manager (REEM) to audit and analyse the total electrical energy consumption in the buildings. As reference in Suruhanjaya Tenaga there are 16 universities in Malaysia that fall under this regulation (EMEER) [35]. For this research, three (3) UiTM's branches have been chosen as the case study, which are UiTM Tapah, UiTM Pasir Gudang and UiTM Sg. Buloh. The aims of this research are to evaluate the efficiency of electricity consumption through an energy audit for residential accommodations. All of these residential colleges are not non-air-conditioned and provided with a ceiling fan and a fluorescent lamp.

9. Scope and limitation

There are many research has been done on commercial buildings but limited research on students accommodation. The scope of work for the implementation of energy audit are as per below:

1. To concentrate and study on energy pattern of the buildings
2. To identify and understand any energy saving opportunities that can be implemented
3. The thesis focuses on residential college only as a target building for energy improvement

The limitation in this research are as per below:

- The results and findings will be focus on the significant and best practices of energy audit in Malaysian universities and limited by the perceptions of recognized experts in energy audit.

10. Significant of the study

As recommended by the government, public and private universities should improve their energy efficiency in order to reduce the cost of university's operational. This research can be used as an important guidance and provide useful information for residential colleges in Malaysia particularly on current energy consumption and potential of energy saving improvement through the best practice of energy audit that has been identified through the research.

11. References

[1] Suruhanjaya Tenaga Malaysia, "Efficient Management of Electrical Energy Regulations," p. 19, 2008.
[2] International Organization for Standardization, "Energy Management System," p. 52, 2011.
[3] S. Tenaga, "Efficient Management of Electrical Energy Regulation 2008," *Effic. Manag. Electr. Energy Regul. 2008*, pp. 1–30, 2008.
[4] T. Star, "The Star Online," The Star, Petaling Jaya, 2015.
[5] V. Galis and P. Gyberg, "Energy behaviour as a collectif," *Energy Effic.*, vol. 4, no. 2, pp. 303–319, May 2011.
[6] C. Teh, "Extraordinary mind discuss ideas," 1 July 2016. [Online]. Available: http://www.christopherteh.com/blog/2010/09/electricity-demand/.
[7] KeTTHA, 26 June 2016. [Online]. Available: http://www.kettha.gov.my/portal/index.php.
[8] M. Z. Abd-Razak, N. K. F. Mustafa, A. I. Che-Ani, N. A. G. Abdullah, and M. F. I. Mohd-Nor, "Campus sustainability: Student's perception on campus physical development planning in Malaysia," *Procedia Eng.*, vol. 20, pp. 230–237, 2011.
[9] S. Ahmad, Y. H. Mohammad, A. Hayati, H. Rahman and M. Majida, "Energy Efficiency Measurements in a Malaysia," in *IEEE International Conference on Power and Energy*, Kota Kinabalu Sabah, Malaysia, 2012.
[10] M. B. Raghav, M. S. Srijaa, G. S. Rao, K. N. Bhavya, and Y.Suchitra, "ENERGY CONSERVATION ANDAUDIT-A CASE STUDY," *Int. J. Adv. Res. Electr. Electron. Instrum. Energy*, vol. 2013, 1970
[11] J. E. Piper, Operations and maintenance manual for energy, New York: Sharpe Professional, 1999
[12] V. Galis and P. Gyberg, "Energy behaviour as a collectif," *Energy Effic.*, vol. 4, no. 2, pp. 303–319, May 2011
[13] KeTTHA, 26 June 2016. [Online]. Available: http://www.kettha.gov.my/portal/index.php
[14] Wong, "Energy Conservation and Human Behaviors: The Professional Faculties Building," in *University of Calgary*, Calgary, 1997
[15] M. Haji-Sapar and S. E. Lee, "Establishment of energy management tools for facilities managers in the tropical region," *Facilities*, vol. 23, pp. 416–425, 2005.
[16] G. Keeffe and B. Grimshaw, "Energy Management," *Managing Educational Property: A Handbook for Schools, Colleges and Universities,* pp. 196-209, 1994
[17] A. Loozen and C. D. Moosdijk, "A Consumer Advise on Energy Efficient Use and," *Energy Efficiency in Household Appliances and Lighting,* pp. 468-474, 2001
[18] S. R. Bhawarkar and S. Y. Kamdi, "Electrical energy audit of a electroplating unit - A case study," *2011 Int. Conf. Recent Adv. Electr. Electron. Control Eng. IConRAEeCE'11 - Proc.*, pp. 25–29, 2011
[19] C. Daniel, N. Barron, G. President, D. B. President, and P. S. Association, "the Educational Attributes of Some of the World ' S ' Top 50 ' Universities – a Discussion Paper," no. May, pp. 1–140, 2008
[20] J. P. Meriac, D. J. Woehr, and C. Banister, "Generational Differences in Work Ethic: An Examination of Measurement Equivalence Across Three Cohorts," *J. Bus. Psychol.*, vol. 25, no. 2, pp. 315–324, Jun. 2010
[21] A. A. Jamaludin, N. Z. Mahmood, N. Keumala, A. R. M. Ariffin, and H. Hussein, "Energy audit and prospective energy conservation: Studies at residential college buildings in a tropical region," *Facilities*, vol. 31, pp. 158–173, 2013

[22] G. H. Brundtland, "Our Common Future: Report of the World Commission on Environment and Development," *Med. Confl. Surviv.*, vol. 4, no. 1, p. 300, 1987
] T.Prugh, "The Local Politics of Global Sustainability," p. 196, 2000

[24] S. A. Roosa, *Sustainable development handbook*. Fairmont Press, 2008.
[25] Jayamaha, Energy Efficiency Building Systems, New York: McGraw-Hill, 2006.
[26] Economic Planning Unit, 26 June 2016. [Online]. Available: http://www.epu.gov.my/en/tenaga.
[27] W. W. Choong, A. H. bin Mohammed, and B. Alias, "Energy Conservation Model for Malaysian Higher Learning Institution," 2010.
[28] Wong, "Energy Conservation and Human Behaviors: The Professional Faculties Building," in *University of Calgary*, Calgary, 1997.
[29] Krarti, M. (2011). *ENERGY AUDIT OF BUILDING SYSTEMS. Taylor & Francis Group* (Vol. 2). http://doi.org/10.1017/CBO9781107415324.00.4
[30] B. S. K. Chou, "Performance-based Standards for Energy Efficient Buildings Introduction-Situation and Observations," no. February, 2004
[31] R. Saidur, H. H. Masjuki, and M. Y. Jamaluddin, "An application of energy and exergy analysis in residential sector of Malaysia," *Energy Policy*, vol. 35, no. 2, pp. 1050–1063, Feb. 2007
[32] Kamaruzzaman and Edwards, "Evaluating performance characteristics of electricity use of British historic building in Malaysia," in *Facilities Vol. 24 Nos 3/4*, 2006, pp. pp. 141-52
[33] M. B. A. Aziz, Z. M. Zain, S. R. M. S. Baki, and R. A. Hadi, "Air-conditioning energy consumption of an education building and it's building energy index: A case study in engineering complex, UiTM Shah Alam, Selangor," *Proc. - 2012 IEEE Control Syst. Grad. Res. Colloquium, ICSGRC 2012*, no. Icsgrc, pp. 175–180, 2012.
[34] Ministry of Higher Education, "Ministry of Higher Education," 16 June 2016. [Online]. Available: https://www.mohe.gov.my/en/muat-turun/awam/statistik/2014
[35] Suruhanjaya Tenaga & Pusat Tenaga Malaysia 2007 for Energy Efficiency and Conservation Guidelines

Acknowledgements
I take this opportunity to express my profound thanks and deep sense of gratitude towards my advisor, Dr. Nur Azfahani Ahmad and Dr Haryati Isa who gave their valuable guidance and the Faculty of Architecture, Planning and Surveying whose constant encouragement and expert advice in the completion of this research.

IOP Conf. Series: Earth and Environmental Science **67** (2017) 012028 doi:10.1088/1755-1315/67/1/012028

The study of recirculating aquaculture system in pond and its purification effect

Qu Jiangqi*, Zhang Qingjing, Jia Chengxia, Liu Pan and Yang Mu

Department of Fisheries Ecology and Environmental, Beijing Key Laboratory of fishery Biotechnology, Beijing Fisheries Research Institute, Beijing, 100068, China

*E-mail: quqi20122012@163.com

Abstract. In this paper, a recirculating aquaculture purification system (RAPS) was designed to solve the problems of aquaculture pollution and shortage of freshwater resource according to the characteristic of northern freshwater ponds of China. The system were arranged in series and composed of high density culture pond, deposit pond, floating and submerged plant pond, ecological floating bed pond and biofilm filtrate pond. At the fish density of 20~30kg/m^3 in the high density culture pond, the water quality parameters were monitored seasonally. The results indicated that the removal rate of total nitrogen, total phosphorus, ammonia nitrogen and nitrite nitrogen in the recirculating aquaculture system were 69.59%, 77.89%, 72.54% and 68.68%, respectively. The floating and submerged plant pond and ecological floating bed pond can remove TN and TP obviously, and increase dissolved oxygen and transparency significantly. And the biofilm filtrate pond has good effect of removing ammonium nitrogen and nitrite nitrogen, meanwhile, the microbial communities in the recirculating aquaculture system regulate on the water quality. Therefore, the RAPS show significant effects on water saving and pollution emission reducing.

1. Introduction

Pond culture, as the dominant practice of freshwater aquaculture industry, its production and area is top of the whole world and provide communities with better incomes, stable jobs and safe food [1, 2, 3]. Although aquaculture industry development is positive to promote the social and economic progress, it also causes great pressure on the surrounding water environment [4, 5]. Especially in the north of China, the relative lack of water resource is seriously hampering aquaculture development. Therefore, the reuse and purification of wastewater from the pond has become one of the key technologies for the sustainable development of aquaculture.

In recent years, pond water purification and remediation technology such as physical remediation, chemical remediation and bioremediation technology have quickly developed in China [6, 7]. For achieving the main goal of energy saving, reduction discharge, ecological security and efficient breeding, our laboratory adopted multistage ecological and bioremediation technologies to regulate and maintain the balance and stability of pond ecological environment [8]. We designed a recirculating aquaculture purification system (RAPS) with high density culture according to the characteristic of northern farming pond.

The objective of this paper is (1) to access the purification efficiency in different remediation units of the health aquaculture recirculating system. (2) to understand the dynamic characteristics of microbial community structure in the aquaculture system.

2. Materials and methods

2.1. Test system

The RAPS is constructed in Beijing XinMiao fishery company field. The system was consisted of (1) High density culture pond: fish density is 20~30kg/m^3. (2) Deposit pond: to ensure that most of the fish feces, bait and other suspended solid dissolved matter can be trapped. (3) Floating and submerged plant pond: some aquatic plants such as water *hyacinth, Pistia stratiotes, Potamogeton crispus* and *Potamogetonaceae* were used to purify water quality. (4) Ecological floating bed pond:to plant Iris *wilsonii* and *Lythraceae* on the floating bed to absorb pollutants. (5) Biofilm filtrate pond: use artificial substrates to immobilize biofilm microbial structure. (in series) (Figure1).

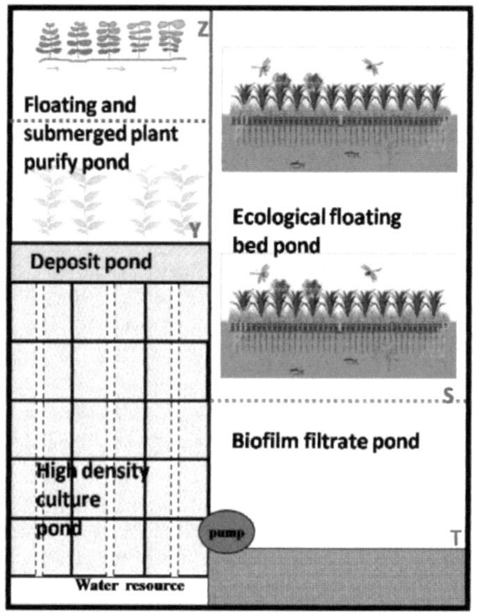

Figure 1. The recirculating aquaculture purification system (sampling site and water flow direction is Y-Z-S-T).

2.2. Water quality monitoring

Water samples were collected from four sampling sites (Y, Z, S, and T) and stored at 500 ml sterile glass bottles at the middle of each quarter. The water quality parameters include water temperature (T), pH and dissolved oxygen (DO) were tested in the field with a water quality analysis machine (AP-7000, USA).The other parameters were tested in the lab according to the standard methods. All data were processed using Microsoft Excel 2007.

2.3. Microbial community monitoring

Water samples were collected, filtered and stored at -70°C. The DNA extraction, PCR amplification and sequencing were performed by Shanghai Major Bio-pharam Technology C.,Ltd using Ilumina Miseq PE 250 strategy.

3. Results

3.1. Water quality changes

The average values of water temperature, pH value and DO concentrations in recirculating aquaculture purification system were15.79±6.5℃, 5.83±0.64, and 8.21±0.41mg/L, respectively. As shown in Figure 2 (a-d), the water quality parameters of TN, TP, NH_4 and NO_2 in the high density culture pond were significantly decreased along the flow direction. The result indicates that wastewater pollutants have been effectively absorption and purification by the ecological engineering facilities. It is also suggested that the system can reduce the pollution to the outside environment, save water and maintain the balance of nutrients in the pond.

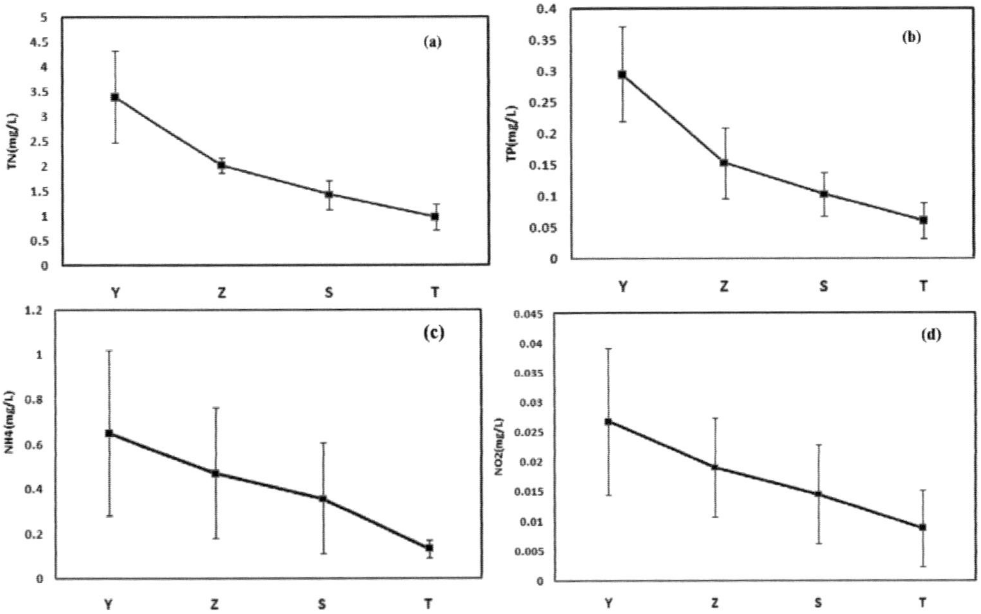

Figure 2. Changes of the water quality parameters in the recirculating aquaculture system.

3.2. The purification efficiency

In total, the removal rate of total nitrogen, total phosphorus, ammonia nitrogen and nitrite nitrogen in the recirculating aquaculture system were 69.59%, 77.89%, 72.54% and 68.68%, respectively. The higher removal rate of total nitrogen (36.87%) and total phosphorus (45.44%) belong to the floating and submerged plants pond. Especially in summer and autumn, the removal efficiency of total nitrogen and total phosphorus in the floating and submerged plants pond were significantly higher than those of other water purification facilities. It means that the purification efficiency gradually increase with the growth of aquatic plants. The result also indicated that there was a significant purification effect of biofilm filtrate pond on ammonium nitrogen and nitrite nitrogen in aquaculture wastewater. In the high nutrient of summer and fall season, the removal rate of ammonium nitrogen and nitrite nitrogen has reached 71.58% and 52.52%, respectively. It is suggested that the biodegradation of biofilm filtrate have regulated on the water quality significantly (Table 1).

Table 1. Water purify efficiency in recirculating aquaculture system

season	site	Removal Rate (%)			
		TN	TP	NH_4	NO_2
	Z	12.00	24.00	20.00	31.11
Spring	S	18.18	36.84	25.00	12.90
(April)	T	33.33	33.33	11.11	33.33
	total	52.00	68.00	46.67	60.00
	Z	53.33	55.00	29.47	14.29
Summer	S	28.57	33.33	23.88	38.89
(June)	T	20.00	50.00	70.59	36.36
	total	73.33	85.00	84.21	66.67
	Z	50.00	76.67	17.58	43.48
Autumn	S	36.84	28.57	17.33	30.77
(August)	T	41.67	60.00	72.58	66.67
	total	81.58	93.33	81.32	86.96

IOP Conf. Series: Earth and Environmental Science **67** (2017) 012028 doi:10.1088/1755-1315/67/1/012028

	Z	32.14	26.09	42.37	22.22
Winter	S	36.84	29.41	38.24	21.43
(November)	T	33.33	33.33	38.10	36.36
	total	71.43	65.22	77.97	61.11

3.3. Microbial community dynamics

The microbial community composition of the recirculating aquaculture system was assessed by Miseq PE250 platform. It is shown that there is a different composition of the microbial community in the purification sections (Figure 3).The main phyla were *Proteobacteria* (41%), *Bacteroidetes* (22%) and *Firmicutes* (20%) in the culture pond, and *Lactococcus* and *Comamonadaceae* were the dominate genera. In the floating and submerged plant pond, sequences belong to *Proteobacteria* (88%) or *Rhodobacter* was found in high abundances. *Actinobacteria* and *Bacteroidetes* were the most group in the ecological floating bed pond with 35% and 28%, respectively, as well as *Alpinimonas* and *Runella* were the main genera. At the biofilm filtrate pond, Proteobacteria (35%) and *Fusobacteria* (28%) were the dominate phyla, and *Cetobacterium* and *Rhodobacte* were the main species at the genus level. The result indicate that the composition of microbial communities in the recirculating aquaculture system regulate on the water quality, as well as the different purification sections contribute to the temporal variations.

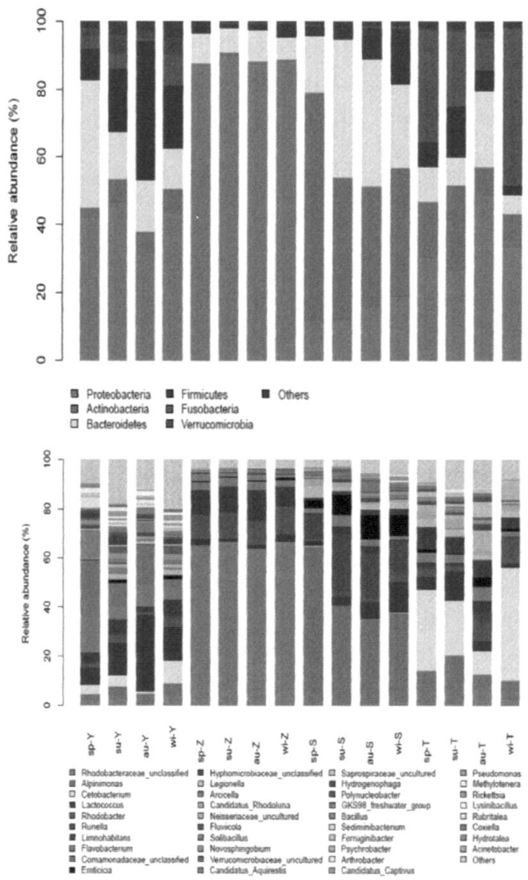

Figure 3. Microbial communities dynamic in the ponds.

Note: sp=spring, su=summer, au=autumn, wi=winter

4. Results

In order to explore the biodegradation efficiency of the whole system of recirculating aquaculture water, the water quality parameters such as the total nitrogen, total phosphorus, ammonia nitrogen, and nitrite nitrogen combined with the dynamic changes of microbial community were monitored seasonally. Our results show that the system purification effect is very significant, especially in summer and fall season. Many reports identified that the recirculating aquaculture system operated well and controlled over the water quality parameters and environments by utilize many different components, thus ensuring optimal conditions for fish culture [9, 10, 11].

In this study, water *hyacinth* and *Pistia stratiotes* were chosen to construct floating plant purification zone. Because of these two aquatic plants have developed root, nitrogen and phosphorus in water and other organic matters were removed effectively by rhizosphere adsorption and species competition mechanism. Previous studies have revealed that rhizoremediation is an environmentally friendly remediation technology which can remove and degrade pollutants and contaminants by microbial activity in the rhizosphere of aquatic plants [12, 13, 14]. The macrophyte allelopathy interact with phytoplankton is considered as one of the mechanisms that contribute to the water environments stabilization [15, 16, 17].In the present study, the submerged macrophytes such as *Potamogeton crispus* and *Potamogetonaceae* can release allelochemicals can inhibit the growth of algae, decrease the algae biomass and improve the water quality and transparency. In the ecological floating bed sections, the aquatic plant roots were colonized to form the microorganism membrane, which increased the contact with surface water area and time, so as to enhance the effect of water purification and promote increased the content of dissolved oxygen in water. Luo et al（2011）also identified that submerged macrophytes combinations with ecological floating beds on in-situ purification and regulation of water quality is quite stable and water quality quite good [18]. Nowadays, Nitrification biofilters are widely used to remove ammonia and other metabolic waste products in recirculating aquaculture systems [19, 20].Our results indicate that biofilms can significantly degrade ammonium nitrogen and nitrite nitrogen in the water, meanwhile, it also can change the structure of microbial commutates and diversity and strengthen the water self-purification function. Moreover, *Lactococcus* and *Comamonadaceae* were the dominate genera in the high density culture pond, which belong to beneficial bacterial benefit to promote the fish growth.

The recirculating aquaculture system is a water-saving, land-saving and high production yield culture mode, which have the advantage of resource demand, ecological protection and production capacity. With the continuous deterioration of the natural environment and more restrictive of environmentally policy frameworks, the recirculating aquaculture will become the future direction of aquaculture development. However, at present in our country, there are very few farmers utilize recirculating aquaculture technology in actual production, because of the technology is more specialized and need to train and improve their management. In order to apply the model to the actual production, it is necessary to explore in the experiment, exploration and practice.

5. Conclusions

During the investigation, the average concentrations of TN in floating and submerged plant pond, ecological floating bed pond and biofilm filtrate pond were 2.205, 1.425 and 0.975 mg/L, respectively. The average contents of TP in the different purification system sections were 0.153, 0.103 and 0.06 mg/L, respectively. According the national fishery wastewater discharge standards, TN and TP pollutants discharge all reached the first class discharge standard($TN \leqslant 2.0$ mg/L, $TP \leqslant 0.3$mg/L).The results showed that the constructed recirculating aquaculture have good degradation effect on nitrogen, phosphorus and other pollutants.

Acknowledge

The work was supported by the founds of Beijing Academy of agriculture and Forestry Sciences Capacity Construction of Science & Technology Innovation (KJCX20151203, KJCX20161502-3); the Agricultural science and technology demonstration and extension project of BAAFS (201664); Beijing

Young Scholar Project (2015000020060G146); Beijing nova program (Z161100004916084). Beijing agricultural Science & technology innovation team (SCGWZJ201705).

References

[1] Tang Q S, Ding X M, Liu S L et al. 2014 Strategy and task for green and sustainable development of Chinese aquaculture *Chin Fish Econ* 32 1 6-14.

[2] Xing L R, Xu X 2016 Economic Profit and Environmental Impact of Aquaculture: Comparison of Different Modes of Tilapia Pond Culture *Ecological Economy* 32 7 143-147.

[3] Dong S L 2011 History, principles, and classification of integrated aquaculture in China *Journal of Fishery Sciences of China* 18 5 1202−1209. DOI=10.3724/SP.J.1118.2011.01202.

[4] Zhang B Z, Ni H G, Guan Y F, Zeng E Y 2010 Occurrence, bioaccumulation and potential sources of polybrominated biphenyl ethers in typical freshwater cultured fish ponds of South China *Environ Pullut*.158 5 1876–1882. DOI=http://dx.doi.org/10.1016/j.envpol.2009.10.043.

[5] Cao L, Wang W M, Yang, Y 2007 Environmental impact of aquaculture and countermeasures to aquaculture pollution in China *Environmental Science and Pollution Research-International* 14 7 452–462. DOI=10.1065/espr2007.05.426.

[6] Deng L F, Jiang X L 2013 The application and research progress of bioremediation in pond aquaculture *Ocenologia et limnologia sinica* 44 5 1270-1275.

[7] Wang W, Chen J, Liu H, He Y P 2010 The overview of aquaculture water purification technology in China *Journal of shanghai ocean university* 19 1 41-49.

[8] Crab R, Defoirdt T, Bossier P, Verstraete W 2012 Biofloc technology in aquaculture: Beneficial effects and future challenges *Aquaculture* 356–357 351–356. DOI=http://dx.doi.org/10.1016/j.aquaculture.2012.04.046.

[9] Pham T A N, Miranda P M, Tru C L, et al. 2016 Adoption of recirculating aquaculture systems in large pangasius farms: A choice experiment .*Aquaculture* 460 90–97. DOI=http://dx.doi.org/10.1016/j.aquaculture.2016.03.055.

[10] Zhang S Y, Li M, Cheng J, et al. 2016 Response and environmental assessment of two Chinese conventional carps to water quality regulation in recirculating aquaculture *Aquaculture Engineering* 74 180–185. DOI=http://dx.doi.org/10.1016/j.aquaeng.2016.09.001

[11] Liu X, Xu H, Wang X, Wu Z, Bao X 2014 An Ecological Engineering Pond Aquaculture Recirculating System for Effluent Purification and Water Quality Control *Clean-Soil Air Water* 42 3 221–228. DOI=10.1002/clen.201200567.

[12] Kristanti R A, Kanbe M, Toyama T, et al 2011. Accelerated biodegradation of nitrophenols in the rhizosphere of *Spirodela polyrrhiza*. *Journal of Environmental Sciences* 24 5 800-807. DOI=10.1016/S1001-0742(11)60839-5.

[13] Toyama T, Furukawa T, Maeda N, et al. 2011 Accelerated biodegradation of pyrene andbenzo[a]pyrene in the *Phragmites australis* rhizosphere by bacteria-root exudate interactions *Water Research* 45 4 1629-1638. DOI=http://dx.doi.org/10.1016/j.watres.2010.11.044

[14] Toyama T, Yu N, Kumada H, et al. 2006 Accelerated aromatic compounds degradation in aquatic environment by use of interaction between *Spirodela polyrrhiza* and bacteria in its rhizosphere *Journal of Bioscience and Bioengineering* 101 4 346 – 353. DOI=http://dx.doi.org/10.1263/jbb.101.346.

[15] Shpigel M, Ben-Ezra D, Shauli L, et al. 2013 Constructed wetland with *Salicornia* as a biofilter for mariculture effluents *Aquaculture* 412–413 52–63. DOI=http://dx.doi.org/10.1016/j.aquaculture.2013.06.038.

[16] Snow A, Anderson B, Wootton B 2012 Flow-through land-based aquaculture wastewater and its

treatment in subsurface flow constructed wetlands *Environ. Rev* 20 1 54–69. DOI=10.1139/A11-023.

[17] Dai Y R, Wu J, Ma X H, et al. 2017 Increasing phytoplankton-available phosphorus and inhibition of macrophyte on phytoplankton bloom *Science of the Total Environment* 74 180–185. DOI=http://dx.doi.org/10.1016/j.scitotenv.2016.11.002.

[18] Luo S T, Zhang Y J, Li J Y 2011 Effect of combination of submerged macrophyte with ecological floating bed on aquaculture pollution controlling *Journal of Ecology and Rural Environment* 27 2 87- 94.

[19] Paula R T, Per B P, Lars F P, et al. 2016 Bacterial activity dynamics in the water phase during start-up of recirculating aquaculture systems *Aquacultural Engineering* DOI=http://dx.doi.org/10.1016/j.aquaeng.2016.09.004.

[20] Christiansona L, Lepinea C, Tsukudaa S, et al. 2015 Nitrate removal effectiveness of fluidized sulfur-based autotrophic denitrification biofilters for recirculating aquaculture systems *Aquacultural Engineering* 68 10–18 DOI=http://dx.doi.org/10.1016/j.aquaeng.2015.07.002.

Assessment of air pollution of settlement areas in Ulaanbaatar city, Mongolia

Ch Sonomdagva, Ch Byambatseren, B Batdelger

Department of Environmental Sciences and Chemical engineering, School of Engineering and Applied Sciences, National University of Mongolia, Ulaanbaatar, Mongolia
Email: ch_sonomdagva@num.edu.mn

Abstract. The purpose of this study is to analyses mass concentration varied by its measurement of air pollution in Ulaanbaatar city, Mongolia. Ulaanbaatar city will have been increasing air pollution due to rapidly expanding vehicular population, growing industrial sector in last 10 years ago. In addition, people use to heat the carbon from 10 month in every year. This becomes a base cause of air pollution in Ulaanbaatar.
We studied a change of mass concentration an air pollution elements in Ulaanbaatar, Mongolia. To research work, we used information that based on data of my measurements of air pollution and Metropolitan air quality agency until 2006 to 2016.
This research important result is air pollution levels are limited to the areas around Ulaanbaatar areas are the most polluted in the center of city are the least polluted areas whereas Tolgoit, Sapporo, 1st Khoroolol, Amgalan, Shar Khad are moderately polluted and the areas around Baruun 4 zam, Factory, Zaisan, Nisekh are normally polluted. The results of pollution are illustrated four zones. By dividing the polluted areas into such zones, we are trying to make it easier to take preventive measures against the pollution itself and protective measures for safeguarding the health of mass population.

1. Introduction
This study area is with basic 6, cortege 3 districts, and only one capital city of Mongolia (Figure 1). The site surrounds the Bayazurkh, Chingeltei, Songinokhairkhan and Khan-Uul mountains and located an along of Tuul and Selbe river's valley in Mongolia center. Ulaanbaatar, the capital city of Mongolia, with a population of 1.1 million is located at an altitude of about 1350 m and in a valley. This study is the first to document the characteristics of the urban heat island (UHI) in Ulaanbaatar. The average UHI intensity is 1.6°C. The UHI intensity exhibits a large seasonal dependence, being strongest in winter (3.3°C) and weakest in summer (0.3°C). The average daily maximum UHI intensity is 4.3°C. The strongest daily maximum UHI intensity occurs in winter with an average intensity of 6.4°C, and the weakest one occurs in summer with an average intensity of 2.5°C. The occurrence frequency of the daily maximum UHI intensity in the night time is 5.6 times that in the daytime [4].
We conducted based on the results and samples from air quality tracking stations around the city. In addition, the research was conducted the measurements of PM_{10} and $PM_{2.5}$ at four different places, by using Dust Trak TM II 8532.
Late years, air, soil and water pollution have been adding from due to growing of population in Ulaanbaatar city.

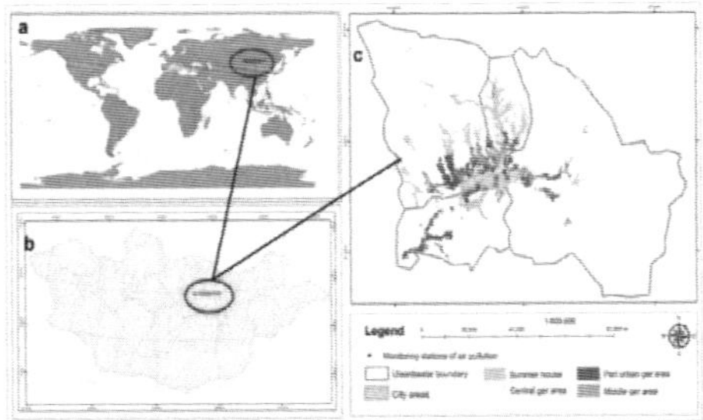

Figure 1. Study area.

Many research studies a change, pollution, assessment and risk of air quality in Ulaanbaatar City. By their studies, The spatial patterns in SO_2 and PM, seasonal and diurnal patterns in $PM_{2.5}$, and high wintertime $PM_{2.5}/PM_{10}$ ratios were consistent with a major impact from coal and wood combustion in the city's low-income traditional housing (ger) areas. The annual average concentration of $PM_{2.5}$ measured at a centrally located government-monitoring site was 75µg/m^3 or more than seven times the World Health Organization's $PM_{2.5}$ air quality guideline, driven by a wintertime average concentration of 148µg/m^3. $PM_{2.5}$ concentrations measured in a traditional housing area were higher, with a wintertime mean $PM_{2.5}$ concentration of 250µg/m^3 [1].
The World Health Organization (WHO) listed the air pollution in Ulaanbaatar, Mongolia among the top 5 cities with the worst air quality in the world [5].
The study concluded that mass concentration of PM $_{2.5}$ and PM $_{10}$ increases during winter and decreases during summer. This variation depends on climate and pollution sources [6]. In our study, we assessed an air pollution of settlement areas of Ulaanbaatar city.

2. Data and study method

The research regarding polluters such as CO, NO_2, SO_2, $PM_{2.5}$, and PM_{10} was conducted based on the results and samples from air quality tracking stations around the city. Also the research was conducted the measurements of PM_{10} and $PM_{2.5}$ at four different places such as Sansar Tunnel, Zaisan Hill, Kyokushyu Tower, Denjiin 1000 by using Dusttrak TM II 8532 TSI (Equation 1).

$$concentration = \frac{3000x\,(filter\,post\,weight - filter\,pre\,weight)}{2x\,dustTrak\,monitor\,flow\,rate}\,x\,total\,sample\,time \quad (1)$$

The concentration of an air pollutant was calculated by dividing the mass of the air pollutant collected by the volume or mass of stack gas sampled (Equation 2). Continuous emission monitoring systems (CEMS) measure gaseous air pollutants directly by fluorescence (SO_2), infrared spectroscopy (CO), and chemiluminescence (NOx). Identifies the concentrations of the criteria air pollutants typically found in stack test and CEMS results. Concentrations were reported on a mass or volume basis. The concentration of nitrogen dioxide (NO_2) reported is assumed to be the difference between total nitrogen oxides and nitric oxide. A stream of sample air was drawn through a cell where it is exposed to pulsed ultraviolet light, resulting in excitation of sulphur dioxide molecules. These molecules subsequently re-emit light but at a different wavelength; they fluoresce. The intensity of the fluorescent light measured by a photomultiplier tube is proportional to the concentration of sulphur dioxide in the sample air.
Sample air was drawn into a cell where a beam of infrared light is passed through it to a photodetector. Some of the light has been absorbed by carbon monoxide in the sample, the amount being proportional to the number of molecules present. By comparing the light intensity received by the photodetector through the sample cell with that received through a similar cell containing reference gas, the concentration of carbon monoxide may be determined. This standard usage is assessing, monitoring the indoor and outdoor air quality in town, apartment, office, public service place, and civil building [7].

$$concentration = \frac{mass\ of\ air\ pollutant\ collected}{Volume\ or\ mass\ of\ air\ sample} \quad (2)$$

Illustration was made by using Interpolation method of IDW (Inverse distance weight) of ArcGIS 10.1 [2] (Equation 3). A model for pollutants plumes was designed through a GIS to calculate the dispersal of total suspended particulate (TSP) loads over the landscape and to estimate its spatial distribution and to analyse its relationship with other geographic land use attributes [3]. The GIS was used for determining the type and area of the natural landscape affected by the pollutants.

$$D = \sum_{g=1}^{G} d_g^{-b},\ \widehat{P_0} = \frac{1}{D} * \sum_{g=1}^{G} d_g^{-b} * P_g \quad (3)$$

3. Research results

The research on air quality of Ulaanbaatar was conducted based on the facts and information we produced and information provided by the air quality tracking stations around the city.

3.1. Research on air quality of ulaanbaatar and dispersion

Contents of SO_2, NO_2, PM_{10} and $PM_{2.5}$ in air of Ulaanbaatar were defined by the average points provided by the air quality tracking stations (Figure 2).

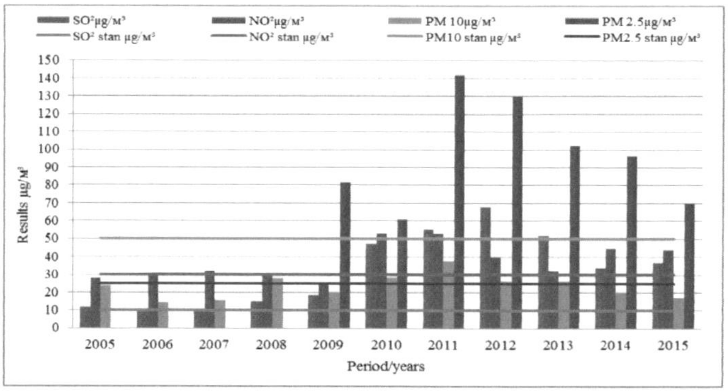

Figure 2. Figure is content of SO_2, NO_2, PM_{10} and $PM_{2.5}$. The average annual content of quite stable in from 2005 to 2015.

The average monthly content of NO_2 in 2012 was at its lowest and November and December were the highest compared to those of other years (Figure 2). Yearly diagrams demonstrate that 2011 was the lowest point whereas the content of NO_2 in air was increasing in 2012-2013 and started to decrease in 2014-2015. The content of SO_2 in air was shown in the figure that SO_2 started increasing from 2009 to 2012. In these years, the content of SO_2 in air slow increased 2009 to 2010. However it was greatly advanced content of SO_2 in air in 2012. The content of SO_2 in air was the highest in 2012 and 2010 and it was shown in the figure that SO_2 started increasing from 2009 and it started decreasing from 2013. In addition, figure shows that during the winters the content of SO_2 in air is about 1.5-1.7 times higher than the standard. The content of PM_{10} in air in Ulaanbaatar city was higher volume contained at spring and autumn. By research, content of PM_{10} is high from standards of Mongolia and World Health Organization in four seasons in year (Figure 2). As we can see in the figure 2, the content of PM_{10} in air was at its peak during 2009, especially in April, May and November. The content of PM_{10} is higher than the standard all through the year. The Content of $PM_{2.5}$ in air was highest in 2011, 2013 which were 3 times higher than the standard then (Figure 2).
The average monthly content in December and January has relatively high than other months of the year whereas during the summer when it has the highest level of rainfall, the content of $PM_{2.5}$ in air is low. Even though it was assumed that there is some seasonal impact on the content of $PM_{2.5}$, the difference in average annual content is insignificantly low.

When we compared the average contents of polluters in air of 2015 to those of 2009, PM_{10} has decreased by 15%, SO_2 by 20%, NO_2 by 19% respectively. During the winter of 2015, the average content of SO_2 has increased by 25%, PM_{10} by 25% compared to the previous year. The $PM_{2.5}$ has decreased by 13% compared to 2009 and relatively similar to other years.

3.2. *Illustration of air pollution dispersion*

The GIS was used for determining the type and area of the natural landscape affected by the pollutants. Next part of our research, we illustrated the picture of distribution of air pollution in Ulaanbaatar city. The program automatically calculated the pollutant concentration result the all pollution, altogether, at each node of a user defined grid covering all sensitive areas vulnerable to the air pollution impact.
We calculated by average measurement of sample in 6 months in winter in 2015 at nine stations that sampling the air quality in Ulaanbaatar city.
The research on air quality of Ulaanbaatar city was shown below according to the air quality tracking stations and the results produced (Figure 3-6).

a)

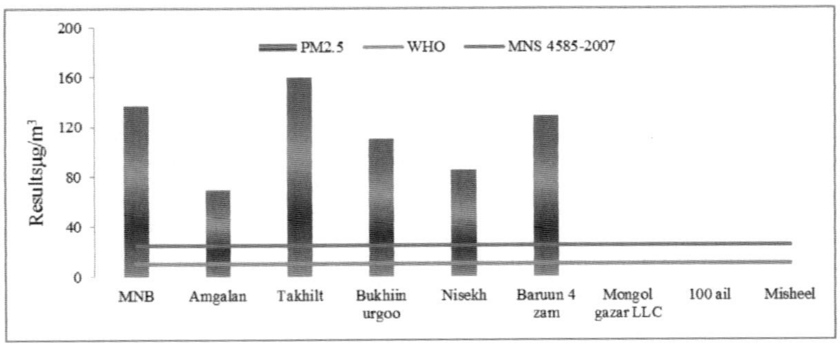

b)

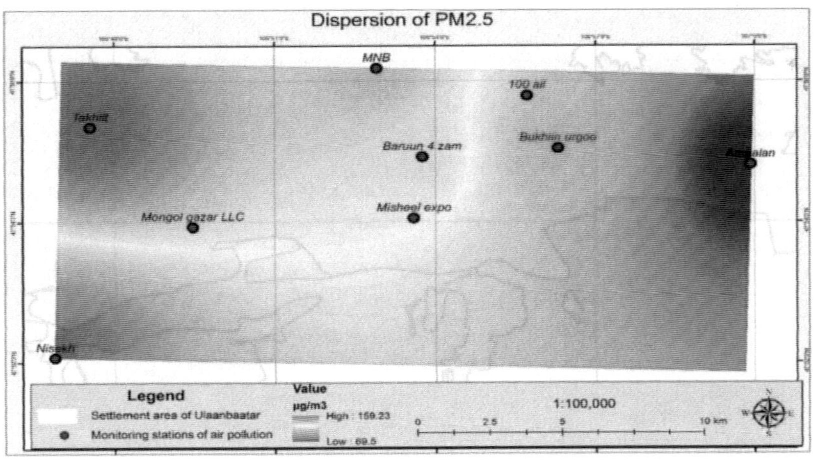

Figure 3. (a, b) Content of $PM_{2.5}$ and its dispersion (by the average of 10, 11, 12 months, 2015 and 1, 2, 3 months, 2016) $PM_{2.5}$ has been recorded to be higher than the standard at all stations. The highest level of pollution by $PM_{2.5}$ was recorded at Takhilt station what was 221.47μg/m^3 or 4.4 times higher than the standard as of November 2015 whereas the lowest mean was 98.2μg/m^3 recorded at Bukhiin Urguu station.

a)

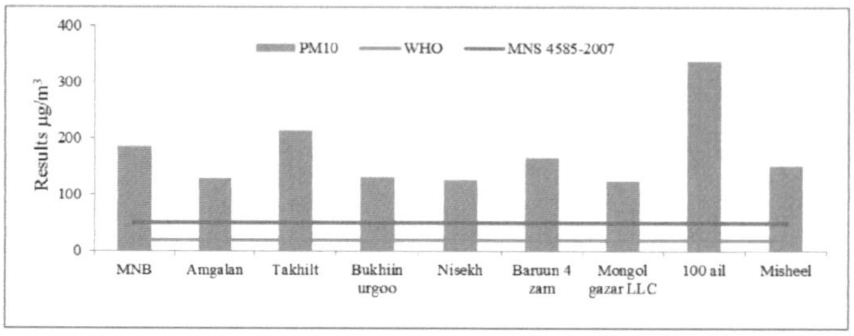

b)

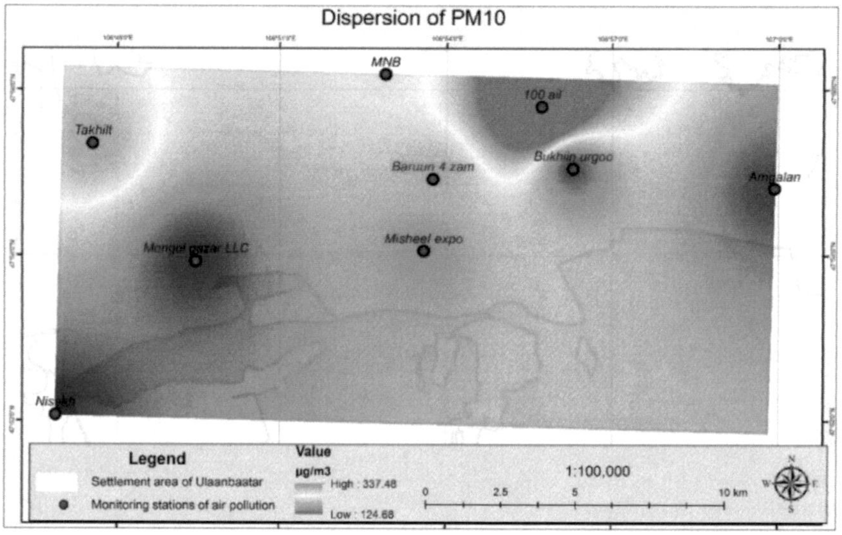

Figure 4. (a,b). Content of PM_{10} (by the average of 10, 11, 12 months, 2015 and 1, 2, 3 months, 2016) PM_{10} has been recorded to be higher than the standard at all stations. The highest level of pollution by PM_{10} was recorded at 100 ail station, which was 291.9μg/m^3 or almost 3 times higher than the standard level as of November 2015 whereas the lowest mean was 102.7μg/m^3 recorded at Misheel Expo station.

a)

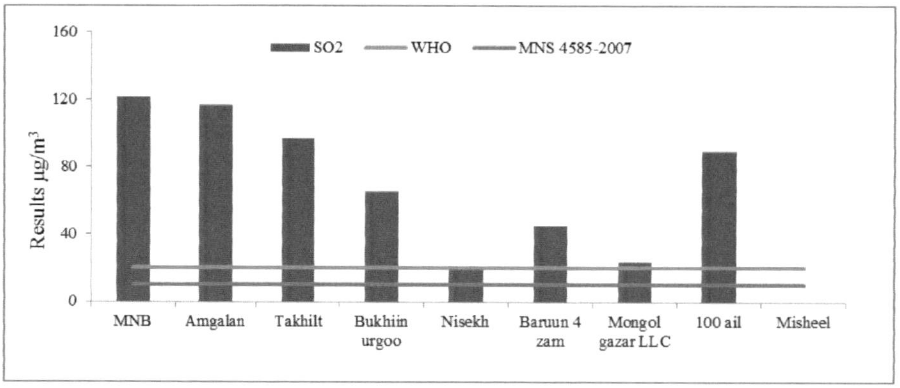

b)

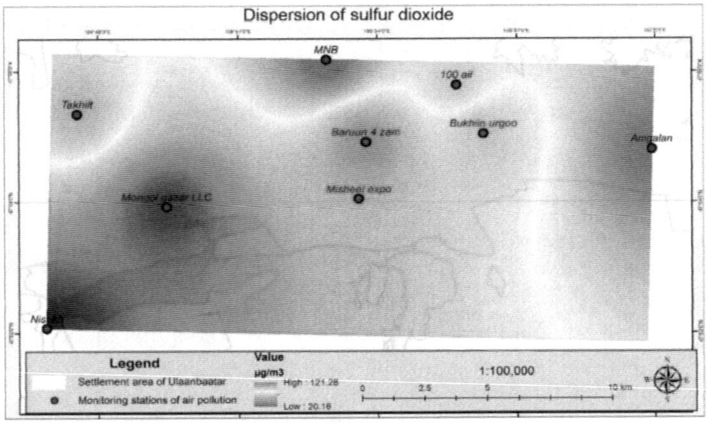

Figure 5. (a,b). Content of SO_2 in air of Ulaanbaatar (by the average of 10, 11, 12 months, 2015 and 1, 2, 3 months, 2016) Content of SO_2 in air of Ulaanbaatar was at the highest level at the areas around the stations Takhilt, MNB and Amgalan which were about 3 or 4 times higher than the standard level. The lowest point was recorded at the areas around Mongol Gazar LLC, Baruun 4-Zam and Nisekh, where was 25.8-34.5 µg/m^3.

a)

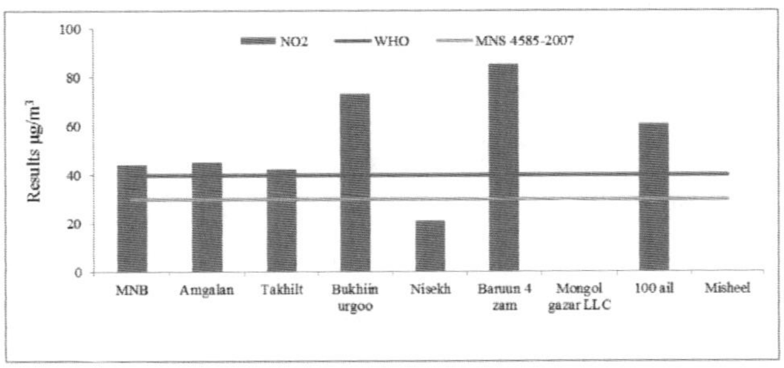

b)

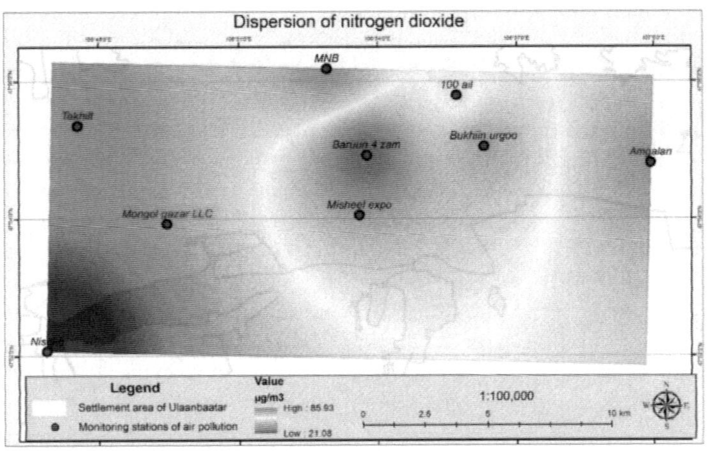

Figure 6. (a,b). Content of NO_2 (by the average of 10, 11, 12 months, 2015 and 1, 2, 3 months, 2016) the content of NO_2 in air of Ulaanbaatar is at the highest level around Baruun 4-Zam, 100 Ail, Bukhiin Urguu and MNB which were 1.2-2 times higher than the standard level. However, Nisekh was polluted less than the other areas and the pollution level was lower than the standard level.

The highest amount of air polluters were recorded around the western part of the city, areas such as Bayankhoshuu, 1st Khoroolol, Tolgoit and Khanyn Material. These numbers were already 2-3 times higher than "the dangerous level" according to the air quality index (AQI) of Mongolia. The areas around the city centre, Nisekh and Amgalan were relatively low in amount of air polluters compared to the other parts of the city; those numbers were still 1-1.5 times higher than the dangerous level of AQI (Figure 7).

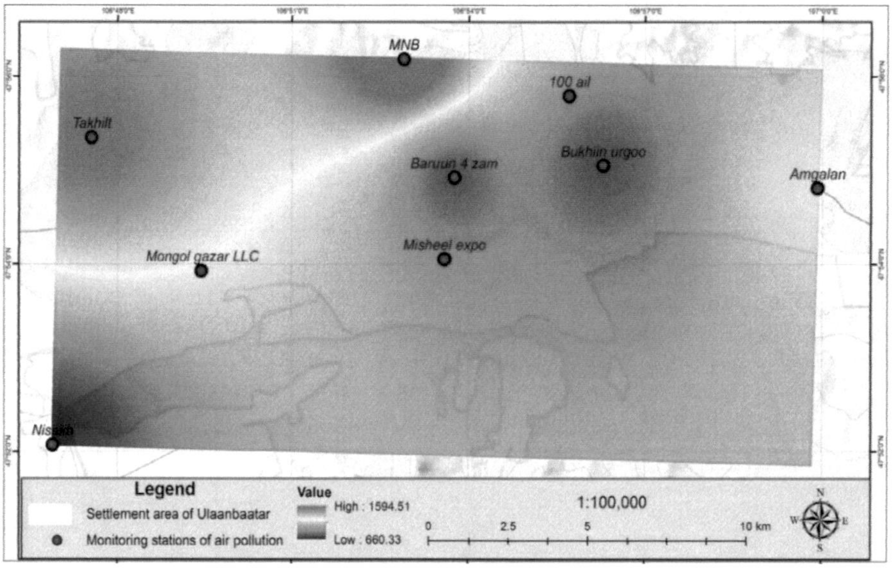

Figure 7. The combined amount of polluters in air of Ulaanbaatar.

4. Conclusion

When air pollution levels in the areas around Ulaanbaatar were measured using the results recorded at the stations, the highest pollution level was recorded during the winter of 2011 and it was shown in the research that air pollution started decreasing in 2013 and began increasing in 2015. SO_2, PM_{10} $PM_{2.5}$ and CO are the dominating factors of air pollution in the outskirts of the city and the stations at Takhilt and MNB have produced the most polluted samples. NO_2 is the most influential polluter around the areas of Baruun 4-Zam, 100 Ail, and 13rd khoroolol whereas Amgalan is the area least polluted by NO_2. PM (Particulate Matter) is the biggest polluting factor during the winters.

When air and soil pollution levels are limited to the areas around Nisekh, Tolgoit, MNB, Amgalan, MNB areas are the most polluted in the city and Sansar Tunnel and 13rd Khoroolol are the least polluted areas whereas Tolgoit, Sapporo, 1st Khoroolol, Amgalan, Shar Khad are moderately polluted and the areas around Baruun 4-Zam, Factory, Zaisan, Nisekh are normally polluted. The results of pollution are illustrated and MNB, Bayankhoshuu and Denjiin 1000 are the first zone, Tolgoit, Baruun Salaa, Zuun Salaa, 1st Khoroolol and Sapporo are the second zone, Yaarmag, 25th Pharmacy, 10th Khoroolol and western area of Factory are the third zone and Sansar, Factory, Zaisan, and Dunjingarav are the fourth zone. By dividing the polluted areas into such zones, we are trying to make it easier to take preventive measures against the pollution itself and protective measures for safeguarding the health of mass population.

5. References

[1] Allen R W, Gombojav E, Barkhasragchaa B, Byambaa T, Lkhasuren O, Amram O, Janes C R, 2011 An assessment of air pollution and its attributable mortality in Ulaanbaatar, Mongolia. *Air Quality, Atmosphere & Health*, *6* (1), p 137–150

[2] Curtis A. J., Mills, J. W., & Leitner, M. (2006). Spatial confidentiality and GIS: re-engineering mortality locations from published maps about Hurricane Katrina. *International Journal of Health Geographics*, *5*(1), p 1–12
[3] Fouda, Y. E. (2001). A GIS for environmental assessment of air pollution impacts on urban clusters and natural landscape at Rosetta City and region, Egypt. *Urban Ecosystems*, *5*(1), p 5–25.
[4] Ganbat, G., Han, J.-Y., Ryu, Y.-H., & Baik, J.-J. (2013). Characteristics of the urban heat island in a high-altitude metropolitan city, Ulaanbaatar, Mongolia. *Asia-Pacific Journal of Atmospheric Sciences*, *49*(4), p 535–541
[5] Guttikunda, S. K., Lodoysamba, S., Bulgansaikhan, B., & Dashdondog, B. (2013). Particulate pollution in Ulaanbaatar, Mongolia. *Air Quality, Atmosphere & Health*, *6*(3), p 589–601.
[6] Sonomdagva, C., Batdelger, B., & Byambatseren, C. (2016). Characteristics of PM_{10} and $PM_{2.5}$ in the Ambient Air Ulaanbaatar, Mongolia. *Environmental Science and Development*, *Vol 7*, p 827–830
[7] “Mongolian National Standard 4585; 2007

6. Acknowledgements
This study has been financial supported by the National University of Mongolia and Science, Technology Foundation of Mongolia and Ministry of Education, Culture, Science and Sport of Mongolia.

IOP Conf. Series: Earth and Environmental Science **67** (2017) 012030 doi:10.1088/1755-1315/67/1/012030

Validation of tsunami inundation model TUNA-RP using OAR-PMEL-135 benchmark problem set

H L Koh[1,2], S Y Teh[3], W K Tan[1,3], and X Y Kh'ng[3]

[1] School of Mathematical Sciences, Sunway University, Bandar Sunway, 47500 Selangor, Malaysia
[2] Jeffrey Sachs Center on Sustainable Development, Sunway University, Bandar Sunway, 47500 Selangor, Malaysia
[3] School of Mathematical Sciences, Universiti Sains Malaysia, 11800 Penang, Malaysia
Email: syteh@usm.my

Abstract. A standard set of benchmark problems, known as OAR-PMEL-135, is developed by the US National Tsunami Hazard Mitigation Program for tsunami inundation model validation. Any tsunami inundation model must be tested for its accuracy and capability using this standard set of benchmark problems before it can be gainfully used for inundation simulation. The authors have previously developed an in-house tsunami inundation model known as TUNA-RP. This inundation model solves the two-dimensional nonlinear shallow water equations coupled with a wet-dry moving boundary algorithm. This paper presents the validation of TUNA-RP against the solutions provided in the OAR-PMEL-135 benchmark problem set. This benchmark validation testing shows that TUNA-RP can indeed perform inundation simulation with accuracy consistent with that in the tested benchmark problem set.

1. Introduction

The 26 December 2004 Andaman tsunami caused the death of a quarter million people worldwide, mainly in Indonesia and Thailand, including 52 deaths in Penang, Malaysia. This catastrophic disaster serves as a wake-up call for all affected countries to develop programs capable of reducing the adverse impacts of tsunamis. These tsunami mitigation programs aim to develop societal resilience to tsunami by cultivating awareness, education and preparedness regarding tsunami hazards, tsunamis risk zones and inundation maps. These tsunami mitigation programs invariably involve a sound understanding about the locations where tsunamis will most likely occur, as well as the extent of run-up and inundation distances. To provide information on run-up and inundation, tsunami run-up and inundation simulation models are normally used to forecast their occurrence. This paper begins with a brief literature review on four commonly used run-up and inundation algorithms. Their relative merits are briefly discussed, leading to a justification for choosing the wet-dry moving boundary algorithm that was incorporated into TUNA-RP. Allowable percentage errors for the three categories of problems listed in OAR-PMEL-135 used for benchmarking are then discussed. Benchmark testing is typically performed by fulfilling the criteria that the percentage errors incurred by a particular algorithm against the three categories of problems must not exceed the prescribed limits [1]. TUNA-RP simulation results are then tested against each of the four benchmark problems chosen for testing, the comparison of which indicates satisfactory performance of TUNA-RP. This paper ends with a conclusion that TUNA-RP is ready for performing run-up and inundation simulations. It is the wish of the authors that this paper will provide useful insights and easy-to-use guidelines on simulation of tsunami run-up.

2. Literature review on wetting-drying algorithms

One of the most challenging problems in tsunami research is modelling the physical process of wave run-up and inundation along shallow beaches. As the wave approaches the beach, it begins to inundate a previously dry area, turning the area into a wet area and continues to penetrate further into the dry land. If the area is flat or has very mild slope, the tsunami waves can indeed inundate further inland up to several km, causing potentially immense damages. The model algorithm must be able to dynamically track the constantly evolving wet-dry areas. As the tsunami wave recedes back to the seaward shoreline after reaching its maximum run-up height, the model must exclude the dry area from simulation. This addition of wet areas and removal of dry areas are named as wetting-drying (WD) algorithm in tsunami modelling terminology. The capability of WD algorithm is determined by the accuracy in simulating inundation distance and run-up height, as these are ultimately the most crucial information needed by tsunami hazard assessment.

A variety of numerical models has been developed to simulate tsunami inundation in one, two and three dimensions. These models employ a diverse set of innovative numerical schemes to solve the governing nonlinear shallow water equations (NSWE) and to optimally discretise the computational domains to suit the selected numerical schemes. This section reviews WD algorithms implemented in various numerical models reported in the literature with the aim of providing useful insights and easy to use guidelines to modellers interested in simulating tsunami run-up and wave inundation. Generally, WD algorithms fall into four broad frameworks: (a) thin film, (b) element removal, (c) depth extrapolation and (d) negative depth algorithms, as elaborated in the sections that follow.

2.1. Thin film algorithm

Thin film algorithm specifies a small artificial sublayer of fluid over the entire computational domain in order to give the computational domain a minimal depth at all time. This allows all nodes and all cells to be computed at each time step. There is typically a minimum threshold depth that defines the categories of being wet or being dry in the model. The Princeton Ocean Model (POM) employs this thin film algorithm into its finite difference scheme where the most landward cells must always remain "dry", to ensure that water can never flow through the dry boundary located on dry land [2]. On the other hand, a 5 cm thin film of water is added on top of the seaward cells to mark it as "wet". This allows the momentum equations to be solved in these seaward "wet" cells. If the depth is less than 5 cm, the velocity is set to be zero; otherwise, the velocity is computed by solving the momentum equations. The Finite Volume Community Ocean Model (FVCOM) is a finite volume coastal ocean model that also employs this similar thin-film WD algorithm [3]. Subsequently, Chen et al. [4] successfully employed the nested global-coastal FVCOM to simulate the 11 March 2011 Great Eastern Japan Earthquake and Tsunami (GEJET) event to assess the inundation in the central Sendai coastal region.

2.2. Element removal algorithm

Element removal algorithm employs a system of checks to determine if a node or cell is wet or dry. The wet nodes are included in the computational domain, while the dry nodes are excluded. For example, the TELEMAC-2D model is a finite element model [5,6] that employs the element removal algorithm by categorizing all elements as one of the three categories: (a) fully wet, (b) fully dry and (c) partially wet-dry. The categorization is performed at the beginning of each time step to include all fully wet and partially wet-dry elements in the computations and to exclude fully dry elements from the computations. Using this concept, Grilli et al. [7] coupled a tsunami generation and propagation model, codenamed FUNWAVE-TVD to the TELEMAC-2D to assess tsunami hazard along the north shore of Hispaniola due to tsunamis from far and near field Atlantic sources. In the TELEMAC-2D inundation simulation, fine-resolution unstructured meshes of 12 m to 30 m are used to accurately simulate run-up heights and inundation distances along the coast. One of the most commonly used model for simulating coastal hydrodynamics, Delft3D-FLOW also employs this element removal algorithm in its finite difference scheme [8]. Apotsos et al. [9] validated the capability of Delft3D-FLOW in simulating tsunami inundation by comparing its simulation results to tsunami wave data obtained from standard benchmark problems.

2.3. Depth extrapolation algorithm

In depth extrapolation algorithm, the depth is extrapolated from a wet cell onto a dry cell if the depth in the wet cell is sufficiently high to inundate the dry cell. Furthermore, if the depth is able to be extrapolated from the wet cell into the dry cells, then the newly extrapolated depth is used to compute the velocity and the cell now becomes wet. For example, COULWAVE is a finite difference free surface wave model which employs depth extrapolation algorithm to simulate wave run-up [10]. The extrapolation process begins with locating the boundary between a wet cell and a dry cell. The free surface of the dry cell is then estimated using a one-dimensional linear extrapolation coupled with an averaging method. If the estimated free surface of the dry cell is above its land height, then the depth at the boundary is interpolated. The interpolated depth is in turn used to compute the velocity at the boundary. This model was adequately validated in both one-dimensional and two-dimensional space using an idealized computational domain and a sinusoidal wave forcing.

2.4. Negative depth algorithm

In the negative depth algorithm, the water surface is conceptually allowed to "exist" below the ground surface, allowing the governing equations to be computed over the entire domain, with cells having negative depth being considered dry. As the water depth gradually increases and eventually become positive, the waves begin the inundation of dry cells, which is then simulated. A finite element model RMA2 [11] employs this negative depth algorithm by introducing a porosity parameter, allowing the water to plunge below the ground surface and to flow in a low porosity medium. Nielsen and Apelt [12] applied the RMA2 model to four test cases and found that the selection of porosity parameter significantly influenced the simulation results. A major problem with this approach of using the concept of being "underground" is the difficulty in selecting an appropriate porosity parameter value as the physics involved is not real.

2.5. Choice of WD algorithm for TUNA-RP

In this subsection, we briefly discuss the relative merits of each of the four WD algorithms presented earlier and provide justification for the choice of the element removal algorithm used in TUNA-RP. The non-physics, thin film algorithm alters the mass conservation by the addition of an artificial sublayer of fluid on top of dry cells in order to allow the governing equations to be computed over the entire computational domain at every time step. Thin film algorithm tends to produce artificially smooth solutions at wetting-drying fronts, contrary to actual oscillations observed in real tsunami surveys. However, it is computationally less expensive at the expense of accuracy. The element removal algorithm that employs a system of checks to determine if a node or cell is wet or dry, thereby removing dry nodes or cells from computational domain, is the most commonly employed algorithm. Element removal algorithm saves computational cost by not computing the governing equations over the entire computational domain at each time step. However, it tends to perform better on advancing wetting fronts than on receding fronts. Depth extrapolation algorithm that extrapolates the water depth from wet nodes onto dry nodes and then computes the velocities using the newly extrapolated water depth also tends to produce artificially smooth solutions at the wetting front. Further, correction schemes are necessary to conserve mass balance because the extrapolation of water depth introduces new mass into the computational domain. Lastly, the negative depth algorithm that artificially allows the simulated water surface to penetrate below ground surface tends to be used exclusively in the finite element models. The inclusion of "appropriate" porosity parameter can capture the actual physical inundation process well. However, the choice of porosity is performed by trial and error, with the accuracy of simulation depends heavily on choice of good porosity parameter. Since the element removal algorithm saves computational cost and performs well in advancing wetting front, the algorithm is chosen for incorporation into TUNA-RP. Benefits and drawbacks in terms of accuracy, robustness, computational efficiency, and conservation properties of these WD algorithms are detailed in [13].

3. Three categories of criteria for standard benchmarking

A group of tsunami simulation experts discussed key issues of accuracy for long-wave run-up models in the 2004 Third International Workshop on Long-Wave Run-up Models held in Wrigley Marine Science Center, California, on 17-18 June 2004, focusing on inundation modelling and underwater landslide-generated tsunami modelling [14]. The workshop highlighted the importance of benchmarking any numerical model in producing practical and usable inundation maps, and proposed a list of benchmark problems for validating a numerical run-up model. Therefore, it is essential that all numerical models used in tsunami hazard assessment, particularly in producing inundation maps, be subjected to benchmark testing involving validation and verification. Validation is the process of ensuring that the model accurately solves the parent equations of motion, while verification is the process of ensuring that the model represents geophysical reality, following Synolakis et al. [15]. For this purpose, a standard set of benchmark problems, known as the OAR-PMEL-135, is developed by the National Tsunami Hazard Mitigation Program (NTHMP [16]. The benchmark problems include four cases: (1) the semi-analytical solution of run-up on a planar beach [17], (2) laboratory experiments of run-up on a circular island [18], (3) run-up on a simple sloping beach [19] and (4) run-up on a piecewise linear bathymetry ending in a vertical wall [20].

It should be noted that there is no absolute assurance that a numerical model that has passed the benchmark validation and verification test will always produce realistic inundation projections in a real tsunami event. The accuracy, spatial resolution and quality of input data driving the numerical model are the main determinant of the model accuracy. However, validated and verified models greatly reduce the level of uncertainty in their simulation results. The uncertainty in the initial conditions (tsunami generation) remains a major concern. Benchmark testing is typically performed by fulfilling the criteria that the percentage errors incurred by a particular algorithm against the three categories of problems must not exceed the prescribed limits as follows: (i) analytical solutions (< 5%) (ii) laboratory experiments (< 10%) and (iii) field measurements (< 20 %). We will perform validation and verification test for our TUNA-RP model in the following section, using the criteria mentioned as a guide for allowable errors for three categories of problems used for benchmarking, as suggested by Horrillo et al. [1].

4. Results of TUNA-RP benchmarking

In this section, we will present TUNA-RP validation and verification test results by comparing its simulation results with four benchmark run-up solutions provided in OAR-PMEL-135 to demonstrate the satisfactory performance of TUNA-RP.

4.1. N-wave run-up on a planar beach

Carrier et al. [17] used the NSWE to simulate tsunami run-up and draw-down dynamics on a planar beach with a semi-analytical solution technique for examining shoreline movement and velocity field. In the 2004 Third International Workshop on Long-Wave Run-up Models held in Wrigley Marine Science Center, California, the semi-analytical solution technique is used to produce the benchmark data for a leading-depression N-wave run-up on a planar beach. The benchmark problem setup and data are available in the Inundation Science & Engineering Cooperative (ISEC) website.

The computational domain of this benchmark problem is a uniformly sloping beach (planar beach) of 50 km with 1/10. The depth (bathymetry) increases linearly from the open sea (-5 km) to mean sea level (0 km), and the dry land (topography) elevates linearly from mean sea level (0 m) to 20 m in height. The initial free surface displacement is an N-wave composed of a leading depression followed by a smaller elevation, typically generated by an underwater landslide. The computation setup is then incorporated into TUNA-RP simulation to validate the capability of TUNA-RP in simulating the maximum run-up height. In this simulation, radiation boundary is employed at the seaward end to allow the wave to propagate out of the computational domain. In addition, a spatial step Δx of 5 m is used, for the simulation to obtain more accurate run-up simulation, with a time step Δt of 0.01 s to ensure numerical stability. The Manning's roughness coefficient n is set to zero, because the semi-analytical solution technique solves the NSWE without taking the friction term into account. It should be noted that the initial N-wave velocity is zero.

Figure 1 compares the simulation results of TUNA-RP with the semi-analytical solution plots available at the ISEC website for the benchmark problem. Simulation snapshots at time t = 160 s, 175 s and 220 s represent initial draw-down, maximum draw-down and maximum run-up, respectively. At t = 160 s, the wave is receding towards the seaward end (east boundary) with a positive velocity, indicating the wave will continue to recede until the maximum draw-down is achieved. At t = 175 s, maximum draw-down is achieved with a negative velocity, indicating the wave will soon propagate towards the dry land (west boundary). At t = 220 s, the simulated maximum run-up of 15.26 m is achieved. Compared to the analytical solution maximum run-up of 15.78 m, the relative error is 3 %, which is within the allowable error. The TUNA-RP simulated free surface displacement and velocity at t = 160 s, 175 s and 220 s agree well with the semi-analytical solution. Therefore, it can be concluded from this comparison that TUNA-RP satisfactorily simulates the run-up on a planar beach.

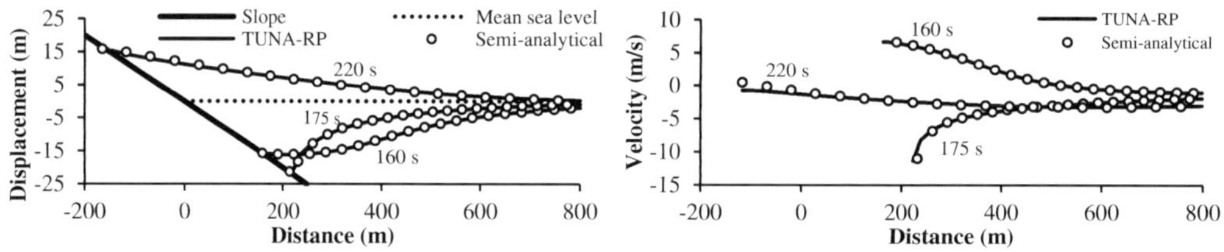

Figure 1. (a) Free surface displacement and (b) velocity at time t = 160, 175, 220 s simulated by TUNA-RP compared to semi-analytical solution.

4.2. Solitary wave run-up on a circular island

On 12 December 1992, an earthquake of magnitude Mw = 7.3 occurred near Flores Island, Indonesia. This tsunami generated a tsunami that struck the circular shaped island, named Babi Island, from the north, but inundated villages located in the southern or lee side of the island with unexpectedly high run-up [21]. Similar phenomenon was also observed on the pear shaped Okushiri island during the 12 July 1993 Hokkaido tsunami event with a measured run-up of 20 m at the lee side of the island [22]. Recognizing the need for a better understanding of this unexpected phenomenon, Briggs et al. [18] conducted a large-scale laboratory experiment that mimics the inundation inflicted on Babi Island.

The experiment was performed at the US Army Engineer Waterways Experiment Station, Coastal Engineering Research Center, Mississippi (USAEWES), in a 30 m wide and 25 m long wave basin. A circular island with 7.2 m base diameter, 0.625 m height and 14.04° slope was positioned in the basin with its center located at x = 15 m, y = 13 m. A Directional Spectral Wave Generator (DSWG) was installed along the x-axis to generate an initial solitary wave in a 0.32 m water depth. This experiment is recognized as one of the most important benchmark problems listed in OAR-PMEL-135 standard set by NTHMP to validate a numerical model. The physical experiment setup is then incorporated into TUNA-RP simulation to validate the capability of TUNA-RP in simulating the incident solitary wave refraction around the circular island and amplification of the run-up at the lee side of the island.

Figure 2 presents the time series of free surface displacement simulated by TUNA-RP in comparison with the experimental data at four wave gauge stations. The experimental data are available in the NOAA website (http://nctr.pmel.noaa.gov/benchmark/), including the locations of wave gauge stations. Good overall agreement is achieved between experimental data and TUNA-RP simulation results. In the front of the island, wave gauges 6 and 9 show an elevation wave followed by a depression wave, resembling the draw-down dynamics after the solitary wave reached the maximum run-up height. Note that, TUNA-RP does not produce the second elevation wave observed in the experimental data between t = 25 s and 30 s at wave gauges 6 and 9, because the second elevation wave is the reflected draw-down wave from the DSWG. In TUNA-RP simulation, radiation boundary is employed to allow the draw-down wave from the island to propagate out of the computational domain without being

IOP Conf. Series: Earth and Environmental Science **67** (2017) 012030 doi:10.1088/1755-1315/67/1/012030

reflected. On the lee side of the island, the reflection of solitary wave from the boundary of the basin at wave gauge 22 is also evident. However, this reflected wave does not affect the maximum run-up height observed in the laboratory experiment, because the first wave often generates the maximum run-up height and the subsequent reflected waves generate significantly smaller run-up height.

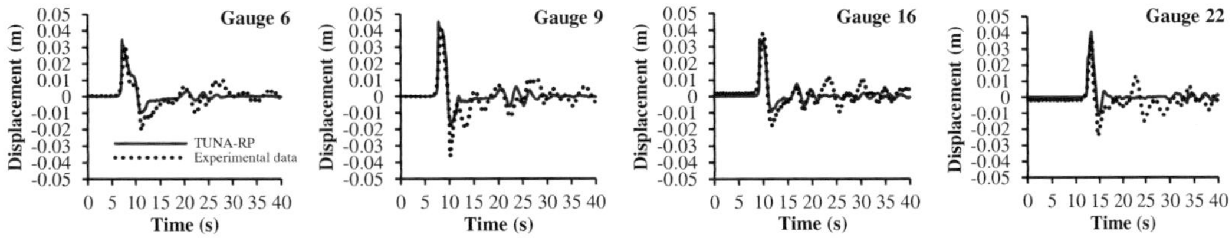

Figure 2. Time series of free surface displacement at four different wave gauges simulated by TUNA-RP compared to experimental data.

4.3. Solitary wave run-up on a simple sloping beach

Synolakis [19] conducted a series of laboratory experiments to investigate the run-up dynamics of a solitary wave on a simple sloping beach. These experiments, which cover a wide range of non-breaking and breaking waves, have been cited extensively as a benchmark testing problem to check the accuracy of numerical solutions. TUNA-RP is used to reproduce this benchmark problem for the run-up of a non-breaking solitary wave. The run-up on a beach with 2.884° and relative wave height of 0.0185, for which experimental results are easily available, is examined. A sequence of three snapshots of the free surface elevation is presented in figure 3, showing the shoaling, run-up and run-down of the solitary wave. The simulated free surface profiles are compared with experimental data from Synolakis [19]. TUNA-RP correctly reproduces both the moving shoreline and the run-up, thereby reinforcing the validity of TUNA-RP in simulating tsunami run-up and inundation.

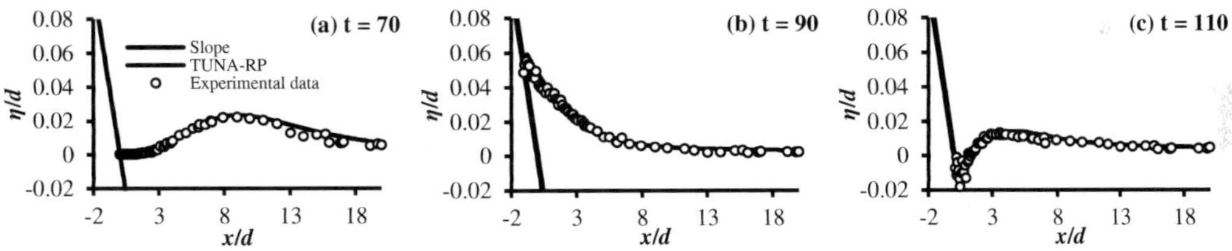

Figure 3. Run-up of solitary wave with $\eta_0/d = 0.0185$ on 1:19.85 slope. The solid line shows the TUNA-RP results whereas circle symbols represent the experimental data from Synolakis [19].

4.4. Solitary wave run-up on a piecewise linear bathymetry ending in a vertical wall

Briggs et al. [20] conducted a laboratory experiment to examine solitary wave propagation and transformation over a compound slope and the subsequent run-up on a vertical wall. This laboratory data allows validation of model capabilities in handling nonlinearity, dispersion, breaking and run-up at the vertical wall. The model beach consists of three piecewise linear slopes of slopes 1:53 (first segment), 1:150 (second segment) and 1:13 (third segment) from seaward to shoreward. The vertical wall is located at the landward end of the 1:13 slope. A water gauge, which is used to measure time dynamics of the water surface height, is installed at the beginning and in the middle of each sloping segment. The locations of the gauges are available at the NOAA website. Three cases denoted by A, B and C with relative wave heights of 0.039, 0.264 and 0.696 respectively are considered in this experimental investigation. The solitary waves will propagate and transform over the three slopes before being reflected from the right wall.

Figure 4 displays the comparison of TUNA-RP simulated and experimentally-measured wave profiles for Case A, which involves a small amplitude non-breaking solitary wave. In both numerical simulation and the experiment, the wave propagates towards the vertical wall and is reflected by the wall without any breaking. The numerical solution shows very good agreement of the amplitude and waveform with the measurements despite slight phase lags in the reflected wave (the second peak in the time series). Validations for Cases B and C are briefly tabulated in Table 1, which shows good agreement between TUNA-RP and other NSWE run-up model simulation results.

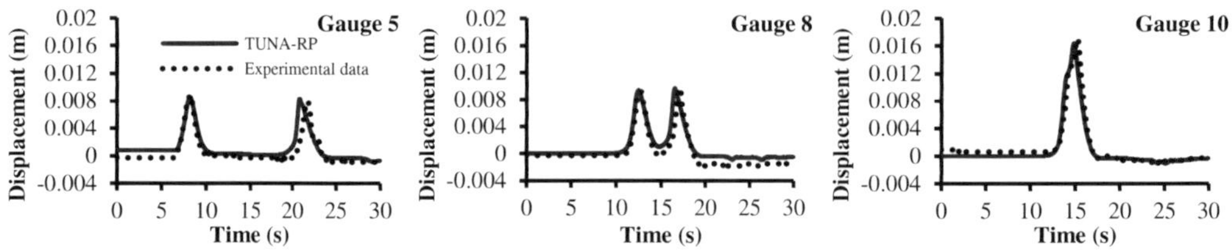

Figure 4. Comparison between the TUNA-RP simulated solutions (dotted line) and laboratory measurements (solid line) for Case A.

Table 1. Validation of TUNA-RP for Cases A, B and C.

Case	Relative wave height	Experimental result		Other NSWE model (Yeh et al. [23])		TUNA-RP result	
	η_0/d	R	R/d	R	R/d	R	R/d
A	0.039	2.74	0.13	2.18	0.10	2.12	0.10
B	0.264	45.72	2.10	8.81	0.40	8.43	0.39
C	0.696	27.43	1.26	12.91	0.59	13.29	0.61

5. Conclusion

TUNA-RP is successfully validated by comparing its simulation results to four well acknowledged international benchmark problems. In the first benchmark problem, TUNA-RP successfully reproduces the semi-analytical solution to correctly demonstrate that the shoreline is drastically drawn-down when a depression wave approaches. Then the wave propagates towards the planar beach and inundates the dry land. Such a drastic wave draw-down accompanying a leading depression wave was also observed in Penang and Langkawi during the 2004 Indian Ocean tsunami. Hence, it is important for the coastal communities to keep in mind that this natural phenomenon of drastic wave draw-down is a sure sign of tsunami arriving within minutes. On the second benchmark test problem to assess the capability of TUNA-RP in simulating solitary wave run-up on a circular island, TUNA-RP correctly shows that the front of the island is exposed to tsunami threats. More importantly, the lee side of the island is exposed to tsunami threats too, due to wave refraction—waves wrapping around the island—and amplification of run-up when waves collide behind an island. This observed phenomenon of wave amplification behind an island, both in real wave basin experiment and in numerical simulations, provides useful insights for authorities in planning evacuation routes and in coastal development around islands. In particular, island residents should always seek high ground whenever a tsunami warning is issued, regardless of the locations along the island, front or lee side. Finally, two additional benchmark validation tests performed on TUNA-RP further confirm its satisfactory performance in simulating run-up and inundation.

6. References

[1] Horrillo J, Grilli S T, Nicolsky D, Roeber V and Zhang J 2015 *Pure Appl. Geophys.* **172** 869–884

[2] Mellor G L 2004 *Users Guide for a Three-dimensional, Primitive Equation, Numerical Ocean Model* (NJ: Princeton University) p 56

[3] Chen C, Liu H and Beardsley R C 2003 *J. Ocean. Atmos. Tech.* **20** 159–186

[4] Chen G -Y, Liu C -C and Yao C -C 2014 *IEEE J. Oceanic Eng.* **40** 37–47

[5] Galland, J -C, Goutal N and Hervouet J M 1991 *Adv. Water Resour.* **14** 138–148.
[6] Hervouet J M 2000 *Hydrol. Process.* **14** 2209–2210
[7] Grilli S T, Grilli A R, David E and Coulet C 2016 *Nat. Hazards* **82** 777–810
[8] Deltares 2012 *User Manual Delft3D-FLOW* (Delft: Deltares) p 676 pp
[9] Apotsos A, Buckley M, Gelfenbaum G, Jaffe B and Vatvani D 2011 *Pure Appl. Geophys.* **168** 2097–2119
[10] Lynett P J, Wu T -R and Liu P L -F 2002 *Coast. Eng.* **46** 89–107
[11] Donnell B P, Letter Jr J V , McAnally Jr W H and Thomas W A 2011 *Users Guide for RMA2 Version 4.5* (Vicksburg: US Army Corps of Engineers)
[12] Nielsen C and Apelt C 2003 *J. Hydraul. Eng.* **129** 628–636
[13] Medeiros S C and Hagen S C 2013 *int. J. Numer. Meth. Fluids* **71** 473–487
[14] Liu P L -F, Yeh H and Synolakis C E 2008 *Advanced Numerical Models for Simulating Tsunami Waves and Runup* (Singapore: World Scientific Publishing)
[15] Synolakis C E, Bernard E N, Titov V V, Kânoğlu U and González F I 2008 *Pure Appl. Geophys.* **165** 2197–2228
[16] Synolakis C E, Bernard E N, Titov V V, Kanoglu U and Gonzalez F I 2007 *NOAA Technical Memorandum OAR PMEL–135 Standards, Criteria and Procedures for NOAA Evaluation of Tsunami Numerical Models* (Seattle: Pacific Marine Environmental Laboratory) p 55
[17] Carrier G F, Wu T T and Yeh H 2003 *J. Fluid Mech.* **475** 79–99
[18] Briggs M J, Synolakis C E, Harkins G S and Green D R 1995 *Pure Appl. Geophys.* **144** 569–593
[19] Synolakis C E 1987 *J. Fluid Mech.* **185** 523–545
[20] Briggs M J, Synolakis C E, Kanoglu U and Green D R 1996 *Benchmark Problem 3 Run-up of Solitary Waves on a Vertical Wall,* Long-wave Runup Models, ed H Yeh, P L -F Liu and C E Synolakis (Singapore: World Scientific Publishing) pp 375–383
[21] Yeh H, Liu P L -F, Briggs M J and Synolakis, C E 1994 *Nature* **372** 353–355
[22] Takahashi T, Takahashi T, Shuto N, Imamura F and Ortiz M 1995 *Pure Appl. Geophys.* **144** 747–767
[23] Yeh H, Liu P L -F and Synolakis C E 1996 *Long-Wave Runup Models* (Singapore: World Scientific Publishing)

7. Acknowledgments

Financial support provided by Grant #305/PMATHS/613418 is gratefully acknowledged. KHL and TWK gratefully acknowledge the financial support provided by Sunway University to present this paper in the ICEII 2017 conference.

IOP Conf. Series: Earth and Environmental Science **67** (2017) 012031 doi:10.1088/1755-1315/67/1/012031

Inactivation disinfection property of *Moringa Oleifera* seed extract: optimization and kinetic studies

M A Idris[1], M S Jami[2] and A M Hammed[1]

[1]International Institute for Halal Research and Training (INHART), International Islamic University Malaysia, 50728, Kuala Lumpur, Malaysia
[2]Biotechnology Engineering Department, International Islamic University Malaysia, 50728, Kuala Lumpur, Malaysia

Abstract. This paper presents the statistical optimization study of disinfection inactivation parameters of defatted *Moringa oleifera* seed extract on *Pseudomonas aeruginosa* bacterial cells. Three level factorial design was used to estimate the optimum range and the kinetics of the inactivation process was also carried. The inactivation process involved comparing different disinfection models of Chicks-Watson, Collins-Selleck and Homs models. The results from analysis of variance (ANOVA) of the statistical optimization process revealed that only contact time was significant. The optimum disinfection range of the seed extract was 125 mg/L, 30 minutes and 120rpm agitation. At the optimum dose, the inactivation kinetics followed the Collin-Selleck model with coefficient of determination (R^2) of 0.6320. This study is the first of its kind in determining the inactivation kinetics of *pseudomonas aeruginosa* using the defatted seed extract.

1. Introduction

Moringa oleifera seed is a biomaterial commonly utilized for drinking water treatment in various part of the developing countries (specifically in Africa). The seed extract has been harnessed for water treatment and purification, documented in various researches [1]. Comparing *Moringa oleifera* with conventional chemical coagulants reveals it advantages such as cost effectiveness, ready availability, production of bio-degradable sludge, lower sludge volume, and production of less hazardous, non-corrosive by-product. As a result of these advantages, the seed extract is eco-friendly and cheap which proves to be a good alternative to chemical coagulants [2], [3]. The seed extract contains bioactive agents with excellent coagulation properties [4] and some previous researches reveal that the coagulation efficiency of *Moringa oleifera* seeds can be greatly enhanced by extracting active agents with salt solution having one valence electron such as NaCl, KCl etc [5]. The salt solution increased the coagulation capacity performance to about 7.4 times better than that when using distilled water. This improved performance according to the research was envisaged to be a protein-protein dissociation caused by the salt solution. Few other researchers report that a polypeptide "flo" present in the seed function as coagulant and also exert antibacterial effect on harmful bacterial strains. 4 αL-rhamnosyloxy-benzyl isothiocyanate has been identified as the bioactive antimicrobial agent which is common to both defatted and normal seed extracts. This antimicrobial agent is about 8-10% of the seed chemical composition [6], [7]. Most studies have focused on the antibacterial effect of the seed extract on different microbes such as *Escherichia coli, salmonella typhi*, however, its application as a disinfectant for water treatment is an emerging area that is worth exploring. Although few researches on the disinfectant property of the seed extract have be done with focus on *E.coli* as the target microbe in water. However, Both *Pseudomonas aeruginosa* and *E.coli* are both gram negative microbes, predominantly present in the environment whose natural habitat is water surface. *E.coli* is identified as a microbial indicator for fecal contamination while *Pseudomonas aeruginosa* is as an opportunistic pathogen whose transmission is through water [8], [9], [10], [11]. This means removal of *pseudomonas aeruginosa* in water is also crucial to maintain the aesthetic state of drinking water making it fit for consumption. Few studies on using *moringa oleifera* seed extract as a disinfectant in water treatment was conducted lately in a research carried out [12] the disinfection process were 31 minutes, 85rpm mixing speed and 3.25 mg/L dosage was concluded as the optimum disinfection conditions. In another similar study, the optimized conditions of the seed extract was utilized against E.coli microbes where the inactivation kinetics of the seed extract was explored. In the research, it was concluded that the seed extract followed a first order reaction and was well fitted with the Chicks-Watson disinfection kinetics model [13]. The optimization of the disinfection process conditions of *Moringa oleifera* seed extract as disinfectant is an emerging area with few researches conducted. Hence, it is essential to study the process condition of the seed extract on other bacterial strains such as *P.aeruginosa* for comparative purposes. It is also crucial to understand the inactivation kinetics of the seed extract during disinfection process which provides an insight into how the microorganisms behave during disinfection process over a period of time. This requires using various disinfection models such as the Chick-Watson, Collin-Selleck and Hom disinfection models to understand the inactivation mechanism of the seed extract. The Chicks-Watson model is limited in various disinfection process because the rate of kill is not constant. Rather, it decreases with time that depends on the form of disinfectant, type of organism to be inactivated and other operating conditions [14]. The Collins-Selleck model developed addressed deceleration rate (like a convex curve) of disinfection [8] which describes the declining rate of inactivation by observing lag in real system otherwise known as tailing. The tailing phenomenon occurs as a result of an ineffective inactivation of the subpopulation of the microbial population due to the disinfectant exposure [15]. The objectives of this study are to determine the optimum inactivation disinfection process conditions of defatted seed extract using three level factorial design under the response surface methodology, to determine the order of reaction and the inactivation kinetics of the seed extract by exploring the different disinfection inactivation models on *Pseudomonas aeruginosa* bacterial strain.

2. Materials and methodology

The following materials and methods were used in this study.

2.1. Sample preparation

Dry *Moringa oleifera* seed were de-husked and two grams of defatted *Moringa oleifera* seed powder were weighed into a liter of distilled water. Mixing rate was maintained at 6000rpm in a centrifuge for 10 minutes. The resulting solution was filtered and stock solution of about 1000mg/L was made [16].

2.2. Water sample

A thousand bacterial cells of *Pseudomonas aeruginosa* were spiked inside distilled water to make the synthetic water which was used throughout this research.

2.3. Inoculum preparation

P.aeruginosa cells obtained from laboratory stock solution at the Department of Biotechnology Engineering (IIUM) were inoculated into 10mL of LB broth with agitation overnight at 37oC. Bacterial cells inoculum density were determined with haemocytometer and about1000cells/mL inoculum density was measured.

2.4. Total viable count (TVC)

The measure of microorganism activity in water is determined by changes in the bacterial population by a process called TVC. The procedure involves taking 1 mL of the sample into a petri dish and molten LB agar was poured. The agar solidifies and the dish incubated at 37 0C for 24 hours. The concentration of the bacteria present are determined by the presence of colonies.

IOP Conf. Series: Earth and Environmental Science **67** (2017) 012031 doi:10.1088/1755-1315/67/1/012031

2.5. Three level factorial Design

The process conditions were determined using the three level factorial design. The optimum concentration of the factors were determined and the experimental region was illustrated for the highly significant factors. The three important factors during disinfection process are dosage, agitation and time and these factors were studied at low, medium and high levels. The factorial design using design expert software for three factors is 32 experimental runs with 6 centre points as shown in Table 1.

Table 1. Experimental design using three level factorial design in response surface methodology and the corresponding output of bacterial count

Run	Dosage (mg/L)	Time (minutes)	Agitation (rpm)	TVC (CFU/mL)
1	100	30	100	29
2	125	30	100	15
3	150	30	100	30
4	100	60	100	28
5	125	60	100	12
6	150	60	100	28
7	100	90	100	31
8	125	90	100	14
9	150	90	100	34
10	100	30	120	27
11	125	30	120	14
12	150	30	120	30
13	100	60	120	30
14	125	60	120	13
15	150	60	120	28
16	100	90	120	32
17	125	90	120	15
18	150	90	120	32
19	100	30	140	29
20	125	30	140	12
21	150	30	140	31
22	100	60	140	25
23	125	60	140	13
24	150	60	140	29
25	100	90	140	34
26	125	90	140	15
27	150	90	140	33
28	125	60	120	12
29	125	60	120	15
30	125	60	120	12
31	125	60	120	12
32	125	60	120	14

The second order polynomial model was used to analyse the observation.

$$Y=\beta_0+\beta_1X_1+\beta_2X_2+\beta_3X_3+\beta_{12}X_1X_2+\beta_{13}X_1X_3+\beta_{23}X_2X_3+\beta_{11}X_1^2+\beta_{22}X_2^2+\beta_{33}X^2_3 \quad (1)$$

where Y is the dependent factor (Bacterial Count); X_1, X_2, X_3 are independent factors (dosage, contact time and agitation); β_0 is the intercept term; β_1, β_2, β_3 are linear coefficients; β_{12}, β_{13}, β_{23} are the interaction coefficients, β_{11}, β_{22}, β_{33} are the quadratic coefficients. The model was evaluated by analysing regression coefficient values, ANOVA (analysis of variance), p-values and F-values that reveals the goodness of fit of the model and coefficient of determination, R^2. The Design Expert 8.0.7.1 trial version (Stat Ease Inc., Minneapolis, USA) a statistical software was used to determine the regression model in predicting parameters that have significant effects on the disinfectant ability of seed extracts.

2.6. Validation of experimental model

The point prediction feature of the statistical software Design Expert 8.0.7.1 trial version predicted the combinations which were used to validate the model. Experimental combinations of four different runs were executed and both experimental and predicted results were compared.

2.7. Inactivation kinetic models

The inactivation models were determined with residual colonies estimated at a given time intervals, optimum dosage and optimum agitation. The data collected from the experimental procedure was used to fit the different disinfection inactivation models such as the Chicks, Watson, Collin-Selleck, Homs models using the following equations.

$$\ln\left(\frac{N}{N_0}\right) = -\Lambda C^n t \tag{2}$$

$$\ln\left(\frac{N}{N_0}\right) = -\Lambda_{CS}\left[\ln(Ct) - \ln(b)\right] \tag{3}$$

$$\ln\left(\frac{N}{N_0}\right) = -\kappa C^n t^m \tag{4}$$

Where equation 2 is the chicks-watson model, equation 3 is the Collin-sellecks model and equation 4 is Homs model. Where N is total of organisms present at time t, N_0 is the total of organism present at time 0, t is time, k is the rate constant, Λ is coefficient of specific lethality signifying the relative potencies of disinfectants, n is Hom dilution coefficient unit less, and m is Hom time exponent unit less. Multiple regression analysis plotting of Hom's model gives values for k, n and m. The model with the highest coefficient of determination R^2 result was selected as the best fitted. Also, multivariate regression statistical analysis with respect to the seed extract was examined.The t-test was used to assess the significance level of the regression coefficient [17].

3. Results and discussion

3.1. Disinfection parameters

The disinfection parameters range used in this optimization research was based on finding from earlier research [12]. In their research, the use of one factor at a time was used to establish the optimum range of the parameters where the disinfectant property of the seed extract was harnessed. The range of *Moringa oleifera* dosage according to the study was between 100mg/L to 200mg/L. The research also concluded that the seed extracts inhibitory action is due to the presence protein in the seed. These proteins are believed to be lipophilic and cationic in nature, acting like a molecular knife that can disrupt the microbial cell wall electrostatically and binds inside the cytoplasmic membrane of the bacterial cell [19], [20], [21] and as a result, for this study the range was fixed at 100-200mg/L. The contact time is also an important factor in the disinfection process. The effect of contact time was also investigated in several articles and most researchers concluded that the seed extract is mostly bactericidal within 30 minutes to 90 minutes. Although there is quite some discrepancies about efficient timing of the extract. In a study by [12], their reports revealed that the longer the contact time, the greater the rate of kill and in some other study, and the rate of kill of the seed extract against microbes was within 60 minutes. In this optimization study, 30 minutes to 90 minutes was selected and for the mixing rate100 rpm to 140 rpm were selected based on reports from earlier studies showing mixing rate increases the rate of kill in higher magnitudes [14] and from previous research, significant reductions in the bacterial cells were achieved were within 100-140 rpm agitation rate [18].

3.2. *Optimization of process conditions using three level factorial design*

3.2.1. *Statistical analysis of results*

A second order regression equation was developed that indicates the dependence of contact time, dosage and mixing rate on the disinfectant property of the seed extract. The multiple regression analysis of the second order model equation reveals the empirical relationship between the process parameters and response, shown in the equation below and the analysis of variance results are reported in Table 2.

$$Y=12.56+0.56A+1.28B+0.000C+16.09A^2+2.26B^2-0.24C^2-0.33AB+0.083AC+0.42BC \quad (2)$$

where the total viable count for *P.aeruginosa* is the response (Y) and A, B and C are dosage, contact time and agitation, respectively.

Table 2. Analysis of variance (ANOVA)

Source	Sum of Squares	DF	Mean Square	F Value	Prob > F	
Model	2237.77	9	248.64	118.32	< 0.0001	
A	5.56	1	5.56	2.64	0.1182	
B	29.39	1	29.39	13.98	0.0011	
C	0	1	0	0	1	
A^2	1852.37	1	1852.37	881.45	< 0.0001	
B^2	36.48	1	36.48	17.36	0.0004	
C^2	0.42	1	0.42	0.2	0.6597	
AB	1.33	1	1.33	0.63	0.4342	
AC	0.083	1	0.083	0.04	0.844	
BC	2.08	1	2.08	0.99	0.3302	
Residual	46.23	22	2.1			
Lack of Fit	38.23	17	2.25	1.41	0.3768	non-significant
Pure Error	8	5	1.6			
Cor Total	2284	31				

Analysis of variance (ANOVA) was utilized to check for the adequacy of the model and the results are reported in Table 2. The F value of 118.32 and p-value of <0.0001 suggest that the model is significant. The not significant lack of fit suggests that the calculated experimental responses satisfactorily fit with the model. The higher values of R^2 (0.9798) and adjusted R^2 (0.9715) also revealed the efficacy of the model. [22]. Based on this, R squared value for the model was significant and much fitted. Adequate precision value of 25.11 shows that the model is able navigate the design space. Coefficient of variation (CV) of about 6.37 indicates a good degree of precision and a low CV shows that the experiment is reliable. The regression equation shows the coefficient values listed in Table 2. The significance of the coefficient is indicated by the p-values revealing the interaction between independent variables. The p-values reveals the significant corresponding coefficient and the responses shows that only the linear coefficients C (time), and one quadratic coefficient were significant ($p<0.05$).

3.2.2. *Three dimensional (3D) surface plot*

The three dimension plots shown in Figure 1 were based on the function of two variables with the other variable being at its optimum level. The saddle or elliptical nature of surface contour plots shows the significant interactions among variables [23].

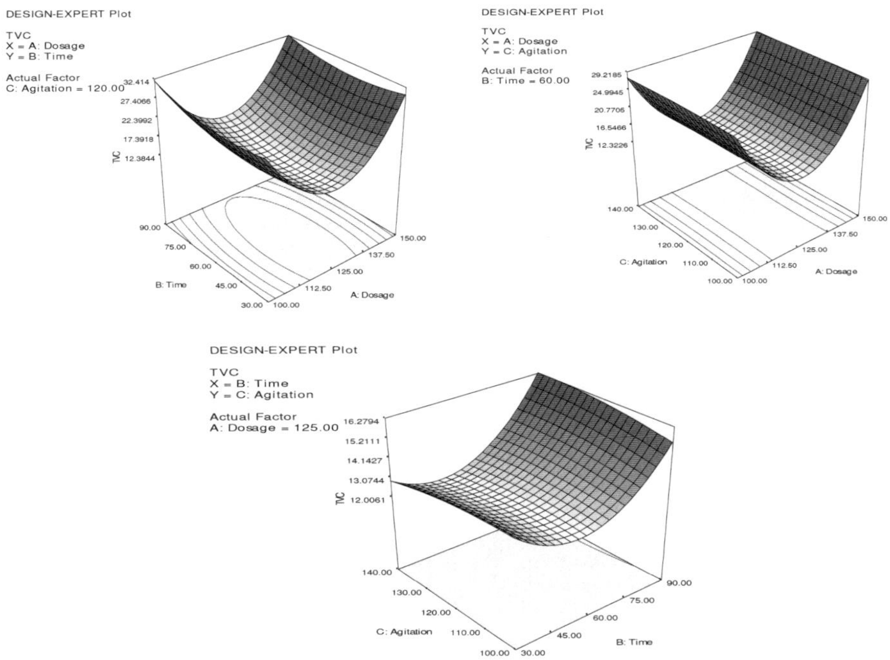

Figure 1. 3D graphical representation of surface plot

The interaction between dosage and contact time is illustrated in Figure 1a. The response surface curves showed an elliptical contour plots shape between the contact time and dosage revealing a significant interaction between these two variables. The predictive optimum range is between 120 mg/L to 125 mg/L for the dosage at a low contact time. In Figure 1b, the elliptical contour plot shows a well define optimum operating conditions between agitations in the range of 116rpm to 120rpm indicating that the effects of interaction between the two factors was significant. In case of dosage and mixing rate (Figure 1c), the response plot was elliptical depicting interaction between them. A good interaction was observed between these two variables and the predictive optimum range is between 30 minutes to 60 minutes for the contact time, 116 rpm to 120 rpm for the mixing rate.

3.3. Validation of experimental model

The second-order regression model was validated with some sets of experiments replicated thrice and were performed according to the point prediction of Design Expert presented in Table 3. The result shows that the optimum disinfection process occurred at 125mg/L, contact time of 30 minutes and 120 rpm mixing rate.

Table 3. Validation of the optimum values

Run order	Dosage (mg/l)	Time (minutes)	Mixing rate (rpm)	TVC (experimental)	TVC (predicted)	Error
1	100	60	120	27	20.07	0.260
2	140	62	100	18	19.01	0.056
3	130	33	120	18	17	0.056
4	125	30	120	17	18.07	0.063

3.4. Disinfection kinetic models

The plot of log survival of the *P.aeruginosa* bacterial colonies against time showed a deceleration of the process called "tail". The tailing as shown in Figure 2 may be explained by a vitalistic hypothesis in which individual bacteria in a population are not identical, and their inherent resistance is distributed in a permanent (time-independent) manner [24]. Also, the presence of tailing revealed in Figure 2 indicates ineffective inactivation of the microbial population when exposed to the seed extract disinfection process [15]. Another possible explanation might be as a result of decrease in the germicidal properties of defatted *Moringa oleifera* extract with time that is observed from the plot. The bacteria population were gradually building up over a period of time which might be as a result of natural heterogeneity in resistance among the microorganisms [8], [15].

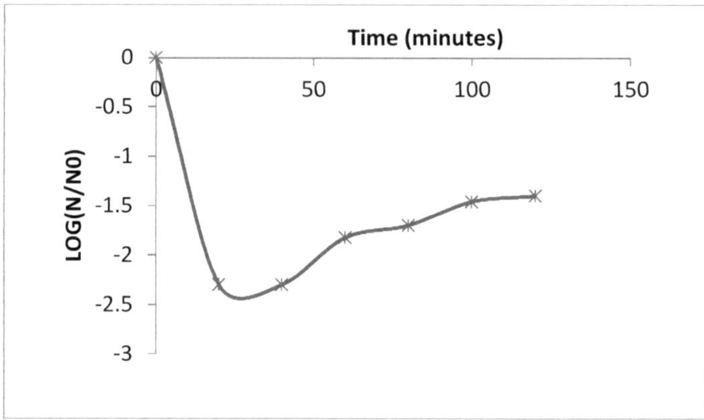

Figure 2. Plot of log survival of *P.aeruginosa* bacterial cells against time

It is also hypothesized that the tailing behaviour indicate that bacterial populations generally referred to as homogeneous are heterogeneous in nature as they comprise of mixtures of cells in varying states such as physiology etc. Hence, from the plot, it is clear that the inactivation kinetics did not follow a first order reaction which shows the deviation from the Chick-Watson model. Other disinfection models such as Hom model and Collins-Selleck empirical model are used in this research to account for deviations from the first order kinetics of the Chicks-Watson model of inactivation kinetics. The three disinfection models (Chick-Watson, Collins, Hom) were determined using linear multivariate regression analysis and results compared for *Pseudomonas aeruginosa* bacterial strain are summarized in Table 4. The coefficient of determination determined showed that the Collins-Selleck model best describe the data with R2 of 0.632 for *Pseudomonas aeruginosa*.

Table 4. Coefficient determination for different disinfection kinetic models for *P.aeruginosa*

Disinfection kinetic model	R^2	Adjusted R^2
Chick-Watson	0.035	-0.158
Collin-Selleck	0.632	0.558
Hom	0.160	-0.008

The results from Table 4 shows the Collin-Selleck model having the highest coefficient of determination of 0.632 and the T-test for the estimated Λ_{CS} shows significance value of 0.033 and standard error of 0.142. This indicates that both the dosage and the contact time are very significant for the inactivation kinetics of *Moringa oleifera* seed extract. The findings in this research is contrary to earlier research conducted by [13] whose finding revealed that the inactivation of the seed extract followed the Chick-Watson model indicating a first order reaction. The possible explanation for the differences in the results obtained even though *Moringa oleifera* seed extract was used as the disinfectant is probably due to different dosage used. In the research conducted by Bichi et al., (2012 a), the seed defatted extract was prepared using distilled water during the extraction at the dosage of 3.25 mg/L while in this study, the defatted seed extract dosage was 125mg/L. Also the bacterial indicator used in [21] was *E.coli* while the bacterial indicator used in this research was *P.aeruginosa*. Hence, it can be concluded that *Moringa oleifera* seed extract inactivates bacteria strains differently and the based on different dosage of the seed used in the preparation of the extract also significantly affects the way bacteria strains are being inhibited [20].

4. Conclusion

A second order polynomial regression model was developed for the optimization process of the seed extract using the response surface methodology. The model generated reveals that dosage, contact time and agitation were highly significant when using the seed extract to treat water. The predicted values obtained from the generated model were compared with the results from the experimental runs. The optimum parameter values for the seed extract as disinfectant for water treatment are dosage of 125mg/L at contact time of 30 minutes and mixing rate of 120 rpm. The inactivation kinetics of the seed extract shows that the Collin-Selleck model fits the experimental data with 0.632 thus, the inactivation kinetics of the seed extract is not first order.

5. Acknowledgement

The authors are grateful to the Ministry of Higher Education (MOHE) Malaysia, for financing the research project under the research initiative grant (RIGS16-075-0239) and to Department of Biotechnology Engineering, International Islamic University Malaysia for their support.

6. References

[1] Muyibi SA, Birima AHM, Mohammed TA, Noor MJMM (2004) Conventional Treatment of Surface Water Using Moringa Oleifera Seeds Extract as a Primary Coagulant. IIUM Engineering Journal **5**:25–35.

[2] Luz L de A, Silva MCC, Ferreira R da S, et al (2013) Structural characterization of coagulant Moringa oleifera Lectin and its effect on hemostatic parameters. International journal of biological macromolecules **58**:31–6.

[3] Sanchez-Martin J, Beltran-Heredia J, Pere JA (2012) Improvement of the flocculation process in water treatment by using Moringa oleifera seeds extract. Brazilian Journal of Chemical Engineering **29**:495–501.

[4] Madrona GS, Serpelloni GB, Salcedo Vieira AM, et al (2010) Study of the Effect of Saline Solution on the Extraction of the Moringa oleifera Seed's Active Component for Water Treatment. Water, Air, & Soil Pollution 211:409–415.

[5] Okuda T, Baes AU, Nishijima W, Okada M (2001) Isolation and characterization of coagulant extracted from Moringa oleifera seed by salt solution. Water Research **35**:405–410.

[6] Oluduroa OA, Aderiye BI, Connolly JD, et al (2010) Characterization and Antimicrobial Activity benzyl thiocarboxamide ; a Novel Bioactive Compound from Moringa oleifera Seed Extract. Folia Microbiology **55**:422–426.

[7] Anwar F, Rashid U (2007) Physico-chemical characteristics of moringa oleifera seeds and seed oil from a wild provenance of Pakistan. Pakistan Journal of Botany **39**:1443–1453.

[8] MWH (2005) Water treatment: Principles and Design, Second Edi. John Wiley & Sons, United States of America

[9] Brahmi M, Hassen A (2011) Disinfection of Wastewater by UV Irradiation : Influence of Hydrodynamics on the Performance of the Disinfection. Environmental Engineering Research **16**:243–252.

[10] Cabral JPS (2010) Water Microbiology: Bacterial Pathogens and Water. International Journal of Environmental Research and Public Health 3657–3703.

[11] Okuda T, Baes AU, Nishijima W, Okada M (1999) Research note improvement of extraction method of coagulation active components from moringa oleifera seed. Water Research **33**:3373–3378.

[12] Bichi MH, Agunwamba JC, Muyibi SA (2012) Optimization of operating conditions for the

application of Moringa oleifera (Zogale) seeds extract in water disinfection using response surface methodology. African Journal of Biotechnology **11**:15875–15887.

[13] Bichi MH, Agunwamba CJ, Muyibi SA (2012) Kinetics of Water Disinfection with Moringa Oleifera Seeds Extract. Journal of Environmental and Earth science **2:**58–69.

[14] Metcalf, Eddy (2004) Wastewater Engineering: Treatment and Reuse, 4th ed. Mc Graw Hill

[15] Pennell KG, Aronson a I, Blatchley ER (2008) Phenotypic persistence and external shielding ultraviolet radiation inactivation kinetic model. Journal of applied microbiology 104:1192–202.

[16] Ali EN, Muyibi S a., Alam MZ, Salleh HM (2012) Optimization of water treatment parameters using processed Moringa oleifera as a natural coagulant for low turbidity water. In: 2012 International Conference on Statistics in Science, Business and Engineering (ICSSBE). IEEE, pp 1–6

[17] Azzellino A, Antonelli M, Canziani R, et al (2011) Multivariate modelling of disinfection kinetics: A comparison among three different disinfectants. Desalination and Water Treatment **29:**128–139.

[18] Muyibi SA, Idris MA, Jami S, et al (2013) Statistical Optimization of Process Conditions for Disinfection of Water Using Defatted Moringa oleifera Seed Extract. International Journal of Environmental Engineering **7**:717–722.

[19] Shebek K, Schantz AB, Sines I, et al (2015) The flocculating cationic polypetide from moringa oleifera seeds damages bacterial cell membranes by causing membrane fusion. Langmuir **31**:4496–4502.

[20] Idris MA, Muyibi SA, Ismail MA, et al (2013) Investigating Antibacterial Activity of Defatted Moringa oleifera Seed Extract on Selected Gram-negative and Gram-positive Microbes. International Journal of Academic Research **5**:101–106.

[21] Viera GHFG, Mourão JAJ, Ângelo ÂM, et al (2010) Antibacterial effect (in vitro) of Moringa oleifera and Annona muricata against Gram positive and Gram negative bacteria. Revista do Instituto de Medicina Tropical de São Paulo **52**:129–132.

[22] Salihu A, Alam M, Ismail MA, Salleh HM (2011) Optimization of lipase production by Candida cylindracea in palm oil mill effluent based medium using statistical experimental design. Journal of Molecular Catalysis B: Enzymatic **69**:66–73.

[23] Alam MM, Alam MZ, Jami MS, et al (2016) Study of the Effects of Process Parameters on Electroforced Sedimentation in Solid--Liquid Separation Using Response Surface Methodology. Waste and Biomass Valorization **7**:583–591.

[24] Lee Y, Nam S (2002) Reflection on Kinetic Models to the Chlorine Disinfection for Drinking Water Production. The journal of Microbiology **40**:119–124.

IOP Conf. Series: Earth and Environmental Science **67** (2017) 012032 doi:10.1088/1755-1315/67/1/012032

Advanced wastewater treatment using microalgae: effect of temperature on removal of nutrients and organic carbon

Mohamad Shurair[(1)], Fares Almomani[(1)], Simon Judd[(2)], Rahul Bhosale[(1)], Anand Kumar[(1)], Ujjal Gosh[(1)], Majeda Khreisheh[(1)]

1) Department of Chemical Engineering, Qatar University, P.O box 2713, Doha, Qatar
2) Gass Processing Center, Qatar University, P.O box 2713, Doha, Qatar

Abstract. This study evaluated the use of mixed indigenous microalgae (MIMA) as a treatment process for wastewaters and CO2 capturing technology at different temperatures. The study follows the growth rate of MIMA, CO2 Capturing from flue gas, removals of organic matter and nutrients from three types of wastewater (primary effluent, secondary effluent and septic effluent). A noticeable difference between the growth patterns of MIMA was observed at different CO2 and different operational temperatures. MIMA showed the highest growth grate when injected with CO2 dosage of 10% compared to the growth for the systems injected with 5% and 15 % of CO2. Ammonia and phosphorus removals for Spirulina were 69%, 75%, and 83%, and 20%, 45% and 75 % for the media injected with 0, 5 and 10% CO2. The results of this study show that simple and cost-effective microalgae-based wastewater treatment systems can be successfully employed at different temperatures as a successful CO2 capturing technology even with the small probability of inhibition at high temperatures.

1. Introduction

In the near future, improving the overall efficiency of the Carbon Capture and Storage (CCS) chain will become increasingly important, as the implementation of the various CO2 existing capture methods from flue gas are not currently economically feasible. On the other hand, it is to be expected that CO2 will become available from a multitude of different sources. For the GCC (Gulf Cooperation Council) this will include power plants, refineries, various petrochemical processes (methanol, ammonia), and will encompass gas separation at processing plants and at LNG production facilities. CO2 has to be removed from flue gas (CO2-N2) and from natural gas (CO2-CH4) mixtures. The increase in CO2 concentration which considered as one of the greenhouse has been associated to several negative environmental impacts, such as climate change and change in ocean water chemistry. Therefore, it becomes urgent for world economies to reduce their carbon emissions [1-3]. Additionally, since this greenhouse gas presents long residence times [4,5], new strategies to reduce CO2 concentration in the atmosphere and in flue gas emissions are required [6].

The current average costs of CO2 capture and storage, based on chemical absorption, are estimated to be at least 55 US$ per ton of CO2 taking a coal-fired power plant as a reference [7]. In the absorption process, the solvent regeneration step accounts for approximately 75% of the energy consumption (and thus costs) [8,9]. The high energy penalty, which for power plants can be as high as 10% (based on the overall power plant efficiency of 35-45%), is one of the main drivers for research into more economically viable, and ultimately sustainable, approaches to be included in the full CCS chain.

At the same time, with the increase on world population the high volumes of wastewater generated by human was always a permanent problem that requires appropriate treatment. While domestic wastewater treatment plants (WWTPs) is considered an effective treatment for removal of biochemical oxygen demand (BOD), suspended solids and bacteria, the removal of inorganic nutrients, still limited and challenging in these processes. With the implementation of the new wastewater discharge regulation (e.g. Canada Gazet, 2012), the removal of nutrients, mainly dissolved nitrogen and phosphorus, is becoming one of the essential requirement for the approval of the WWTP performance. The new regulations were based on the fact that the presence of these nutrients in wastewater will leads to eutrophication by stimulating the growth of unwanted plants such as algae and aquatic macrophytes [10-12], and increase the effluents toxicity to fishes and aquatic organisms [13]. Moreover, untreated nutrients in wastewater effluent reduce the performance of disinfection step and increase the chlorine demand [14]. As a result, there is a big need to find a treatment process that can remove these nutrients before the effluents are discharged [15, 16].

Different studies have used algae for treatment of different types of wastewater. Most recently, Zhu et al. [17] reported a removal efficiency of 65-76% of COD, 68-81% of TN and 90-100% of TP from piggery wastewater by fresh water microalgae Chlorella zofingiensis. Wang et al. [13] reported COD removal of 50.9%and 56.5% from primary and secondary effluents by green algae Chlorella sp. The study showed significant nutrient removals efficiencies. A total nitrogen and Phosphorus removal efficiencies of 68.4%, and 50.8%, and 82.8%and 85.6 % were reported for the same previous wastewater samples, respectively.

So far, limited attention has been paid to the dual role of microalgae as advanced technology for treatment of different kinds of wastewater and as CO2 bio-fixation technology. Moreover, there are knowledge gaps in understanding the performance of different microalgae strains in the treatment of wastewater and their capacity for CO2 capturing. In particular, according to our literature review, no studies were carried out using mixed indigenous microalgae cultures as opposed to lab-grown pure culture strains. The objective of this study was to investigate and compare the treatment efficiency and CO2 bio-fixation capacity of lab-grown mixed indigenous cultures of microalgae in different wastewaters. Primary effluent, secondary influent and septic tank effluent was used in the study to determine possible applications and locations for the proposed treatment systems. The study was carried out in 1.50 L-Erlenmeyer flasks and CO2 was provided to microalgae mixed with air though air diffuser.

2. Materials and methods

2.1. Culture system

Algae growth experiments were carried out in 1 L- sterilized Erlenmeyer flasks at 36°C. The growth experiments were performed using an incubator equipped with thermostat and two cool white lamps of 1500 lux each. Temperature control was maintained at 8, 20 and 36 ° C by a fan connected to thermostatic control. Two fluorescent —daylight lamps were used as light sources. The photon flux intensity inside the incubator was measured by NIST Traccable Radiometer (International light. Model IL 400A) and found to be 1.5 μ W/cm2.

A volume of 850 mL of wastewater (primary, secondary or septic) was mixed with algal strain and incubated at constant temperature and continuous lighting for a period of 28 days. No carbon dioxide (CO2) was supplied to stored cultures other than diffusion of CO2 present in the ambient atmosphere. The flasks were stirred manually 4 times a day.

2.2. Characteristics of wastewaters

Wastewaters were collected from three different sources (1) effluent of primary treatment, (2) effluent of secondary treatment, and (3) septic tank effluent. All wastewaters were filtered using glass microfiber filters (934-AH, Whatman, USA) to remove large particles and indigenous bacteria. Table 1 presents the chemical characteristics of the three wastewaters used in this study.

Table 1: Properties of the Wastewaters

Parameter	**Primary effluent**	**Secondary effluent**	**Septic effluent**
COD_s (mg/L)	245	65	165
$NH_4^{+(1)}$(mg-N/L)	44	40	45
NO_2^- (mg-N/L)	0.006	1.1	5
NO_3^- (mg-N/L)	0.33	23	12
Total Phosphorus (mg/L)	15	32	10

3. Results and discussion

3.1 Algae growth at different temperatures

The growths of mixed indigenous microalgae (MIMA)reported as function of optical density (OD) at 690 nm and at different temperatures (20, 25 and 36 ◦C) and different CO2 dosage in primary effluent, secondary effluent and Septic effluent were plotted as function of time in Figure 1. The growth curve of cultivated algae follows atypical growth curves consisting of lag phase, exponential growth phase and stationary phase. The length of the lag phases accounts for the algae capacity to adapt to the new wastewater environment. Figure 1 shows that the lag phases of MIMA depend on the type of wastewater and the cultivation temperatures.

The growth trends of MIMA in primary wastewater effluent has rapid growth. It can be seen in the figure 1a that the lag time for MIMA range from 1 to 3 days, and the best growth trend was obtained at a temperature of 25◦C. The maximum optical density reported for MIMA at 20, 25 and 36 ◦C were 0.74, 1.05 and 0.88 A.U. The measured OD was converted into dry cell weight using pre-determined conversion factors. The final biomass concentrations for experiments carried out with MIMA in primary wastewater effluent after 24 days of cultivation were found to be 1.91, 2.71 and 2.27 g/L at 20, 25 and 36 ◦C, respectively.

The growth rates of MIMA were almost comparable for secondary and primary wastewaters and slightly higher in septic tank effluent. Moreover, it was observed that the mixed algal culture reached a stationary phase within 17 days of cultivation time in Septic tank effluent while it requires 22 days in secondary and primary wastewaters. The average specific growth rates in the first 3 days of growth phase (days 3, 4 and 5) were found to be 0.22, 0.20, and 0.40 day-1 for primary effluent, secondary effluent and Septic tank effluent , respectively.

3.2 CO2 capturing

The effect of increasing CO2 concentrations on its fixation rate showed that the lowest CO2 fixation rates were achieved when MIMA was fed with CO2 at atmospheric concentrations. Higher CO2 dosage increased CO2 but is was followed by a decrease in MIMA growth. The obtained results suggest that MIMA can be effective technology for CO2 capturing. In addition, experimental results showed that the evaluation of Ph can be used as an indicator of the CO2 capturing efficiency. Picardo et al. [18] showed that for the pH values in the range of 6.0 to 9.0, bicarbonate (HCO3−) is the most common form of inorganic carbon present in solution, suggesting that external can utilized by microalgae in cell formation. On the other hand, for the condition of low pH 5.0 to 7.0, CO2 utilization occurs through mass transfer. Thus, CO2 up take by algal culture is mass transfer controlled process. It was obseved also that when the algal culture aerated with atmospheric air enriched with CO2-enriched air the pH of water solution decrease suggesting that CO2 uptake in these conditions may be performed through diffusion.

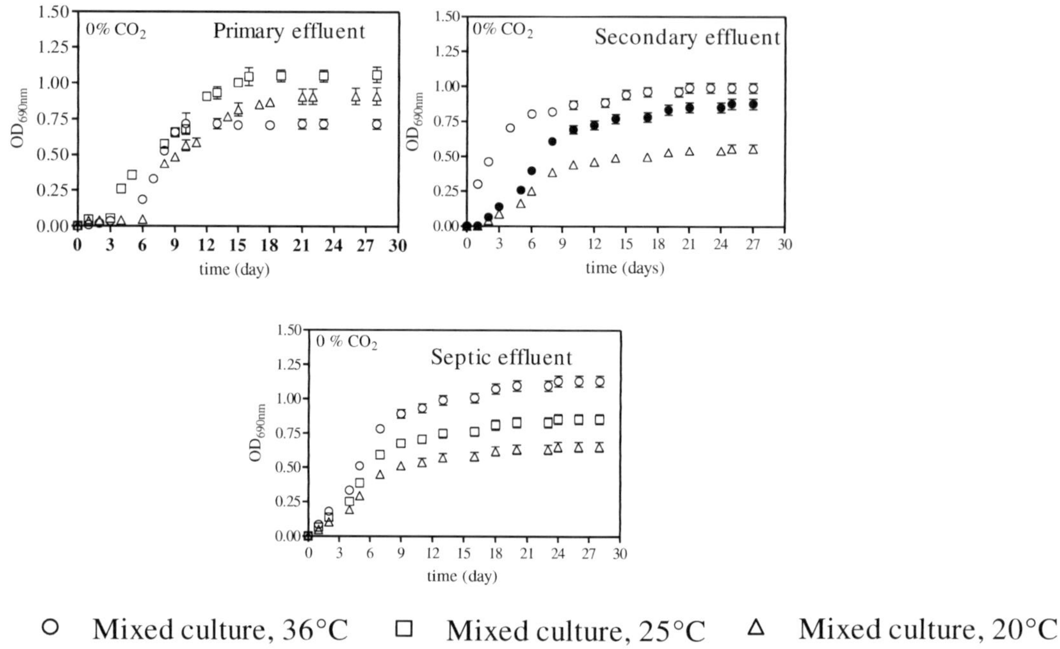

Figure 1: The change in optical density of MIMA as function of temperature and cultivation time in primary effluent, secondary effluent and Septic effluent.

COD removal efficiency also varied among different wastewaters. The results indicate that cultivation times can be shortened based on the optimum COD removal and desired final effluent qualities. Rapid decrease in COD was achieved during the first 7 days of cultivation for all wastewater samples. Significant COD removals were achieved with the mixed indigenous culture, which were 64.9%, 70.3% and 69.3%, respectively.

3.4 Removal of nitrogen compounds

MIMA showed very good efficiencies in removing ammonia (NH4–N) from wastewater (Figure 2b). ammonia removal efficiencies were 63.2%, 67.5% and 71.7% for primary effluent, secondary effluent and septic tank effluent. In addition to ammonia, nitrate (NO3–N) and nitrite (NO2–N) removals were also measured. MIMA 64%, 66.7% and 60% of nitrate from primary effluent, secondary effluent and septic tank effluent, respectively. The removal of nitrate from the effluents corresponded to the generation of nitrite (Figure 2d). The initial nitrite in primary effluent, secondary effluent and Septic tank effluent were 0.005, 0.89 and 0.14 mg-N/L. The concentration of nitrite did not exceed 0.11 mg-N/L for primary effluent, 0.93 mg-N/L for secondary effluent and 0.99 mg-N/L for Septic tank effluent. Wang et al. [13] reported a similar increase in nitrite and a decreasein nitrate during the cultivation of MIMA. Table 2 provides a summary of the percent removals of ammonia, nitrate as well as total inorganic nitrogen (TIN). TIN removal efficiencies were around 60% on average and above 50% could be achieved by MIMA. The highest TIN removal of 80.8% was achieved in Septic tank effluent by using the mixed culture. The results show that algal strains can be used to effectively remove inorganic nitrogen from different types of wastewater. Some of this removal could be due to the volatilisation of ammonia at elevated pH levels as a result of microalgae activity. Experiments carried out with adding CO2 showed also promising results of nitrogen compounds removals.

3.5 Removal of phosphorus

The percent removal of total dissolved phosphorus in primary effuent, secondary effluent and septic tank effluent with MIMA. Similar to COD and nitrogen removal, phosphorus removal was different in different wastewater samples and also varied with the type of microalgae. The general trend indicated that the highest phosphorus removal was achieved in primary effluent and septic tank effluent, and limited removal was observed in secondary wastewater. Reported phosphorus removal efficiencies were 79%, 49% and 21.9% in primary effluent, secondary effluent and septic tank effluent, respectively. Some of the removal could be caused by the precipitation of phosphorus through the formation of hydroxyapatite due to elevated pH levels occurred as a result of microalgae activity.

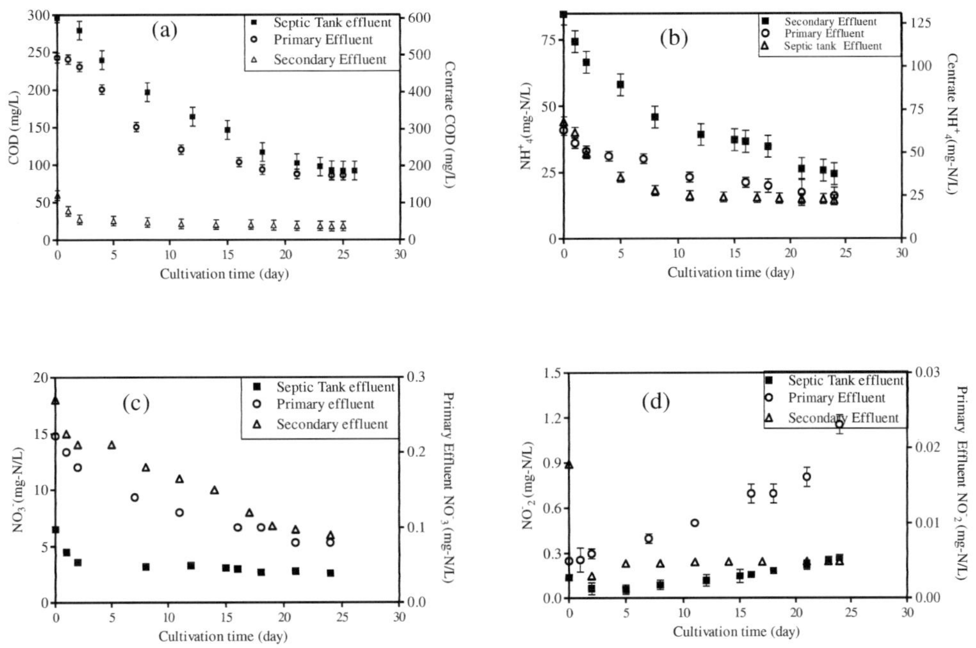

Figure 2: Organic matter and nutrient removals in primary effluent, secondary effluent and septic tank effluent by mixed indigenous microalgae culture at 25 ◦C and without addition of CO2.

Table 2: Percent removal of COD, NH4+, NO3-, total inorganic nitrogen (TIN) and total dissolved phosphorus (TDP) by mixed indigenous microalgae in primary effluent, secondary effluent, and septic tank effluent.

% Removal	Primary effluent	Secondary effluent	septic tank effluent
COD	64.9	70.3	69.3
NH^+_4	63.2	67.5	71.7
NO_3^-	64	66.7	60
TIN	63.2	67.3	80.8
TDP	70.0	30.8	50

4. Conclusions

The suitability of algae process in treating wastewater from different location in municipal wastewater treatment plant at different temperature was evaluated. Growth rate, COD removal and nutrient removals (N and P) were followed. The results of this study show that MIMA is effective in capturing CO2 and removing ammonia and phosphorus from wastewater. This simple and cost-effective microalgae-based wastewater treatment systems can be successfully employed in different regions in spite of some potential inhibition to microalgae due to high temperatures. MIMA showed a maximum growth for media injected with 10% of CO2 at different temperatures. More than 70% of ammonia and 75 % of Phosphorous was removed by MIMAin the temperature range of 20-30°C. Growth rates of microalgae were different not only between the two temperatures but also for the same microalgae injected with different CO2, which indicates that the operational condition play an important role in determining the potential of algae as CO2-capturing technology and this knowledge can be used to optimize the microalgae based treatment systems.

5. References

[1] Abdel-Rauf, N., Al-Homaiden, A. A., and Ibraheem, I. B. M. 2012. Microalgae and wastewater treatment. Saudi Journal of Biological Sciences, 19: 257-275.
[2] Hamilton, S. K., S. J. Sippel, D. F. Calheiros, and J. M. Melack. 1997. An anoxic event and other biogeochemical effects of the Pantanal wetland on the Paraguay River. Limnology and Oceanography 42:257-272.
[3] Hamilton, S. K., and J. W.M. Lewis. 1990. Basin morphology in relation to chemical and ecological characteristics of lakes on the Orinoco River floodplain, Venezuela. ArchivfürHydrobiologie 119:393-425.
[4] Li, W. K. W. 1980. Temperature adaptation in phytoplankton: Cellular and photosynthtic characteristics. In: Falkowski, P. G. (ed.) Primary productivity in the sea. Plenum Press.
[5] Li, Y., M. Horsman, B. Wang, N. Wu, and C. Q. Lan. 2008. Effects of nitrogen sources on cell growth and lipid accumulation of green alga Neochlorisoleoabundans. Applied Microbiology and Biotechnology 81:629 – 636.
[6] Melo, S. d., and V. L. M. Huszar. 2000. Phytoplankton in an Amazonian flood-plain lake (Lago Batata, Brasil): Diel variation and species strategies. Journal of Plankton Research 22:63-76.
[7] Moheimani, N. R. and Borowitzka, M. A. 2006. Limits to productivity of the algae Pleurochrysiscarterae (haptophyta) grown in outdoor raceway ponds. Biotechnology and Bioengineering 96 (1): 27-36.
[8] Park, J. B. K., Craggs, R. J., and Shilton, A. N. 2011. Wastewater treatment high rate algal ponds for biofuel production. Bioresource Technology, 102:35-42.
[9] Pittman, J. K., Dean, A. P., and Osundeko, O. 2011. The potential of sustainable algal biofuel production using wastewater resources. Bioresource Technology, 102: 17-25.
[10] Ratkowsky, D. A., Lowry, R. K., McMeekin, T. A., Stokes, A. N. and Chandler, R. E. (1983) Model for bacterial culture growth rate throughout the entire biokinetic temperature range. J. Bacteriology 154: 1222-1226.
[11] Salvucci, M. E., Crafts-Brandner, S. J. 2004. Inhibition of photosynthesis by heat stress: the activation state of Rubisco as a limiting factor in photosynthesis. Physiology Plant 120(2): 179-186.
[12] Mahapatra, D. M., Chanakya, H. N., Ramachandra, T. V., 2013. Treatment efficacy of algae-based sewage treatment plants. Environmental Monitoring and Assessment, 185,7145-7164.
[13] Wang, L., M. Min, Y. Li, P. Chen, Y. Chen, Y. Liu, Y. Wang, and R. Ruan. 2009. Cultivation of Green Algae Chlorella sp. in Different Wastewaters from Municipal Wastewater Treatment Plant. Applied Biochemistry and Biotechnology 162:1174-1186.
[14] Martin, C., De la Noae, J., Picard, G., 1985. Intensive cultivation of freshwater microalgae on aerated pig manure. Biomass, 7, 245-259.

[15] Mulbry, W., Kondrad, S., Pizarro, C., Kebede-Westhead, E., 2008. Treatment of dairy manure effluent using freshwater algae: Algal productivity and recovery of manure nutrients using pilot-scale algal turf scrubbers. Bioresource Technology, 99, 8137-8142.

[16] Pouliot, Y., Buelna, G., Racine, C., De la Noüe, J., 1989. Culture of cyanobacteria for tertiary wastewater treatment and biomass production. Biological Wastes, 29,81-91.

[17] Zhu, L., Wang, Z., Shu, Q., Takala, J., Hiltunen, E., Feng, P. and Yuan, Z. (2013) Nutrient removal and biodiesel production by integration of freshwater algae cultivation with piggery wastewater treatment. Water Research 47(13), 4294-4302.

[18] M.C. Picardo, J.L. Medeiros, F.A.O. Queiroz, R.M. Chaloub, Effects of CO2 enrichment and nutrients supply intermittency on batch cultures of Isochrysisgalbana, Bioresour. Technol. 143 (2013) 242–250.

Acknowledgements

This publication was made possible by the NPRP grant (NPRP No.: 6 - 1436 - 2 - 581) from the Qatar National Research Fund (a member of Qatar Foundation). The statements made herein are solely the responsibility of authors. The authors gratefully acknowledge the support provided by the Qatar University to do this work.

IOP Conf. Series: Earth and Environmental Science **67** (2017) 012033 doi:10.1088/1755-1315/67/1/012033

Rainfall-runoff modelling of Ajay river catchment using SWAT model

Subhadip Kangsabanik and Sneha Murmu
Department of Civil Engineering, Indian Institute of Engineering Science and Technology, Shibpur, Howrah 711103, India.

Abstract. The present study is based on SWAT (Soil and Water Assessment Tool) Model which integrates the GIS information with attribute database to estimate the runoff of Ajay River catchment. Soil and Water Assessment Tool (SWAT) is a physically based distributed parameter model which has been developed to predict runoff, erosion, sediment and nutrient transport from agricultural watersheds under different management practices. The SWAT Model works in conjunction with Arc GIS. In the present study the catchment area has been delineated using the DEM (Digital Elevation Model) and then divided into 19 sub-basins. For preparation of landuse map the IRS-P6 LISS-III image has been used and the soil map is extracted from HWSD (Harmonized World Soil Database) Raster world soil map. The sub basins are further divided into 223 HRUs which stands for Hydrological Response Unit. Then by using 30 years of daily rainfall data and daily maximum and minimum temperature data SWAT simulation is done for daily, monthly and yearly basis to find out Runoff for corresponding Rainfall. The coefficient of correlation (r) for rainfall in a period and the corresponding runoff is found to be 0.9419.
Keywords: Runoff; Rainfall; SWAT;

1. Introduction

In water resources studies runoff is the most important hydrological variables used. Estimation of direct runoff for ungauged river basins is difficult and time consuming. The conventional models for predicting runoff requires considerable hydrological and meteorological data. Many models were developed for watershed hydrology but the availability of temporal and spatial data was the main constraint hindering the implementation of these models especially in developing countries. Remote sensing and Geographic Information Systems (GIS), in combination with appropriate rainfall runoff models, provide ideal tools for the estimation of direct runoff volume, peak discharge and hydrographs. However, the development of remote sensing techniques and Geographic Information System (GIS) capabilities has influenced and enhanced the vast use of these models worldwide.

The model will be capable of forecasting the runoff for any given event of rainfall for the required duration. ERDAS IMAGINE 9.1, ArcGIS 10 and ArcSWAT2012 are the main software used in the present study. In the present study an attempt has been made for estimation of runoff using SWAT model utilizing GIS.

2. Study area

The present study area taken is the catchment of Ajay River. The river Ajay takes its rise from the eastern fringe of the Chotanagpur the Chotanagpur Plateau in Munger District of Bihar and after traversing the Gneissic terrain for a length of 162 km. Figure 1 shows the location map of the Ajay river catchment.

It encounters the Gondwana sedimentary of the Raniganj-Andal Coalfields. Subsequently it flows over the alluvial tract of the Bengal basin area. For the rest of the course up to the confluence point with the Bhagirathi at Katwa in West Bengal. It is a major river in Jharkhand and West Bengal. Total length of the Ajay is 288 km, out of which 152 km is in West Bengal. The catchment area of Ajay River is 5813.75 km2. The important tributaries of the Ajay are Pathro and Jayanti in Jharkhand and Tumuni and Kunur in Bardhaman district of West Bengal. The study area latitudinally extends from 23°26' N to 24°35' N and longitudinally between 86°16' E to 88°10' E.

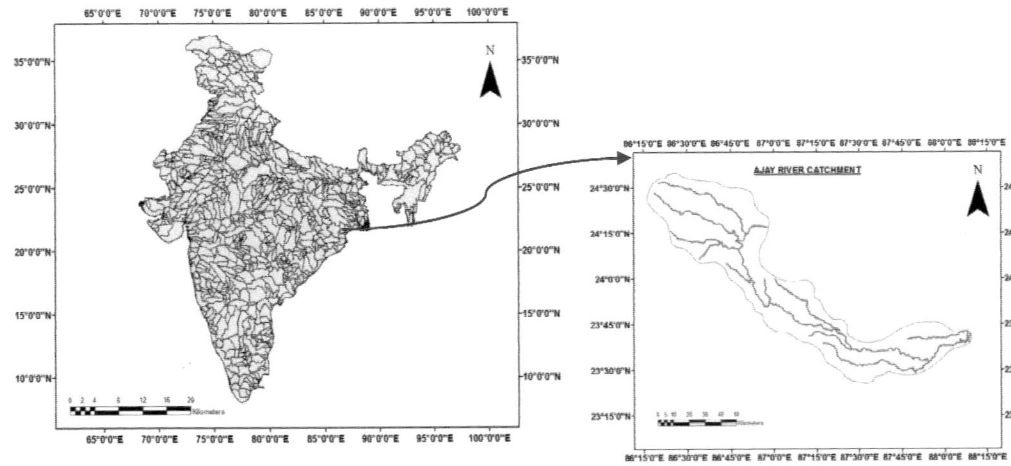

Figure 1. Study area.

3. Description of SWAT model

Soil and Water Assessment Tool (SWAT) is a physically based distributed parameter model which has been developed to predict runoff, erosion, sediment and nutrient transport from agricultural watersheds under different management practices. The SWAT works in conjunction with ArcGIS with an extension ArcSWAT a graphical user interface for SWAT tool. The hydrologic cycle as simulated by SWAT is based on the water balance equation (1). Here the runoff volume is estimated by using the Soil Conservation Service (SCS) curve number technique (USDA, 1972).

$$SW_t = SW_0 + \sum_{i=1}^{t} \left(R_{day} - Q_{surf} - E_a - w_{seep} - Q_{gw} \right) \qquad (1)$$

Where, SWt is the final soil water content (mm H2O), SWo is the initial soil water content (mm H2O), t is time in days, Rday is amount of precipitation on day i (mm H2O), Qsurf is the amount of surface runoff on day i (mm H2O), Ea is the amount of evapotranspiration on day i (mm H2O), wseep is the amount of percolation and bypass exiting the soil profile bottom on day i (mm H2O), Qgw is the amount of return flow on day i (mm H2O).

For simulation, a watershed is sub-divided into a number of sub-basins or sub watersheds. The use of sub basins in a simulation is particularly beneficial when different areas of the watershed are dominated by land uses or soils dissimilar enough in properties to impact hydrology. The total runoff mainly depends on the actual hydrologic condition of each land cover soil type and slope present in the watershed. Therefore, the impact of each type of land use is considered in this modelling to calculate runoff of the basin. After the overlay of the land-use, soil maps and slope, the distributions of the Hydrological Response Units (HRUs) were determined. HRUs divide the sub basin in to the area of similar land use, soil type and slope. Runoff is predicted separately for each HRU and routed to obtain the total runoff of the watershed. This increases accuracy and gives a much better physical description of the water balance.

4. Methodology

SWAT requires many sets of spatial and temporal input data. As semi-distributed model, SWAT has to process, combine and analyze spatially these data using GIS tools. Therefore, to facilitate the use of the model, it was coupled with GIS software as free additional extension ArcSWAT for ArcGIS. The methodology for the runoff modelling at the basin outlet using SWAT is depicted in flow chart Figure. 2.

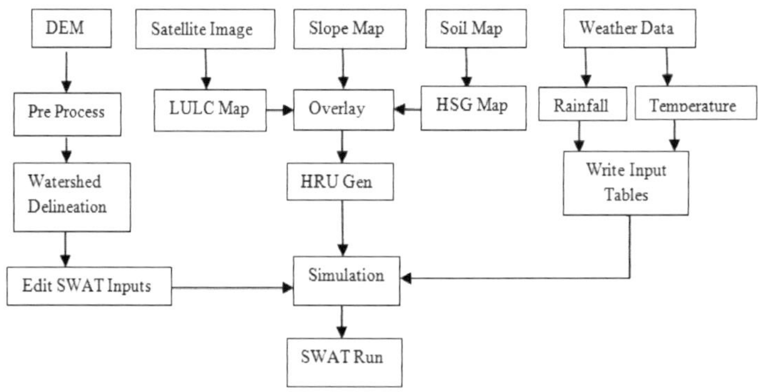

Figure 2. Methodology for rainfall- runoff modelling.

4.1 Degital elevation model(DEM)

DEM (Figure 3a) is extracted from the website of NASA. It is used to delineate the watershed and sub-basins as the drainage surfaces, stream network and longest reaches. The topographic parameters such as terrain slope, channel slope or reach length were also derived from the DEM.

4.2 Land use map

Land use/ land cover map (Figure 3b) for the study area is prepared using satellite IRS-P6 LISS-III image (acquired on January 2006) with spatial resolution of 23 m. The Supervised classification technique has been carried out to derive and distinguish the most present land use classes in the Ajay basin. Eight major classes were so identified. The dominant categories are fallow land (66.31%) and water (22%) and urbanized areas (4%). The forest and agriculture represents 2% and 1% of the watershed respectively.

4.3 Soil map

The soil map is extracted from the Harmonized World Soil Database (HWSD) raster world soil map. In the context of a complete update of the global agro-ecological zones study, FAO and IIASA combined existing regional and national updates of soil information worldwide and incorporated these with the information contained within the 1:5,000,000 scales FAO-UNESCO Soil Map of the World (FAO, 1971-1981). There are six types of textural classes present in the study area. The complete soil map is shown in Figure 3c.

4.4 Hydro meteorological data

SWAT requires daily values for precipitation, maximum and minimum temperature, solar radiation, relative humidity and wind speed for modeling of various physical processes. Rainfall data and Temperature data of Ajay river catchment has been downloaded from Indian Meteorological Department, Pune and http://swat.tamu.edu respectively. Here 30 years of 0.5 degree gridded daily rainfall data and also 30 years of 1 degree gridded daily maximum and minimum temperature data is used from the year 1976 to 2005 in this study area. The location map of the raingauge stations as well as temperature stations has been shown in Figure 3d.

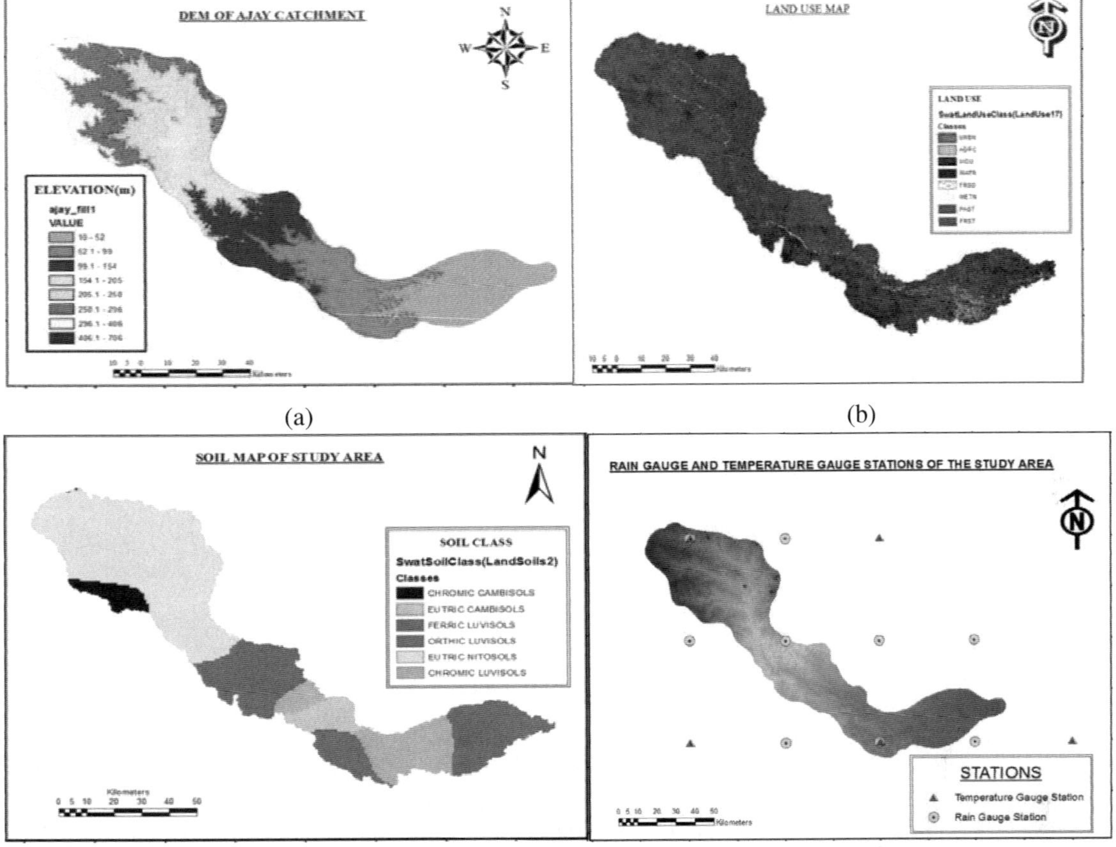

(a) (b)

(c) (d)

Figure 3. Spatial and weather data of the study area (a) DEM (b) Landuse Map (c) Soil Map (d) Location of weather stations.

5. Model setup

The entire database required by the SWAT model is developed for the study area and the model has been setup. ArcSWAT2012 allows us to delineate sub-watersheds based on an automatic procedure using DEM. DEM is imported in the model and the mask is manually created in the model in order to extract out the ajay sub-catchment area. The outlet is defined for the catchment area and Watershed was delineated for the present study (Figure 4.) and all the parameters are calculated for each sub basins. The total area of the catchment is found to be 5813.75 km2. The minimum and maximum elevations of the study area are 0 m and 705 m respectively. The catchment is divided into 19 sub-basins.

SWAT allows us to import the land use and soilmap to the model, then evaluate slope characteristics and determine the land HRUs for each sub-watershed. Land use category is used for specifying the land use layer and soil look up table is used for specify the type of soil to be modeled for each category. The soil map reclassified the database in 4 hydrological soil group (HSG) named A, B, C and D based on their infiltration rate. The LULC map is also reclassified into 8 different categories. The slope map is reclassified in 5 classes i.e. 0-3%, 3-8%, 8-20%, 20-45% and above 45%. Next the land use, soil and slope data layers have been overlaid. The distribution of hydrologic response units (HRUs) within the catchment has been determined. A total of 223 HRUs were generated in the Ajay river catchment. To eliminate minor land use, soil and slope, a threshold percentage of 10% has been adopted for all the land use, soil and slope classes.

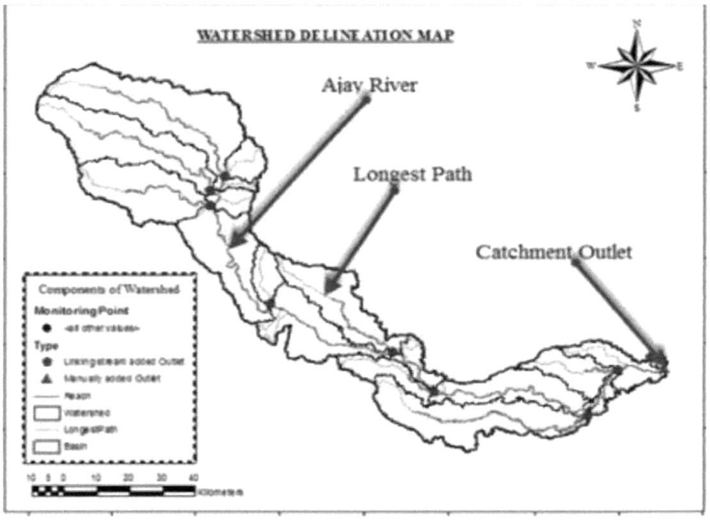

Figure 4. Delineation of sub-basins of Ajay catchment.

The model requires daily data for precipitation and temperature. SWAT allows the user to load weather station locations into the current project and assign weather data to the sub-watersheds. Here 30 years (1976 to 2005) daily rainfall data for 9 raingauge stations and daily maximum and minimum temperature data for 5 stations the same years has been utilized. Weather data is loaded using the 'Write Input Tables' menu on the ArcSWAT toolbar. This tool allows user to load weather station locations into the current project and assign weather data to the sub-watersheds. For each type of weather data loaded, each sub-watershed is linked to one gauge. Before SWAT can be run, the initial watershed input values must be defined. These values are set automatically based on the watershed delineation and Land use\ soil\ slope characterization or from defaults. After this step, the model was run to simulate the surface runoff.

6. SWAT simulation

The SWAT Simulation menu allows user to finalize the setup of input for the SWAT model and run the SWAT model, perform sensitivity analysis and perform auto-calibration.

7. Results and discussions

This study shows the structure of the SWAT-based model used in modeling of the Rainfall Runoff process. SWAT simulation is done for daily, monthly and yearly basis. Average runoff for average yearly rainfall is shown in Figure 5 from which it can be seen that the maximum runoff occurred in the year 1984. The rainfall runoff correlation has also been done for 30 years data and a good correlation is found with r2 value 0.8872.

SWAT also gives daily Runoff for corresponding daily Rainfall value throughout the year. Here the graphical representation of daily maximum Rainfall-Runoff values for each year for 30 years period has been shown in Figure 6.

8. Conclutions

SWAT model for the catchment produced good simulation results for daily, monthly and yearly runoff values as for the other water balance components. In this context, the observed correlation coefficient(r) is 0.9419. The evaluation of the model performance was carried out successfully with the recommended statistical coefficients. These performances can be enhanced furthermore using more accurate input data especially for the soil, landuse and DEM data that were estimated in this study with global data. The integration of climatic data such as rainfall data, temperature data also helps to compute accurate rainfall-runoff correlation. With the help of observed runoff data of the catchment, model validation can be done.

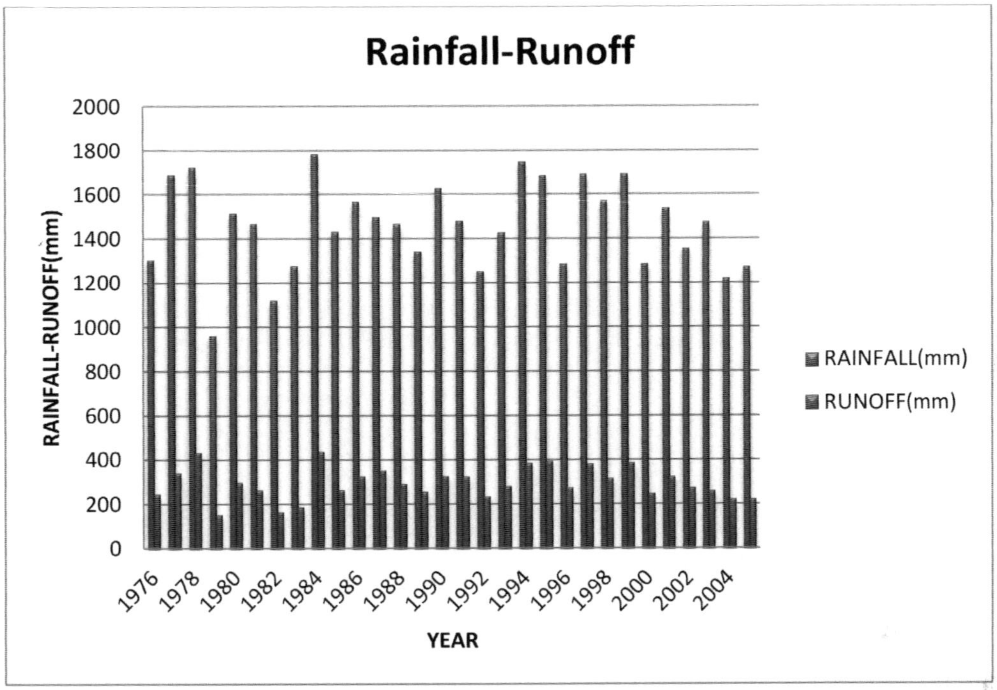

Figure 5. Yearly average Rainfall-Runoff .

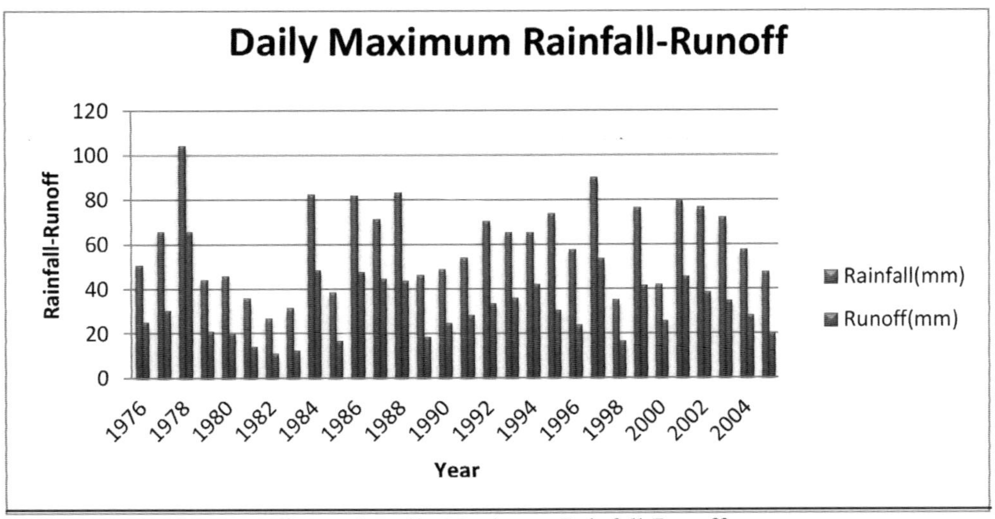

Figure 6. Daily Maximum Rainfall-Runoff.

9. References

[1] Khare D, Singh R and Shukla R 2014 *Int. J. of Geology, Earth & Environmental Sciences* **4** (1) 224-235

[2] Singh V, Bankar N, Salunkhe S S, Bera A K, and Sharma J R 2013 CURRENT SCIENCE, 104(9) 1187

[3] Winchell M, Srinivasan R, Luzio M D, and Arnold J 2013) "ArcSWAT Interface For Swat2012"

[4] Fadil A, Rhinane H, Kaoukaya A, Kharchaf Y and Bachir O A 2011 *J. of Geographic Information System* **3** 279-289

[5] Easton Z M, Fuka D R, White E D, Collick A S, McCartney A, Biruk B, Awulachew M, Ahmed S B and Steenhuis T S, 2010 *Hydrol. Earth Syst. Sci.* **14** 1827–41

[6] Simić Z, Milivojević N, Prodanović D, Milivojević V and Perović N 2009 *J. of the Serbian Society for Computational Mechanics* **3(1)** 38-63

[7] Setegn S G, Srinivasan R, and Dargahi B 2008 *The Open Hydrology Journal* **2** 49-62

[8] Arnold J G and Fohrer N 2005 *Hydrol. Process.* **19** 563–572

[9] Singh V P and Woolhiser D A 2002 *Journal of Hydrologic Engineering* **7(4)** 270-292

[10] Chong Y X 2002 "TEXT BOOK OF HYDROLOGIC MODELS", Uppsala

[11] Bellal M, Sillen X, Zeck Y, 1996 (Proceedings of the Vienna Conference, April 1996

[12] Kovar K, Nachtnebel H P *Int. Association of Hydrological Sciences, Series of Proceedings and Reports,* **235** 99-106

[13] Miloradov M, Marjanovic P, 1991 3rd Rhine-Danube Workshop Proceedings

The effect of anthropogenic activities to the decrease of water quality

N V Sidabutar[1], I Namara[2], D M Hartono[3], T E B Soesilo[4]
[1]School of Environmental Science Universitas Indonesia
[2] Department of Civil Engineering Universitas Indonesia
[3]Department of Civil Engineering Universitas Indonesia
[4] School of Environmental Science Universitas Indonesia
Email: noni.valeria90@gmail.com

Abstract. The raw water in Jakarta is supplied from Jatiluhur Dam, which is distributed pass through West Tarum Canal with an open canal about 70 km long. This water quality does not meet the standard set by the government and heavily polluted by anthropogenic activities along its river. This research uses a quantitative research approach with the mix-method. This research did an in-depth interview with inhabitants along the riverbank about their daily activity. The water along the riverbank is polluted by anthropogenic activities, such as: first: domestic activities (washing, cooking, and bathing), second: littering into the river, and third: discharging waste water from households into the river. This present research measures water quality for parameters pH, temperature, Dissolved Oxygen (DO), Chemical Oxygen Demand (COD), Total Dissolved Solid (TDS), Total Suspended Solid (TSS), and Fecal coliform. In this social segment, it is shown that pH, DO, TDS and Fecal coliformin the downstream part are worse than in the upstream.

1. Introduction

In Sustainable Development Goals (SDG)'s goals, water is as the 6th goal, namely: "Ensure availability and sustainable management of water and sanitation for all"[1]. It shows that clean water is one of human basic needs, and it is processed from raw water to clean water by using any treatment. To reach the standard of living, water is a necessary for humans [2]. The 1945 Constitution of the Republic Indonesia also described that every natural resource is used to meet citizens needs i.e water. Jakarta as the largest city in South East Asia with 9.7 million inhabitant [3], has just about 58 percents of needs that can be fulfilled. Government of Jakarta's treat and distribute water to inhabitants are supported by two private companies (PT Lyonnaise Jaya (Palyja) and PT Aetra Air Jakarta). The raw water from Jatiluhur Dam has flown through West Tarum Canal since 1997, which uses open canal of ±70 km. It divided into three drinking water units, namely: Buaran, Pulogadung and Pejompongan area.

Concentration of BOD in the intake drinking water unit is between 14-26 mg/ l while the standard by the government is 2 mg/ l. The COD is between 17-61 mg/l while the standard is 10 mg/l. Solid is between 109-122 mg/l with the standard is 1,000 mg/l. Total coliformis measured 34,000-4,800,000/ml with the standard of 100 mg/l[4]. Water quality is strongly influenced by anthropogenic activities pollutants that enter to the body of water as the result of human activities, such as domestic activities (households), urban activities, as well as industrial activities [5].As the consequence of those, the raw water should fulfill the first's class water quality based on the Government Regulation No. 82/2001 regarding Water Quality Management and Water Pollution Prevention.

The existence of population growth is followed with the number of settlements and the growth of the industry, particularly along watersheds that dump waste into surface waters, causing a decline in the quality of surface water [6,7]. The uses of riverbanks as the area to conduct a bath, wash, and draw the waste water to the river give an impact in the quality of raw water in Jakarta's drinking water unit [8].Human activities exploit water and soil resources that reduce water quality [9]. In other, the citizens in the riverbank, who have lived more than 20 years along the river, could become the effect of the decreasing water quality. This present research aims to analyze the anthropogenic activities along the riverbank, i.e. domestic activities (washing, cooking, and bathing), littering into the river, and waste water from households discharged into the river. This thesis contributes to find the proper strategy for water quality cleaning from upstream to downstream. This is different from the current literature, that pays attention to the intake of Water Treatment Process (WTP), not the condition of raw water[4,10,11].

Formulation of the problem in this research is the quality of the raw water of Jakarta's drinking water unit being contaminated due to anthropogenic activities along the riverbank, whereas raw water to be used should meet the applicable standards. The research question posed is how do the anthropogenic activities in the river affect water quality of the river? This research aims to analyze anthropogenic activities on the riverbank which affect river water quality deterioration. The academic benefit from this research is to analyze the status of river water quality and identify which anthropogenic activities could worsen the water quality of the river. As a practical benefit, this research is expected to provide information and feedback on the raw water pollution of Jakarta's Drinking Water Company. By this research, the government shall do urban planning in that city. To keep the water quality along the riverbank, the government shallbuild a good environment for citizens, motivate them to give responsibility of a collective effort to do bathing-washing in their own toilet not in the river's body, clean river regularly, manage solid waste, and other efforts. The contribution of all the communities makes the raw water adequate to be used.

2. Method

This research fits within the framework of water sustainability which has concern in three main aspects: economic, social, environmental. It focusesits analyses on raw water problems affected by anthropogenic activities in the riverbank as an urban area. This research uses a quantitative approach, whereas the method is used the combined qualitative and quantitative method. Interviewees and topics were identified through sampling between September 2015–January 2016. This research was conducted by using primary and secondary data. Primary data are the data obtained by researchers using interviews, observations of anthropogenic activities, and the water quality in the segments of citizens. Human activities were performed at the segment B. Tb. 7-11 (about 7 km from Dam), and it was selected because this area is near upstream and about 5 kmalong this riverbank shows anthropogenic activities. Most questions were open-ended to provide flexibility to express their experience and opinion about issue. Interview lasted about 20 minutes, the majority were recorded, transcribed, and coded. To examine the effect from inhabitants, the water qualityin upstream and downstream is measured, using parameters: temperature, pH, COD, DO, TSS, TDS, and *Fecal coliform.*Secondary data are from the governance and drinking water company. The anthropogenic activities from interviews will described descriptively.

3. Result and discussion

3.1 Description area

Jatiluhur Dam as the source of raw water for the drinking water treatment unit is about 81% from Citarum River, in West Java. This area has several purposes, such as to support agriculture, fisheries, public water supply, industry, hydroelectric power, recreation and others.

Figure 1. Jatiluhur dam

The increasing population in West Tarum Canal affects the river conditions. The river should be bordered by inspection space, but now it becomesa residential settlement for urban people along the riverbank. It makes the river contaminated and makes the river become narrower. Based on the interview with citizens, it is discovered that they have already lived there more than 20 years. About 54% citizens had low income, namely 0-Rp. 2,000,000, so it makes them use the riverbank for their daily activities, which is shown on Table 1.

Table1. Activity Along Riverbank West TarumCanal

No	Activities	Kind of waste
1	Residential	1. Black water from bathroom
		2. Solid waste
		3. Waste from defection in the riverbank
2	Industrial	1. Polluted air from burning
		2. Waste Water
3	Small shop	1. Black water from cleaning the dishes
		2. Organic solid waste
4	Mosque	1. Black water from toilet
		2. Water from 'wudhu'
5	Public Bathroom	1. Black water from bathroom
		2. Waste from defection in the riverbank
6	Littering into river	1. Leachate from solid waste
		2. Solid waste in the river
7	Construction well	1. Construction Waste
		2. Sediment in the river

Figure 2 shown the flow of raw water from Jatiluhur Dam to intake of Drinking water unit. "BangunanTarum Barat (B.Tb.) mentioned the distance from Cibeet Dam.

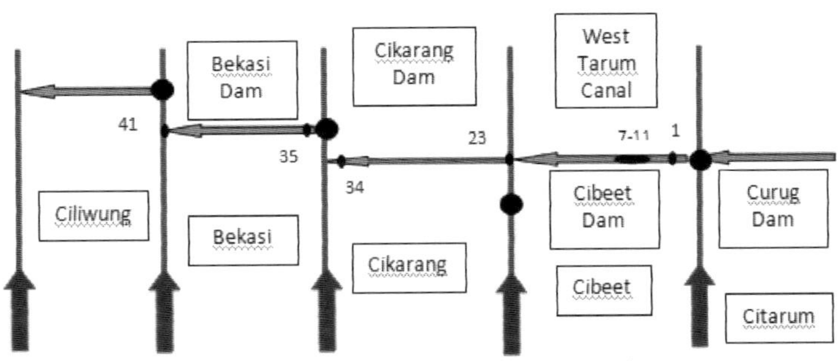

Figure 2. A long west tarum canal
source: PJT II (2015)

3.2 Anthropogenic activities

Water that has decreased in quality is affected by anthropogenic activities along the river bank, such as defection in the riverbank, water dispose of domestic waste water intothe river, taking out the trash and others. Firstly, the anthropogenic activities arebathing, washing, and latrine activities in the river. In the interviewing, 37% have no toilet and 63% of the interviewer have toilet in their own house, although 55% of them prefer to do bathing, washing, and defection in the river than in their own toilet. Along the riverbank, each house has toilet in the body of the river. However, the researched by Parlinto (2012) stated the usage of the riverbank for domestic needs affects the river have bad smell and become blackened.

Secondly, the activities discharge domestic wastewater into the river body such as grey water from bathing, washing and feces. By observing, the researchers found that many houses disposedomestic waste water into the river. About 54% of houses have no septic tank and throw waste water directly into West Tarum Canal.About 17% of houses have septic tank only for defection but throwwaste water directly into West Tarum Canal. The waste water directly flows into water bodies through PVC pipes, culverts and ditches.Based on the results of the observations on the segment of society, most people in the riverside neighborhood dispose domestic waste water from households directly into the river. It is mostly done in the morning (6:00 to 9:00 a.m.) and afternoon (13:50 to 17:30 p.m.), because those are the peak hours.

The third islitteringinto the river. All houses produce any rubbish or garbage or solid waste, and the handling is just making a hole in the ground, sacks, plastic containers, and others. Most of the inhabitants collect and then burn the garbage. The bins are put near the river, and it is possible to be carried over by water when it rains. From the interview, it is found that each house produce about 3.2 kg/day. Based on the interview, about 69% of the people never throw solid waste into river, but 31% do, such as diapers, organic waste, and others. The interviewers also collect the solid waste in the riverbank about 1-2 days before burn it. However, it is possible to come into the river's body because of runways. It is also supported by the inclined surface sloping riverbank, so that the waste can easily be carried into water bodies. The source of spreading pollutants gets into the water as runoff or runoff of surface soil or agricultural settlements. In addition, the existing trees along the riverbank also contribute in additional organic waste that is carried away by the water, because the leaves are falling directly into water bodies.

In addition, there are also other anthropogenic activities that allegedly support the degradation of water quality of the river, such as the construction of overpasses near to Buaran Area (intake of the Drinking Water Unit provided by AETRA). In the other words, many people who have small-scale livestock such as goats, chickens, and ducks placed at the riverbank, can also increase pollutants in the river.

3.3 Water quality

To examine the effects of anthropogenic activities contributing to the decline in water quality, water quality testing was done on the downstream and upstream of society segments. For upstream point in B. Tb. 7 with coordinates 6023'38.2"S and 107020'22.4"E, while the downstream point as B. Tb. 11 with coordinates 6021'53.9"S and 107019'12.3"E. It obtained the test results for the parameter: temperature, pH, COD, DO, TSS, TDS, and Fecal coliform in those sampling points shown in Table 2. Anthropogenic activities on the segments of society B. Tb. 7-11 show the effects in the decrease of water quality, although the result is not significant. Visually, it can be said that the waters become smelly. The temperature is still appropriate with the standard. The temperature could be affected by the height of the above mean sea level (AMSL), radiance of the sun, stream of the river, and depth of the river.

Parameter pH declared the Hydrogen ion concentration is an important quality parameter in water. The low pH affects the mobility of toxic heavy metal compounds and the high value of pH disturb the balance of a biological life. The measurement results in this present research show that the water meetsthe standard, namely: 7-8. It stated that the condition of the wateris still normal.

Table 2. Water quality in citizens segment

No	Parameter	Unit	Up	Down
1	Temperature	^oC	29	29
2	pH		7.21	7.17
3	COD	mg/l	11.3	9.72
4	DO	mg/l	8.71	6.47
5	TDS	mg/l	272	271
6	TSS	mg/l	83	51.5
7	*Fecal coliform*	Colony/100mL	360	930

Chemical Oxygen Demand (COD) has indicated the amount of Oxygen required to oxidize organic materials in the water. In the upstream, it is above of the standard set by the government, but in the downstream it meets the standard. This concentration of COD showed chemical compounds arising from industrial activities. The concentration of COD shows affected by contaminants coming into water bodies, because of the number of residents who live along the West Tarum Canal.
Dissolved Oxygen (DO) is useful for aquatech. The high DO shows the quality of water as well, but it is opposite with the concentration of COD and BOD. The number of COD and BOD explains the number of Oxygen needed to decompose organic and inorganic compounds. The existence of Oxygen in water showsthat the quality of water is good. In the upstream and downstream it meets the standard ($\geq$6 mg/l), but the quality decreases. It can be caused by the influx of domestic waste water. The concentration of DO in water should be increased, because the DO is needed by microorganisms to decompose pollutants in the water body.
Total Dissolved Solid (TDS) is a physical parameter of raw quality and size of the solute (both organic and inorganic substances), which is generally considered not as a major pollutant. The concentration of TDS in water must be controlled, because if the high TDS concentrations can reduce the clarity of the water, contributing to a decrease in photosynthesis, compounds react with heavy metal, and cause an increase in water temperature. The TDS content of ions and minerals contained in the water can accumulate in the body and cause kidney disorders. TDS value is used as an indicator of the aesthetic characteristics of the water and as an indicator of the number of very small particles. In the upstream and downstream, the concentration of TDS is almost similar, not significant, but both are still below the quality standard.
Total Suspended Solid (TSS) measurement based on the dry weight of the particles are trapped by the filter, usually with a specific pore size. Generally, the filter used has a pore size of 0.45 lm. The suspended solids may include organic and inorganic materials. The existence of the suspended solids will block the penetration of the light that enter the waters so that the relationship between TSS and brightness will indicate that the relationship is inversely proportional. Based on the measurements of parameters of TSS in the segment of society, it shows that it is larger in the upstream than in the downstream, but both had passed the quality standard. The factor affecting the suspended solids is rainfall and water runoff is relatively small in the downstream.
The number of *Fecal coliform* is an indicator of water quality changes due to human and animal fecal contamination. The number of *Fecal coliform* is expected to meet the quality standard, since most people use water along the riverbank. The number of *Fecal coliform*are affected by poor sanitation toilets, waste water charging into the river, the inclusion of human and animal feces, and other factors. The data of *Fecal coliform* in the upstream and downstream on the quality standard (100 colonies/100 mL for first class function). The increasing number of *Fecal coliform*is caused by those people who still do domestic activities in the river. The existence of coliform in water shows the pollution inside water, so inhabitants should not be able to directly use the river water without pre-treatment.

The decline is seen in the quality parameters of pH, DO, TDS, and Fecal coliform. It shown that the water should not be used for purposes designated as first class's function based on Government Regulation No. 82/2001 regarding Water Quality Management and Water Pollution Prevention.

3.4 Proper strategy

The anthropogenic activities on the riverbank is examined in this study, and those consist of domestic activities (washing, cooking, and bathing), littering into the river, and dischargingwaste water from households into the river. This present research extends this literature by providing information about raw water quality of this river by determining proper strategies to avoid water contamination from anthropogenic activities based on the reality. Proper strategies are keeping the upstream area, keeping along the riverbank such as releaseanthropogenic activities, treatingdomestic and industrial waste before discharging into the river body, having integrated solid waste management, keeping the river physical wide, and regular monitoring about the anthropogenic activities.

4. Conclution

The conclusion is that the water in the West Tarum Canal has been contaminated, due to anthropogenic activities along the riverbank, namely domestic activities (washing, cooking, and bathing), littering into the river, and discharging waste water from households into the river.Some suggestions by this research arethat it is necessary to learn and observe the influence of industry on water quality in the canal because the industry is located directly and indirectly on the riverbank.

5. References

[1] UN-DESA 2015Sustainable development goals, accessed 12 September 2015, 12.30
[2] Gleick P H 2007The human right to water*20 Pacific Institute Established 1987*
[3] BadanPusatStatistik (BPS) DKI Jakarta 2015 *StatistikdaerahProvinsi DKI Jakarta 2015*, BPSProvinsi DKI Jakarta
[4] FuadiA 2012 Pengaruh residual klorinterhadapkualitasmikrobiologipadajaringandistribusi air bersih (Studikasus: jaringandistribusi air bersih IPA Cilandak)*Final Thesis Universitas Indonesia, Depok*
[5] Effendi H 2003 Telaahkualitas air, bagipengelolaansumberdayadanlingkunganperairan.Kanisius, Yogyakarta
[6] MoersidikS Sand HartonoD M 2009Pendekatanstatistikuntukmenentukan parameter dominandalampengelolaankualitas air baku*LingkunganTropis*Vol. **3**Number 1p 23-32
[7] SugandaE YatmoA Y and ParamitaA 2009Pengelolaanlingkungandankondisimasyarakatpadawilayahhilirsungai*Jurnal Makara, SosialHumaniora*
[8] ParlintoB 2012Peningkatankualitas air SaluranTarum Barat secaramandirimelaluidiversifikasienergi*ProsidingSnaPP 2012: Saints, Teknologi, danKesehatan*
[9] PeilinW andMinghongT 2012Challenges for sustainable urbanization: a case study of water shortage and water environment change in Shandong China*Procedia Environmental Sciences 13. Elsevier* p 919-927
[10] Piri IHomayoonnezhadIand AmirianP 2016 Investigation on Optimization of Conventional Drinking Water Treatment Plant, *International Conference on Chemical and Environmental Engineering*
[11] PratamiM W R 2011 Perencanaansistempengolahanlumpur IPA Pejompongan I dan II Jakarta*Final Thesis Universitas Indonesia, Depok*

Acknowledgments
The authors would like to acknowledge funding support from the Universitas Indonesia PITTA grant in this research.

Numerical thermal performance study in a heat exchanger tube with inclined elliptical rings

W Sodsri[1], S Tamna[2], C Thianpong[1] and P Promvonge[1]

[1] Department of Mechanical Engineering, Faculty of Engineering, King Mongkut's Institute of Technology Ladkrabang, Bangkok 10520, Thailand
[2]Applied Mathematics and Mechanics Research Laboratory (AMM) Department of Automotive Engineering, Faculty of Engineering, TNI Bangkok, Thailand
Email: kpongjet@gmail.com

Abstract. The paper deals with a numerical study on the effect of inclined elliptical ring (IER) on heat transfer augmentation in a uniform heat-fluxed heat exchanger tube. In the present work, the 60° IER was mounted repeatedly in the tube with six different eccentricity ratios (E_R= *b*/*a* = 1, 0.9, 0.8, 0.7, 0.6 and 0.5) at a single ring-pitch ratio P_R= 1.0. Air as the test fluid flows into the tube for Reynolds number ranging from 4000 to 20,000. To find the optimum thermal performance, the effect of E_R values on the heat transfer and pressure loss is investigated. The study indicates that the use of IER can induce higher turbulent intensity imparted to the flow leading to higher heat transfer in range of about 237 to 461% above the smooth tube.

1. Introduction

Heat exchanger is a device that facilitates the exchange of heat between two different temperature fluids while keeping them from mixing with each other. Heat exchanger is widely applied in production process such as drying furnace, ventilation, air conditioning, refrigeration, electronic device, engine of machine and equipment in process industry, etc. The high performance heat exchanger is required in industry field, no matter small or large industrials take heat exchanger application in their process. At present, due to high competition and advanced development of those processes, improving their thermal performance is needed to increase efficiency. Principles used to design the device or thermal system for transfer or exchange of heat must be selected with aware of the features such as work according, laminar or turbulent flow [1], production cost, heat transfer efficacy, structure, compact size, maintenance & repair, reliability and safety. To achieve all features required, the basic key is the great allocation of resources, fuel and materials, so for allocation of resources we need to develop the capacity and potential of the heat transfer system to the best efficient, to minimize the power consumption. The design of compact size can reduce material and installation cost. Those are important that will lead to sustainable development in the future. Turbulators (turbulence promotors) are devices that can be described as static mixers. It can generate the secondary flows. Typically located inside the tubes of heat exchange equipment, turbulators can help increase the tube-side heat transfer efficiency of these units. Custom-made to fit the inside diameter of the tube, turbulators are designed to create and maintain turbulent swirl flow. There are many types of turbulators such as wire coils [2], vortex ring [3], conical rings [4], twisted tapes [5], baffles [6] and wing/ribs [7,8]. Promvonge [3] investigated the heat transfer, pressure drop and thermal performance in a round tube fitted with 30° inclined vortex rings (IVRs) with various geometry parameters such as three relative ring width ratios and four relative ring pitch ratios. Air was employed as the test fluid for the Reynolds number from 5000 to 26,000. The IVR yields the highest TEF of about 1.4 at BR=0.1 and PR=0.5. According to the literature review above, several experimental works on heat transfer enhancement in tubes by using turbulators have been extensively reported. However, only a few numerical works on that issue have been done. Also, the application of inclined elliptical rings in a circular tube has never found in the literature. Therefore, numerical investigations for six different eccentricity ratios of 60° inclined elliptical rings (IER) are conducted to study the flow structure and thermal behaviours and to understand the mechanism of heat transfer enhancement in the tube with IER insert. In the present work, air as the test fluid enters the tube inserted with IER elements for Reynolds number ranging from 4000 to 20,000.

2. Computational model

2.1. Physical flow model

A schematic of the IER tube in this work is shown in Fig. 1. The tube has inner diameter, D = 50 mm while the IER having outer diameter of 48 mm has a ring pitch, P = D. For the elliptical ring, it has a major radius, a = 23 mm which is kept constant and a minor radius, b to be varied (see Fig.1). The inner ring shape is changed to be elliptic with eccentricity ratio, $E_R = b/a$ (= 0.5, 0.6, 0.7, 0.8, 0.9 and 1) where $E_R = 1$ means that the ring shape is circular. The heat is transferred to air through the tube wall due to different temperatures between air and tube wall. The heat transfer coefficient and pressure drop are calculated by numerical computations. The tube flow model of this work is expected to attain a periodically fully developed flow where the velocity field repeats itself from one module to another [9]. Therefore, only one module is used for the current simulation. Due to tube symmetry, only left or right half of the tube flow model is employed as computational domain as shown in Fig. 1.

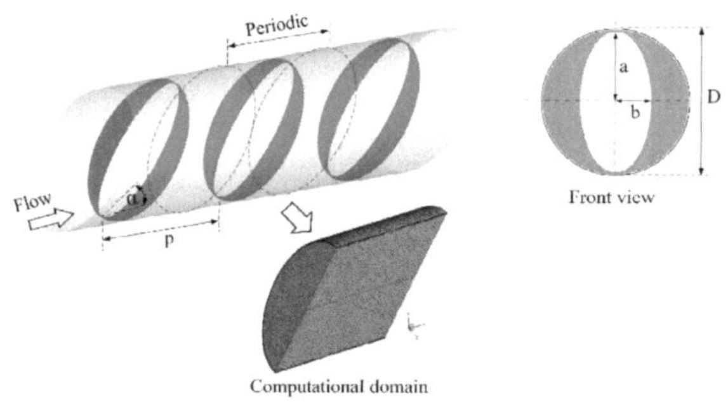

Figure 1. Tube flow model and computational domain.

2.2. Governing equations

The numerical tube flow model in the tube is developed with assumptions as follows: 1) Steady three-dimensional, turbulent and incompressible flow. 2) Constant fluid properties and 3) omitting body forces, viscous dissipation and radiation heat transfer. Based on assumptions above, the tube flow model is governed by the Reynolds averaged Navier-Stokes (RANS) equations and the energy equation. In the Cartesian tensor system these equations can be expressed as:

Continuity equation:

$$\frac{\partial}{\partial x_i}(\rho u_i) = 0 \tag{1}$$

Momentum equation:

$$\frac{\partial}{\partial x_j}(\rho u_i u_i) = -\frac{\partial p}{\partial x_i} + \frac{\partial}{\partial x_j}\left[\mu\left(\frac{\partial u_i}{\partial x_j} + \frac{\partial u_j}{\partial x_i}\right) - \rho\overline{u'_i u'_j}\right] \tag{2}$$

where ρ is fluid density, u_i is a mean component of velocity in direction x_i, p is the pressure, μ is dynamic viscosity, and u' is a fluctuating component of velocity. Repeated indices indicate summation from one to three for 3-dimensional problems.

Energy equation:

$$\frac{\partial}{\partial x_j}(\rho u_i T) = \frac{\partial}{\partial x_j}\left((\Gamma + \Gamma_t)\frac{\partial T}{\partial x_j}\right) \tag{3}$$

Where Γ and Γ_t are molecular thermal diffusivity and turbulent thermal diffusivity, respectively and are given by

$$\Gamma = \mu / \mathrm{Pr}, \text{ and } \Gamma_t = \mu_t / \mathrm{Pr}_t \tag{4}$$

The Reynolds-averaged approach to turbulence model requires that the Reynolds stresses, $-\rho\overline{u'_i u'_j}$ in (2) needs to be modeled. The Boussinesq hypothesis relates the Reynolds stresses to the mean velocity gradients as seen in (5) below:

$$-\rho\overline{u'_i u'_j} = \mu_t\left(\frac{\partial u_i}{\partial x_j} + \frac{\partial u_j}{\partial x_i}\right) - \frac{2}{3}\left(\rho k + \mu_t\frac{\partial u_i}{\partial x_i}\right)\delta_{ij} \tag{5}$$

Where k is the turbulent kinetic energy, as defined by $k = \frac{1}{2}\overline{u'_i u'_i}$, and δ_{ij} is the Kronecker delta. An advantage of the Boussinesq approach is the relatively low computational cost associated with the computation of the turbulent viscosity, μ_t given as $\mu_t = \rho C_\mu k^2/\varepsilon$. The realizable k–ε turbulence model is an example of the two-equation models that use the Boussinesq hypothesis. The steady state transport equations are expressed as:

$$\frac{\partial}{\partial x_j}(\rho k u_j) = \frac{\partial}{\partial x_j}\left[\left(\mu + \frac{\mu_t}{\sigma_k}\right)\frac{\partial k}{\partial x_j}\right] + G_k + G_b + \rho\varepsilon - Y_M + S_k \tag{6}$$

$$\frac{\partial}{\partial x_j}(\rho\varepsilon u_j) = \frac{\partial}{\partial x_j}\left[\left(\mu + \frac{\mu_t}{\sigma_\varepsilon}\right)\frac{\partial \varepsilon}{\partial x_j}\right] + \rho C_1 S\varepsilon - \rho C_2\frac{\varepsilon^2}{k + \sqrt{\nu\varepsilon}} + C_{1\varepsilon}\frac{3}{k}C_3 G_b + S_\varepsilon \tag{7}$$

In the above equations, σ_k and σ_ε are the inverse effective Prandtl numbers for k and ε, respectively. C_1, C_2, C_3, and $C_{1\varepsilon}$ are constants. All the governing equations are discretized by the QUICK numerical scheme, coupling pressure-velocity with the SIMPLE algorithm and solved using a finite volume approach. For closure of the equations, the realizable k–ε model is used in the present study. The solutions are converged when the normalized residual values are less than 10^{-6} for all variables but less than 10^{-9} only for the energy equation. More details on boundary conditions are similar to the case in [8]. There are four parameters of interest in the present work, namely, the Reynolds number, friction factor, Nusselt number and thermal enhancement factor. The Reynolds number is defined as

$$\mathrm{Re} = \rho u_0 D/\mu \tag{8}$$

The friction factor, f is computed by pressure drop, ΔP across the length of the tube, L as

$$f = \frac{(\Delta P/L)D}{2\rho u_0^2} \tag{9}$$

The local heat transfer is measured by local Nusselt number which can be written as

$$Nu_x = h_x D/k_a \tag{10}$$

where k_a is thermal conductivity of air. The area-average Nusselt number can be obtained by

$$Nu = (1/A)\int Nu_x dA \tag{11}$$

The thermal enhancement factor (TEF) is defined as the ratio of the heat transfer coefficient of an inserted tube, Nu to that of a smooth tube, Nu_0, at an equal pumping power and given by

$$TEF = (Nu/Nu_0)/(f/f_0)^{1/3} \tag{12}$$

2.3. Grid independence

Grid independence method is tested to make sure that numerical solutions are free from grids by using four levels of grid systems performed at Re = 12,000 as shown in Fig 2. Comparing the solution of Nu obtained from four different grid levels, 90,749, 187,527, 261,275 and 385,239. It is seen that the Nu difference of the two grids between 90,749 and 187,527 is 1.021% while those between 187,527 and 261,275 and between 385,239 and 261,275 were less than 1.00%. In terms of convergence time and solution precision, the grid size of 187,527 is adopted in the current work.

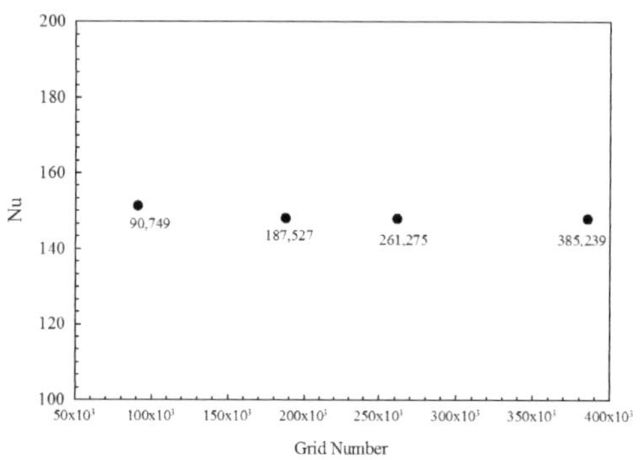

Figure 2. Grid independence test

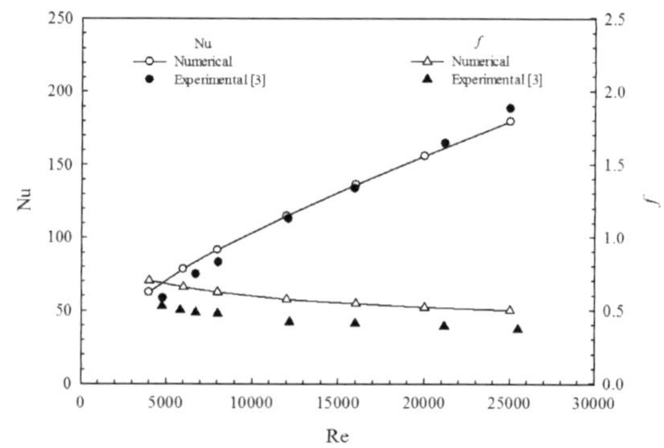

Figure 3. Validation Nu and *f* with measurements of inclined ring insert.

2.4. Model validation

It is necessary to verify the tube flow model and the method adopted in the numerical simulation. The validation of simulated Nu and *f* with Re of the tube with measurements for using 30° inclined ring insert [3] is depicted in Fig. 3. In the figure, it is found that the present numerical results and experimental data are in good agreement. The deviation of both numerical and measured data is within ± 8% for Nu and ± 18% for *f*.

3. Results and discussion

3.1. Flow structure and heat transfer

The flow structure and vortex coherent in the flow model fitted with 60° IER at E_R = 0.5, Re = 12,000 can be displayed in the form of the streamlines in transvers planes and the local Nu_x contours as shown in Fig. 4(a) and 4(b), respectively. In Fig. 4(a), the use of IER can induce two main vortices flow appearing on the left and right sides of the tube. Closer examination reveals that both are counter-rotating vortices with common-flow-down to the bottom wall. The local Nusselt number contours on the tube surface for E_R = 0.5 and Re = 12,000 are presented in Fig. 4(b). It is apparent that high Nusselt number values appear on large areas over the tube wall, especially for the area around the IER and on the bottom wall. The peaks can be observed at the impingement areas on surface close to the

IER. Also, larger regions of higher heat transfer values can be found on the bottom wall area of the tube due to common-flow-down vortices as can be seen in Fig. 4(b). This indicates that the IER enhances the heat transfer rate effectively.

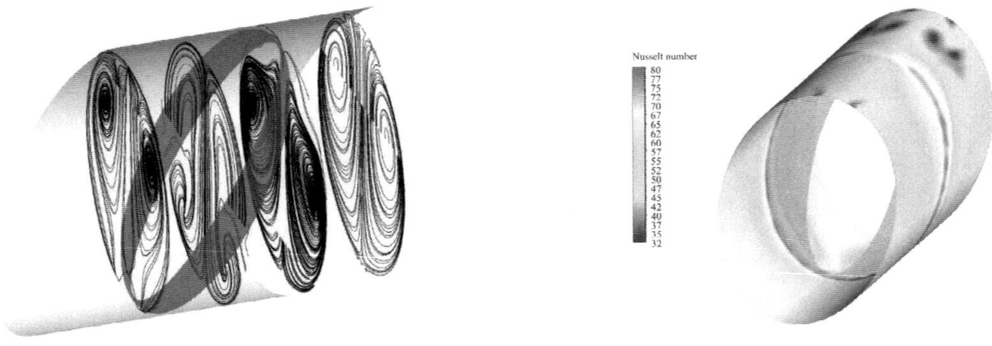

(a) (b)

Figure 4. (a) Streamlines in transverse planes and (b) local Nu_x contours for the tubes with 60° IER at E_R=0.5.

3.2. *Heat transfer result*

Effect of using IER turbulators with different six ER values on the heat transfer enhancement in the form of Nu/Nu_0 is depicted in Fig. 5. In the figure, Nu/Nu_0 decreases steeply with increasing Re and ER especially for Re = 4000 to 8000 but shows a gradual decrease for further Re. The maximum and minimum Nu/Nu_0 is about 6.7 and 3.6 times for E_R = 0.5, Re = 4000 and E_R = 1.0, Re = 20,000, respectively. This is because the presence of IER turbulator can interrupt the development of thermal boundary layer and increases the degree of turbulence level apart from inducing two counter-rotating vortices. The smaller E_R yields the considerable increase in the heat transfer because of higher flow blockage leading to higher vortex strength. The E_R = 0.5 provides the Nusselt number in the range of 4.7 to 6.7 times higher than the smooth tube.

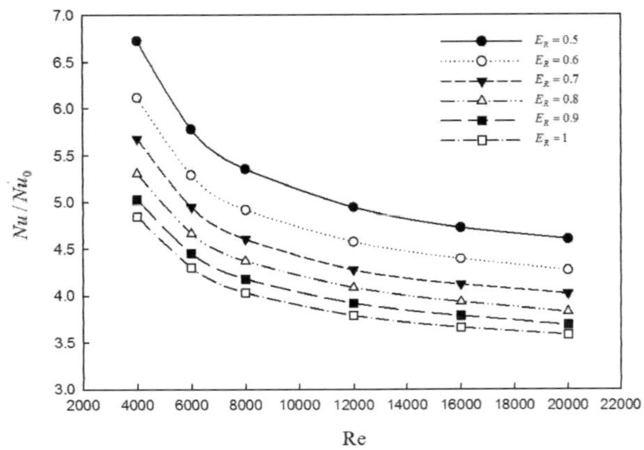

Figure 5. Nusselt number versus Reynolds number.

3.3. *Friction factor result*

The employ of IER insert comes with higher pressure drop in terms of friction factor. Fig. 6 displays the variation of f/f_0 with Re for various E_R values. In the figure, it is noted that f/f_0 tends to increase with the increment of Re but with decreasing E_R. The presence of IER provides the f/f_0 in the range of

36 to 130 times above the smooth tube. The f/f_0 for E_R=0.5 is around 82 – 130 times while that for E_R=0.9 is about 38 to 57 times that is nearly the same as that for circular ring (E_R=1.0). The friction losses mainly come from the dissipation of the dynamical pressure of the air due to high viscous losses near the wall, to the extra forces exerted by reversing flow and to higher friction of increasing surface area.

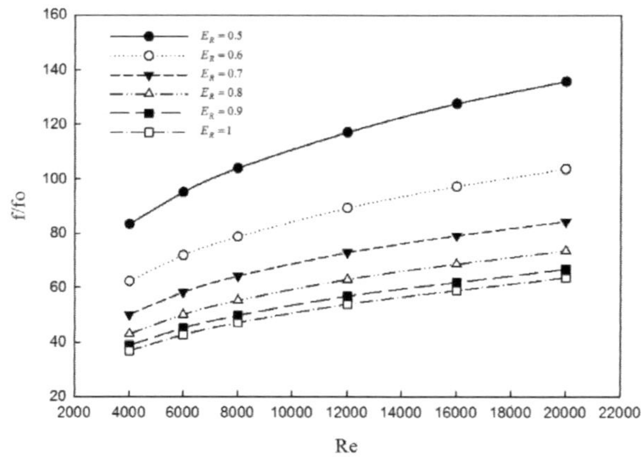

Figure 6. Friction factor against Reynolds number.

3.4. Thermal performance

Fig. 7 depicts the variation of TEF with Re for using the IER. In the figure, it is observed that TEF shows the decreasing trend with increasing Re and E_R values. The IER with E_R = 0.5, 0.6, 0.7, 0.8, 0.9 and 1 at Re=4000 provides the maximum TEF of about 1.54, 1.54, 1.54, 1.51, 1.49 and 1.46, respectively. This indicates the merit of using the IER for enhancing the heat transfer in the heat exchanger tube.

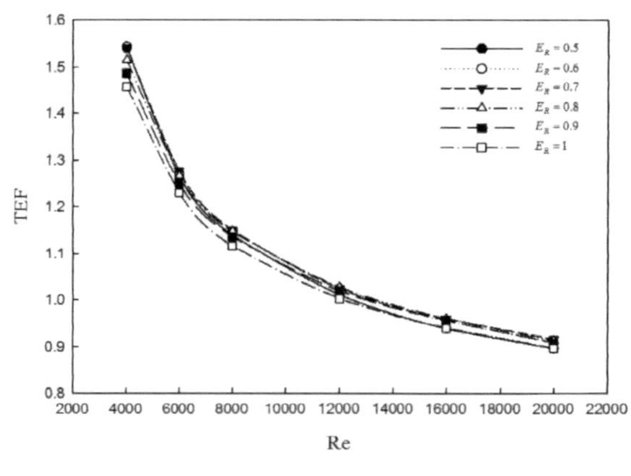

Figure 7. TEF vs Reynolds number.

4. Conclusions

Heat transfer and flow friction characteristics in a round tube with 60° inclined elliptical rings with various E_R values has been numerically investigated for the turbulent regime, Re = 4000 – 20,000 under a uniform heat-flux condition. The findings in the present study can be drawn as follows: The insertion of 60° inclined elliptical rings yields a considerable heat transfer enhancement at about 3.6 to 6.7 times above the smooth tube while the friction factor is 36 to 130 times. The Nu/Nu_0 decreases with increasing Re and E_R while f/f_0 has a similar trend except for Re that shows the reversing trend. The highest TEF of the 60° IER is found to be about 1.54 for a wide range of E_R= 0.5 – 0.7. Thus, the application of IER insert can be considered as a promising device for augmenting heat transfer in a heat exchanger tube.

5. References

[1] Esparza D M and Rojas E S April 2011 Numerical simulations of the laminar flow in pipes with wire coil inserts *Computers & Fluids* vol 44 pp 169–77

[2] Promvonge P 2008 Thermal augmentation in circular tube with twisted tape and wire coil turbulators *Energy Conversion and Management* vol 49 pp 2949–55

[3] Promvonge P, Koolnapadol N, Pimsarn M and Thianpong C 2014 Thermal performance enhancement in a heat exchanger tube fitted with inclined vortex rings *Applied Thermal Engineering* vol 62 pp 285–92

[4] Sripattanapipat S, Tamna S, Jayranaiwachira N and Promvonge P 2016 3D Numerical Heat Transfer Investigation in a Heat Exchanger Tube with Hexagonal Conical-ring Inserts *Energy Procedia* vol 100 pp 522–5

[5] Eiamsa-ard S, Thianpong C, Eiamsa-ard P and Promvonge P 2009 Convective heat transfer in a circular tube with short-length twisted tape insert *Int. Commun. Heat Mass Transfer* vol 36 pp 365–71

[6] Skullong S, Thianpong C, Jayranaiwachira N and Promvonge P 2016 Experimental and numerical heat transfer investigation in turbulent square-duct flow through oblique horseshoe baffles *Chem. Eng. Process. Process Intensif* vol. 99 pp 58–71

[7] Skullong S, Promvonge P, Jayranaiwachira N and Thianpong C 2016 Experimental and numerical heat transfer investigation in a tubular heat exchanger with delta-wing tape inserts *Chem. Eng. Process. Process Intensif* vol 109(3) pp 164–77

[8] Promvonge P, Changcharoen W, Kwankaomeng S and Thianpong C 2011 Numerical heat transfer study of turbulent square-duct flow through inline V-shaped discrete ribs *International Communications in Heat and Mass Transfer* vol 38 pp 1392–9

[9] Patankar S V , Liu C H, and Sparrow E M 1977 Fully developed flow and heat transfer in ducts having streamwise-periodic variations of cross-sectional area *ASME, Journal of Heat Transfer* vol 99(2) pp 180–6

IOP Conf. Series: Earth and Environmental Science **67** (2017) 012036 doi:10.1088/1755-1315/67/1/012036

Numerical heat transfer study in a round tube with 60° V-shaped rings

Y Kaewkohkiat[1], S Tamna[2], W Pairoj[1] and P Promvonge[1]

[1] Department of Mechanical Engineering, Faculty of Engineering, King Mongkut's Institute of Technology Ladkrabang, Bangkok 10520, Thailand
[2] Applied Mathematics and Mechanics Research Laboratory (AMM) Department of Automotive Engineering, Faculty of Engineering, TNI Bangkok, Thailand
Email: kpongjet@gmail.com

Abstract. Turbulent flow and heat transfer behaviors in a 3-dimensional isothermal tube mounted repeatedly with 60° V-shaped rings (V-rings) are numerically investigated. The computation based on the finite volume method was conducted for the airflow rate in terms of Reynolds number (Re) in the range of 4000 to 20,000. The ring parameters include three flow blockage ratios, (BR=e/D=0.1, 0.15 and 0.2) and a single ring pitch ratio, PR=1. The computed result shows that friction factor and Nusselt number (Nu) increase with the increment in BR and the maximum thermal enhancement factor (η) of about 1.73 with $Nu/Nu_0 = 4.5$ is found at BR=0.1 and Re=4000. For the need of maximum heat transfer enhancement (Nu/Nu_0), the case of BR=0.2 V-ring is selected and modified since this case provides the highest heat transfer. The modification is made by cutting off some area of both ends of the BR=0.2 V-ring and the effect of cutting off (called end-cut ratio, ER=c/e) on thermal performance is also examined. The study reveals that for the end-cut V-ring, the friction factor is considerably decreased while the heat transfer rate reduces moderately with increasing ER. The end-cut V-ring yields the maximum η of 1.64 at ER=0.9, Re= 4000 with $Nu/Nu_0 = 5.4$. Thus, the Nu/Nu_0 of the end-cut one is about 20% higher than that of the BR=0.1 V-ring.

1. Introduction

The application of vortex generators (VGs) in a heat exchanger system is commonly known as enhancement devices for augmenting the convective heat transfer coefficient leading to the compact heat exchanger and increasing thermal performance of such a system. The outstanding characteristic of the compact heat exchanger with VGs is considerably higher heat transfer rate, leading to reduced space, weight, energy requirements and costs [1]. The series of baffle/ring/rib is often used in the design of heat exchanger systems in order to increase the degree of cooling/heating levels. Although the heat transfer rate is increased through the baffle/ring arrangement, the pressure drop of the tube is also increased to the decreased flow area effects. Therefore, ring geometry and arrangements such as ring spacing, angle of attack and height/width are among the most significant parameters in the design of heat exchanger tubes. Promvonge et al. [2] studied the effect of flow blockage ratio and pitch ratio of using inclined rings inserted into a round tube on thermal performance behaviors. They found that the ring with larger BR provided higher heat transfer and pressure loss while the one with larger PR yielded the reversing trend. Chingtuaythong et al. [3] examined the influence of relative ring-pitch and blockage ratios at a constant attack angle of the V-shaped rings on heat transfer and flow resistance characteristics. They found that the heat transfer rate was up to 5.8 times above the plain tube whereas the friction factor was up to 82 times. Jedsadaratanachai et al. [4] investigated numerically the laminar periodic flow and heat transfer characteristics in a circular tube fitted with 45° V-baffles and reported that the optimum thermal enhancement factor was about 3.20 at BR=0.2 for the V-upstream and BR= 0.25 for the V-downstream baffles. Poomsalood et al. [5] numerically studied the turbulent flow and heat transfer characteristics in a round tube fitted with angled rings and indicated that the maximum η was about 1.35 at PR=1.75. In general, the experimental investigation on thermal performance in a heat exchanger system is very costly and takes time due to lots of parameters involved. The numerical investigation is considered to be a suitable technique for predicting a wide range of parameters, due to its advantages of geometric change and flexibility. Numerical study has a vital role in developing a heat transfer enhancement technique for application purposes. Nevertheless, thermal behaviors of turbulent flow is quite complicated and difficult to understand the heat transfer mechanism occurring in a heat exchanger tube, especially for experimental work. In the present work, a numerical computation for three dimensional turbulent flows through V-shaped rings mounted repeatedly in a round tube is conducted to examine the flow structure and heat transfer characteristics for turbulent regime.

2. Flow description

2.1. Geometry and configuration of V-shaped rings

The flow system under consideration is a circular tube fitted repeatedly with 60° V-rings as depicted in Fig. 1. The flow model is considered as fully periodic flow by using the concept of periodically fully developed flow [6]. In the figure, the ring pitch, *P* is a distance between the adjacent V-rings and set to *P=D* in which *P/D* is defined as the pitch ratio, PR. The air enters the tube having inner diameter of *D* set to 0.05 m. *e* is the ring width and *e/D* is known as the blockage ratio, BR. To investigate the effect of V-ring inserts on thermal performance of the tube, BR is varied between 0.1 and 0.2 while the PR and attack angle (α) are kept constant at 1.0 and 60°, respectively. To obtain the maximum heat transfer rate, further investigation is made by cutting off both ends of the BR=0.2 V-ring to reduce the pressure loss. Then, the effect of end cut-off ratio (ER=*c/e*) for the BR=0.2 ring on thermal performance is examined by varying ER values set to 0.0 (no cut), 0.1, 0.5 and 0.9 in which *c* is the cut-off length at both ends of the V-ring as seen in Fig. 1.

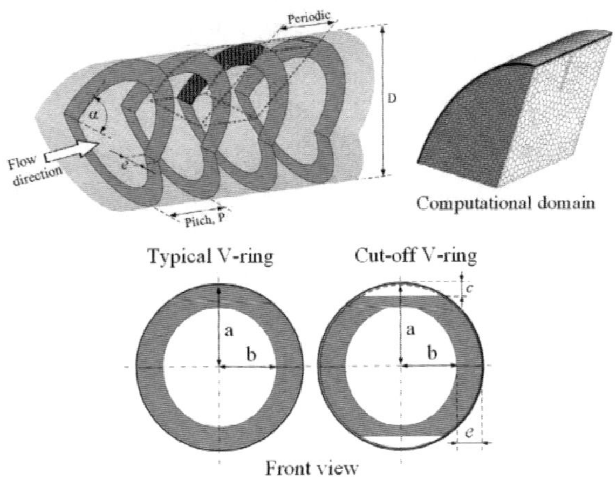

Figure 1. Tube geometry and computational domain of periodic flow.

Due to ring shape symmetry, only one-fourth of the tube flow model is used as the computational domain as shown in Fig. 1. For grid independence test, the variations in Nu and *f* for PR=1.0, BR=0.1 and Re=10,000 are found to be less than 0.5% when increasing the number of cells from 110,156 to 227,274, hence there is no such advantage in increasing the number of cells beyond this value.

Considering both convergence time and solution precision, the grid system of 110,156 is employed for the current computation.

2.2. Boundary conditions

Periodic boundaries are used for the inlet and outlet of the flow domain. Constant mass flow rate of air at 300 K is assumed in the flow direction due to fully periodic flow conditions. The physical properties of the air have been assumed to remain constant at the average air bulk temperature.

Table 1. Range of geometric parameters and operation conditions.

Parameters	Range
Flow Conditions	
Reynolds number (*Re*)	4000 to 20,000
V-ring geometry and tube	
Diameter of tube, D (mm)	50
Ring Pitch, P (mm)	50
Pitch ratio, PR= P/D	1.0
Ring width, e (mm)	5, 7.5 and 10
Blockage ratio, BR=*e/D*	0.1, 0.15 and 0.2
Cut-off length, c (mm)	0 (No cut), 0.5, 2.5 and 4.5
End cut-off ratio, ER=*c/e*	0 (No cut), 0.1, 0.5 and 0.9
Attack angle, α	60°

3. Mathematical foundation and key parameters

In the present study, the numerical simulation of fluid flow and heat transfer in a round tube is developed under the following assumptions: steady three-dimensional, turbulent and incompressible flow; omitting all the body forces, viscous dissipation and radiation heat transfer. Based on the above assumptions, the tube flow model is governed by the continuity, the Navier-Stokes equations and the energy equation [7]. In the Cartesian tensor system these equations can be written as follows:

Continuity equation:

$$\frac{\partial}{\partial x_i}(\rho u_i)=0 \quad (1)$$

Momentum equation:

$$\frac{\partial}{\partial x_j}(\rho u_i u_j)=-\frac{\partial p}{\partial x_i}+\frac{\partial}{\partial x_j}\left[\mu\left(\frac{\partial u_i}{\partial x_j}+\frac{\partial u_j}{\partial x_i}\right)-\rho\overline{u_i' u_j'}\right] \quad (2)$$

Energy equation:

$$\frac{\partial}{\partial x_j}(\rho u_i T)=\frac{\partial}{\partial x_j}\left((\Gamma+\Gamma_t)\frac{\partial T}{\partial x_j}\right) \quad (3)$$

For the Realizable k–ε turbulence model, its transport equations are expressed as:

$$\frac{\partial}{\partial x_i}(\rho k u_j)=\frac{\partial}{\partial x_i}\left[\left(\mu+\frac{\mu_t}{\sigma_k}\right)\frac{\partial k}{\partial x_j}\right]+2\mu_t S_{ij}.S_{ij}-\rho\varepsilon \quad (4)$$

$$\frac{\partial}{\partial x_j}(\rho k u_j)=\frac{\partial}{\partial x_i}\left[\left(\mu+\frac{\mu_t}{\sigma_\varepsilon}\right)\frac{\partial \varepsilon}{\partial x_j}\right]+\rho C_1 S\varepsilon-\rho C_2\frac{\varepsilon^2}{k+\sqrt{\nu\varepsilon}} \quad (5)$$

In the above equations, σ_k and σ_ε are the turbulent Prandtl numbers for k and ε, respectively.

All the governing equations are discretized by the QUICK numerical scheme with the SIMPLE algorithm for handling the pressure-velocity coupling and are solved by a finite volume method. For closure of the equations, the realizable k–ε turbulence model is used in the present study. The solutions are converged when the normalized residual values are less than 10^{-6} for all variables but less than 10^{-9} only for the energy equation.

There are four key parameters of interest in the present work, namely, the Reynolds number, friction factor, Nusselt number and thermal enhancement factor. The Reynolds number is defined as

$$\mathrm{Re} = \rho u_0 D/\mu \tag{6}$$

The friction factor, f is computed by pressure drop, ΔP across the length of tube flow model, L as

$$f = \frac{(\Delta P/L)D}{(1/2)\rho u_0^2} \tag{7}$$

The local heat transfer is measured by local Nusselt number which can be written as

$$Nu_x = h_x D/k_a \tag{8}$$

in which k_a is thermal conductivity of air. The area-average Nusselt number can be obtained by

$$Nu = (1/A)\int Nu_x dA \tag{9}$$

The thermal enhancement factor (η) defined as the ratio of the dimensionless heat transfer coefficient of an inserted tube, Nu to that of a smooth tube, Nu_0, at an equal pumping power is given by

$$\eta = (Nu/Nu_0)/(f/f_0)^{1/3} \tag{10}$$

Where Nu_0 and f_0 stand for Nusselt number and friction factor for the smooth tube, respectively.

4. Results and discussion

4.1. Validation

In the present simulation, the numerical results are validated with the previous experimental data taken from [3]. The comparison of the predicted friction factor and Nusselt number is presented in Fig. 2. The predicted friction factor and Nusselt number obtained from the numerical solutions are in reasonable agreement with the experimental data. The maximum deviations of the predicted friction factor and Nusselt number with measured data are, respectively, within ±19% and ±15% for the 30° V-ring at PR=2.0 and BR=0.1.

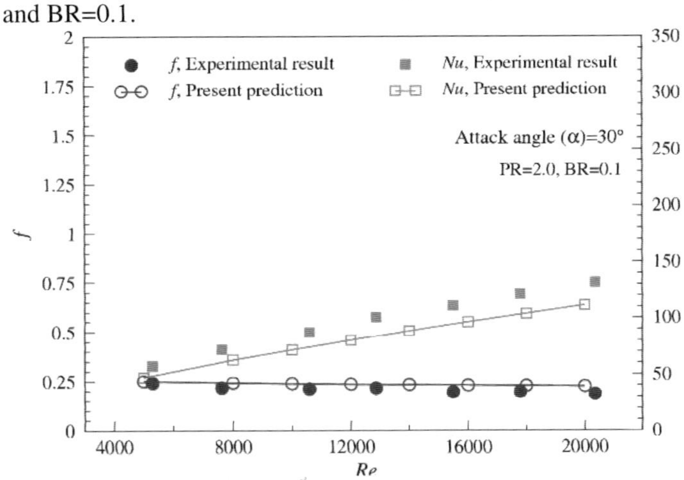

Figure 2. Verification of Nu and f with measurements [3].

4.2. Flow structure

Fig. 3(a) and (b) displays the temperature contours with streamlines in transverse planes at different locations for (a) typical (no end-cut) and (b) ER=0.9, end-cut V-rings, respectively. It can be observed that there are four main longitudinal vortex flows along the tube, two counter-rotating vortices appear on the upper and lower tube due to V-ring symmetry for both the typical V-ring and end-cut V-ring at ER=0.9. The scrutiny of the figure reveals that two pairs of common-flow-down vortices appear on both sides of the tube. These vortices are responsible for the increase in heat transfer rate along both sidewalls of the tube, lower temperature regions can be observed along both sidewalls. Higher temperature areas (red color regions) are found on the upper and lower parts of the tube.

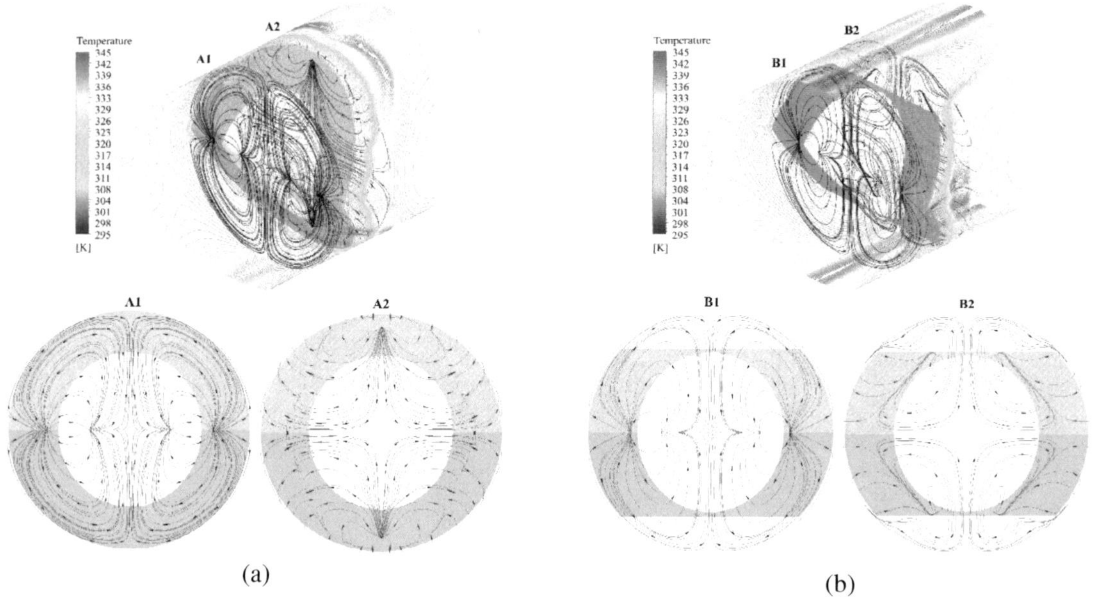

Figure 3. Temperature contours with streamlines in transverse plane for 60° V-ring at Re=4000, PR=1, BR=0.2; (a) typical V-ring, (b) ER=0.9.

4.3. Heat transfer

Temperature contours of using the V-rings at Re=4000 for the typical and cut-off V-rings, ER=0.9 are also displayed in Fig. 3(a) and (b), respectively. In Fig. 3(a) temperature become poorer around the top and bottom areas where the end of V-rings attached. In Fig. 3(b), the high temperature area on the tube wall is pretty linear strip in axial direction over the rings on the top and bottom. It is visible that the vortex flows provide a significant influence on the temperature field. When the end-cut ratio (ER) increases, the high temperature zone appears. This can be explained that the end-cut with larger ER allow the fluid flows through the cut area more easily, resulting in weaker turbulence and fluid mixing and consequently less efficient heat transfer. The local Nusselt number (Nu_x) contours on the tube wall for the V-ring with BR=0.1, 0.15 and 0.2 at Re=4000 are presented in Fig. 4(a), 4(b) and 4(c), respectively. The intensity of the Nu_x contours is displayed by different colors. The peaks (red color) can be observed at the impingement areas on the tube wall where the rings attached. In the cases studied, it is apparent that the typical V-ring with BR=0.2 gives the highest Nu as seen in Fig. 4c. For all cases, the heat transfer rate is intense around the ring, except for a small region on the top and bottom.

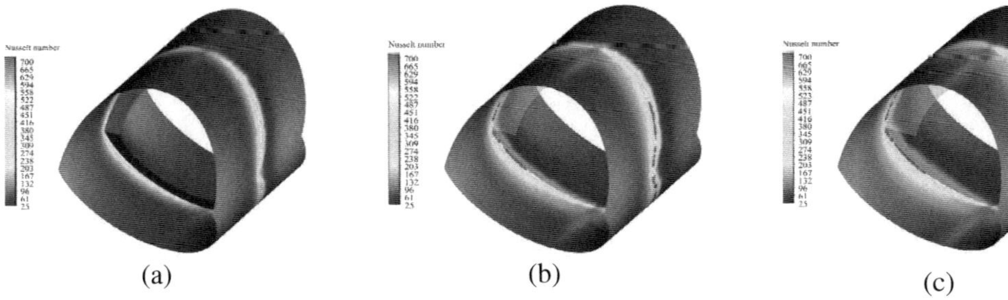

Figure 4. Nu_x contours for 60° typical V-ring at Re=4000, PR=1; (a) BR=0.1, (b) BR=0.15, (c) BR=0.2.

4.4. Effect of BR

Fig. 5 presents the pressure drop of V-rings in terms of friction factor ratio, f/f_0 against Re for various BRs. In the figure, it is noted that f/f_0 tends to increase with the rise of Re and BR values. The use of the V-rings leads to considerable increase in friction factor in comparison with the smooth tube with no ring. The larger BR yields higher pressure loss across the tube. According to numerical results, f/f_0 is found to increase considerably with the increment in BR. The lowest and highest f/f_0 are, respectively, found to be 24 and 235 times for BR=0.1, Re=4000 and BR=0.2, Re=20,000. The decrease in BR gives rise to the reduction of friction factor. The f/f_0 for BR=0.1 and BR=0.15 is, respectively, about 24–39 and 65–102 times while that for BR=0.2 is highest around 150-235 times depending on Re values. Thus, the use of larger BR should be avoided due to extreme pressure drop increase.

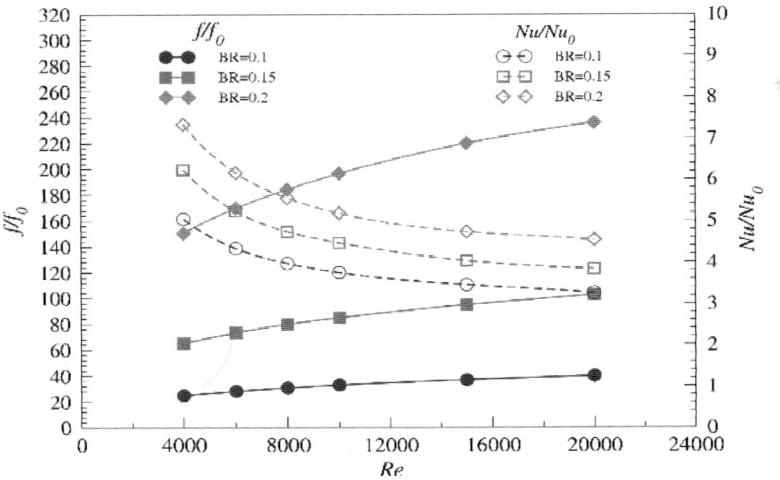

Figure 5. Variations of f/f_0 and Nu/Nu_0 with Re for V-rings.

The variation of heat transfer in terms of Nu/Nu_0 against Re is also plotted in Fig. 5. Nu/Nu_0 tends to decrease with the rise in Re for all BRs and the lower BR leads to the decrease of Nu/Nu_0. It is observed that Nu/Nu_0 increases with the increment of BR while the minimum and maximum Nu/Nu_0 values are, respectively, about 3.25 and 7.33 times at BR=0.1, Re=20,000 and BR=0.2, Re=4000. The use of the V-ring yields the heat transfer rate of about 3–7 times higher than the smooth tube.

IOP Conf. Series: Earth and Environmental Science **67** (2017) 012036 doi:10.1088/1755-1315/67/1/012036

4.5. Performance evaluation

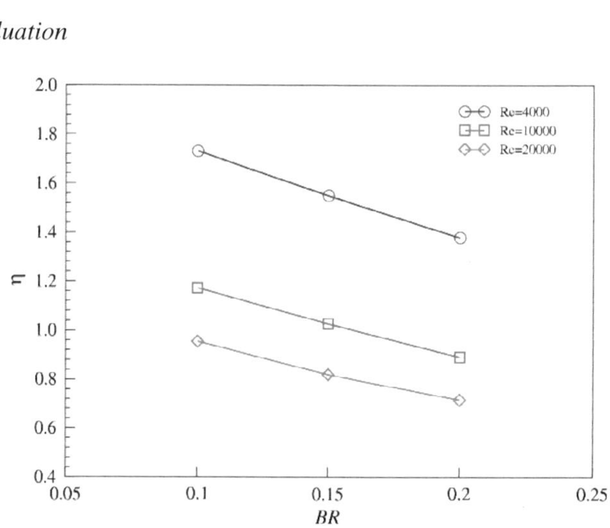

Figure 6. η versus BR for V-ring.

Fig. 6 displays the variation of η with BR at different Re values. It is visible that η shows the decreasing trend with the increase in BR and Re. The highest η is 1.73 at BR=0.1, Re=4000 while the lowest η is 0.71 at BR=0.2 and Re=20,000. Hence, to obtain higher thermal performance the smaller BR should be used.

4.6. Thermal performance improvement for lager BR

Apart from optimum thermal performance as mentioned in section 4.5, the possibly highest heat transfer enhancement is required to reduce the heat exchanger size. In this section, only the BR=0.2 case is considered to be modified because it gives the maximum Nu/Nu_0 as can be seen in Fig. 5. The modification is made by cutting off both ends of the V-ring as mentioned earlier and the numerical results are described next. The variation of f/f_0 and Nu/Nu_0 for the end-cut ring with Re at various ERs is depicted in Fig. 7(a). At the same ER, f/f_0 increases with the increase of Re whereas the larger ER provides lower resistance to the flow. For the range investigated, the use of typical V-ring (BR=0.2) leads to the increase in f/f_0 from 150-235 times and it can be reduced considerably for ER=0.9 (37-62 times). This indicates that the use of end-cut V-ring can reduce the friction factor around 3.7-4 times below the ring with no cut. The Nu/Nu_0 decreases with increasing ER and the highest Nu/Nu_0 is about 5.4 compared with 7.3 times for the no cut ring or around 30% lower.

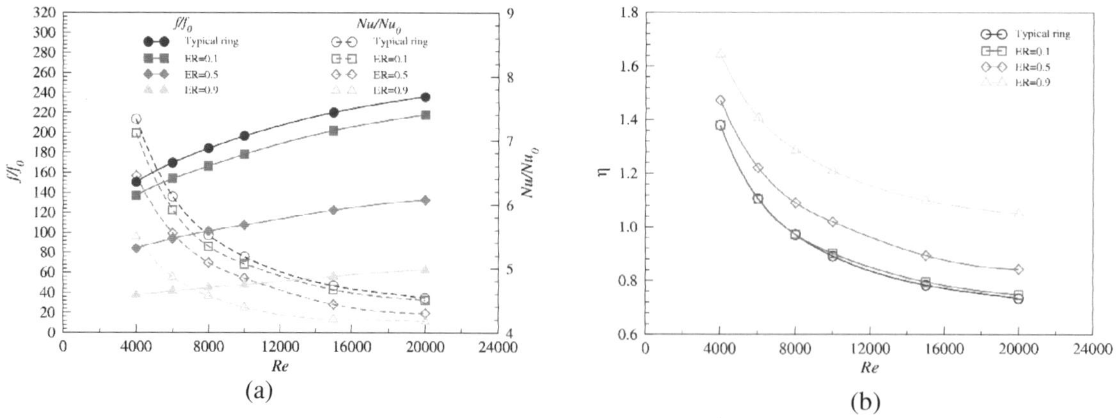

Figure 7. Comparison of (a) f/f_0 and Nu/Nu_0 (b) η with Re for various ER.

Fig. 7(b) exhibits the variation of η with Re for various ERs. The larger ER causes extremely lower friction loss and also gives lower heat transfer enhancement. The effect of ER on η signifies that the influence of decreasing friction loss is more significant than that of decreasing heat transfer. The maximum η of 1.64 is achieved by the end-cut V-ring with ER=0.9 at Re=4000 while the typical BR=0.2 V-ring is 1.37 as seen in Fig. 7(b).

5. Conclusions

A numerical study has been conducted to investigate the effect of BRs of using the V-ring insert on turbulent flow, heat transfer, flow resistance and thermal performance characteristics in a round tube. The computed result for typical V-rings shows that the friction factor and Nusselt number increase with the increment of BR. The lower BR leads to the lowest friction factor and Nusselt number. The maximum thermal performance for the V-ring is about 1.73 at BR=0.1 and Re=4000 with Nu/Nu_0 around 4.5. The improvement on thermal performance for using larger BR by allowing some of air flow through the space above the V-rings in order to reduce the flow resistance. The increase of end-cut ratio (ER) results in the decrease of heat transfer and friction loss. Among the end-cut rings, the larger ER gives the highest thermal enhancement factor of 1.64 at BR=0.2 and Re=4000 with Nu/Nu_0 around 5.4.

6. References

[1] Shah R K, Heikal M R, Thonon B, and Tochon P April 1955 Progress in the Numerical analysis of Compact heat exchanger surfaces *Advances in heat transfer* vol 34 pp 363–443

[2] Promvonge P, Koolnapadol N, Pimsarn M, and Thianpong C 2014 Thermal performance enhancement in a heat exchanger tube fitted with inclined vortex rings *Applied Thermal Engineering* vol 62 pp 285–92

[3] Chingtuaythong W, Promvong P, Thianpong C, and Pimsarn M 2017 Heat transfer characterization in a tubular heat exchanger with V-shaped rings *Applied Thermal Engineering* vol 110 pp 1164–71

[4] Jedsadaratanachai W, Jayranaiwachira N, and Promvonge P 2015 3D numerical study on flow structure and heat transfer in a circular tube with V-baffles *Chinese Journal of Chemical Engineering* vol 23 pp 342–9

[5] Poomsalood R, Promvonge P, and Sripattanapipat S August. 2013 3D simulation of turbulent flow and heat transfer in a round tube fitted with angled orifices *Int. Conf. on Engineering Applied Sciences and Technology* pp 315–9

[6] Patankar S V, Liu C H, and Sparrow E M 1977 Fully developed flow and heat transfer in ducts having streamwise-periodic variations of cross-sectional area *ASME, Journal of Heat Transfer* vol. 99(2) pp 180-6

[7] Shih T. H, Liou W W, Shabbir A, and Zhu J 1995 A New k-ε eddy viscosity model for high Reynolds number turbulent flows-model development and validation *Computer Fluids* vol 24 no.3 pp 227–38

IOP Conf. Series: Earth and Environmental Science **67** (2017) 012037 doi:10.1088/1755-1315/67/1/012037

Experimental study on evaporation characteristics of a hydrogen peroxide droplet at elevated temperature

Jonghan Won[1], Hongjae Kang[1], SeungWook Baek[1]*, and Sejin Kwon[1]

[1] Department of Aerospace Engineering, Korea Advanced Institute of Science and Technology (KAIST), 291 Daehak-ro, Yuseong-gu, Daejeon 34141, Republic of Korea

swbaek@kaist.ac.kr

Abstract. In this study, evaporation characteristics of a hydrogen peroxide droplet, 90% purity, was experimentally investigated at elevated temperature (between 400 and 800 ℃) and atmospheric pressure under normal gravity. Elevated temperature atmosphere was provided by electric furnace inside the chamber. The range of a droplet size was 1.1 mm to 1.3 mm. The evaporation process of a droplet was recorded by high speed CCD camera. As analysing the image extracted from the camera using the program, evaporation rate of a single droplet was calculated at each ambient temperature. After thermal expansion period, evaporation rate of a hydrogen peroxide droplet followed d2-law but thermal expansion period didn't clearly separate at 400℃. The evaporation rate increased with increase in ambient temperature. Also thermal decomposition of hydrogen peroxide increased at high temperature.

1. Introduction

In the recent year, hydrogen peroxide has received attention to monopropellant as well as oxidizer in the space rocket system because of lower toxicity than conventional propellants. That's why lots of researchers have investigated the hydrogen peroxide to apply in space thruster system [1]. Hydrogen peroxide has decomposition process under normal condition. However the decomposition goes very slowly so catalysts have been used to promote. When hydrogen peroxide of high concentration was decomposed, it can be described by a single step reaction of the exothermic process as (1) where Q is generated heat [2]. But this single step reaction could not spontaneously occur at low temperature except for adding a catalyst and high temperature.

$$H_2O_{2\,(l)} \rightarrow H_2O_{(l)} + \frac{1}{2}O_2 + Q \qquad (1)$$

Many previous researchers have studied on hydrogen peroxide for using monopropellant but researches related to evaporation characteristics of a hydrogen peroxide droplet are insufficient so evaporation rate of hydrogen peroxide has used as speculated value based on theoretical equation [3, 4].

Evaporation research of a liquid droplet at high temperature is one of basic studies in spray combustion for applying to various fields such as industrial furnace, incinerator, internal engine, etc. What evaporation characteristics of a single droplet investigate is essential for understanding complicated process of spray evaporation and combustion [5].

This research is experimental study on evaporation characteristics of a hydrogen peroxide droplet which is a basic study for applying to theoretical equation in numerical simulation and spray combustion researches.

2. Experimental apparatus

1.1. Experiment procedure

Fig. 1 is the schematic diagram of experimental apparatus that has been used in previous researches [5, 6, 7]. As using the temperature controller, atmospheric temperature inside electric furnace was uniformly increased. The temperature variation inside the furnace was about ±5 ℃ at targeted temperature. When the electric furnace reached at targeted temperature, a droplet was suspended at the tip of quart fiber using a injector. The diameter of injector hole is very small size (100 μm). The diameter of quartz fiber is 0.125 mm and it has spherical shape bead (0.25 mm) at the tip. The size of suspended droplet was from 1.1 mm to 1.3 mm. When a droplet suspended at the tip of quartz fiber, it existed with a side effect. Heat conduction between quartz fiber and a droplet happened during evaporation process of a droplet. This effect was experimentally and theoretically investigated from previous study [8]. Heat conduction through quartz fiber enhanced evaporation of a droplet and this effect increased when thickness of a fiber was thicker and lower ambient temperature. In this study, used quartz fiber was thin in high temperature environment so this effect can be negligible.

The electric furnace fell down to downstairs along the two guide bars as using the lever after a droplet was suspended at the tip of quartz fiber. When the furnace arrived at the bottom, a droplet started to vaporize and the process of droplet evaporation was recorded by high speed camera (operating at a rate of 500 frames per second) through the quartz window. Acquired images were analyzed for calculation of droplet size using the program based on Visual Basic which was validated from previous study [7].

Table 1. Thermochemical properties of hydrogen peroxide

Density (g/cm^3)	1.41
Boiling point (°C)	141
Specific heat (J/g·K)	2.619 (liquid)
Heat of Vaporization (kJ/g)	1.517
Viscosity (cP)	1.245
Vapor pressure at 20°C (kPa)	0.2

1.2. Fuel

Used fuel was hydrogen peroxide (90% purity) made by HABO chemical. For protecting the decomposition, fuel was storage in the fridge at 5℃. Table I is the thermochemical properties of hydrogen peroxide. It has characteristics of strong surface tension and hard to vaporize because of hydrogen bond comparing with conventional fuel.

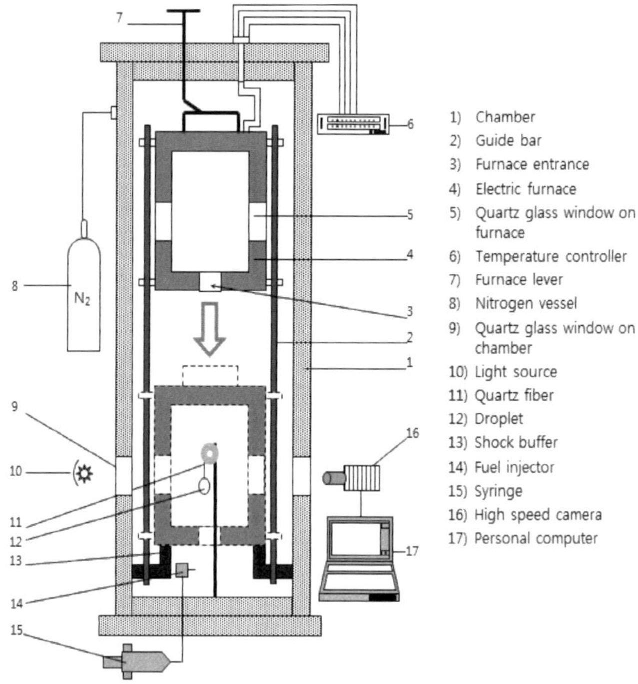

Figure 1. Schematic diagram of experimental setup

3. Results and discussion

Fig. 2 is the normalized temporal variation of squared diameter of a droplet. At relatively high temperature, the diameter of a droplet was increased because of thermal expansion. The thermal expansion happened when the amount of evaporation of liquid fuel was relatively lower than reduced density resulted from increasing ambient temperature. Density is mass divided by volume so droplet size related to volume is increased. The thermal expansion effect was larger in case of higher ambient temperature. Table II is normalized maximum droplet size pass a thermal expansion process at each temperature. At 400℃, a droplet slightly increased a diameter. The droplet diameter increased as ambient temperature was higher. From this result, the effect of reducing density was dominant comparing with an amount of liquid evaporation in thermal expansion period. After this period, a balance between thermal expansion and liquid evaporation determined a droplet size. A balance made a droplet reached a quasi-steady state. In this period, the temperature of a droplet was uniform but a size of a droplet was shrunk because of evaporation. But quasi-steady state was assumed that a droplet had uniform temperature and same size in infinitesimal time. At high temperature, these periods easily distinguished but it was hard to separate between thermal expansion and quasi-steady state at low ambient temperature. At 400℃, a droplet was slowly decreased at beginning evaporation process and then it was linearly decreased. In this temperature, thermal expansion didn't almost happen because the effect of volume increase was not lager than an amount of evaporation. At linear evaporation period, a droplet followed the d2-law expressed as (2) where do is the initial diameter and Cv is the evaporation rate [9]. The linear slopes in fig. 2 are evaporation rates at each ambient temperature. The best slope line was estimated using the least square method.

$$d^2 = d_o^2 - C_v t \tag{2}$$

Each evaporation rate of a droplet is shown as Table III at elevated temperature. In case of higher ambient temperature, evaporation rate increased. Especially, the rate of increase sharply grew from 600°C to 700°C.

Fig. 3 is temporal variation of droplet images at each ambient temperature. At 400°C, a droplet was verified not to almost expand. In case ambient temperature was higher, the thermal expansion effect was bigger and a droplet was not transparent. During the evaporation process, small water vapors boiled inside a droplet because of different boiling temperature comparing with hydrogen peroxide. Due to thermal decomposition, an amount of vapors of water and oxygen increased and then they blocked the back light that's why droplet image changed to dark. At high temperature, thermal decomposition of hydrogen peroxide more actively happened. As soon as the droplet started evaporation process, thermal decomposition of hydrogen peroxide actively occurred so center image of a droplet was fast opaque.

Fig.4 is the micro explosion phenomenon inside a droplet. The micro explosion phenomenon makes a fuel droplet to smaller particles so a droplet can be well atomized. The combustion of atomized droplet is similar to complete combustion that's why this improves combustion efficiency and reduces air pollutant emission [10, 11]. The major cause of micro explosion is the sudden explosion of water droplets which is superheated inside a droplet. Beyond 700°C of ambient temperature, micro explosion phenomenon occurred. The red boxes in fig.4 show that small droplet particle came out from an original droplet. As lots of small droplets went out consistently during evaporation process, a droplet didn't have perfect spherical shape. This phenomenon could be observed at high ambient temperature which was related to thermal decomposition of hydrogen peroxide. Thermal decomposition actively occurred at higher temperature and then water and oxygen gases were produced by single step reaction with heat. Through continuous thermal decomposition reaction, lots of water were produced and superheated. Because the large amount of superheated water was explosive, micro explosion happened in a droplet.

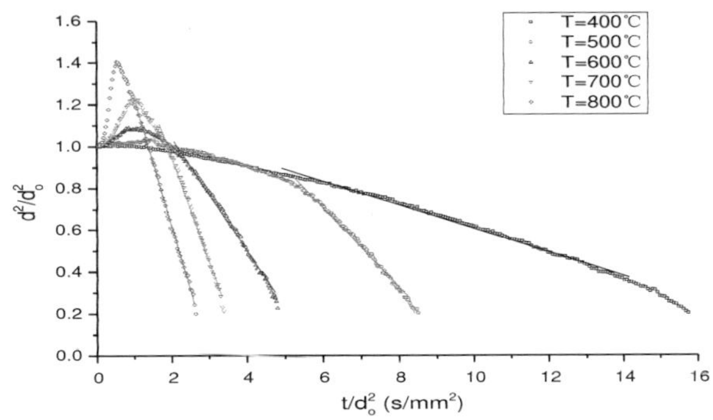

Figure 2. Normalized temporal variation of the diameter squared divided by initial diameter squared of a droplet

Table 2. Normalized maximum diameter during thermal expansion period

Ambient temperature (°C)	400	500	600	700	800
Maximum droplet diameter ($\mathbf{d^2/d_o^2}$)	1.007	1.034	1.090	1.225	1.404

IOP Conf. Series: Earth and Environmental Science **67** (2017) 012037 doi:10.1088/1755-1315/67/1/012037

Table 3. Evaporation rate of a droplet in linear decrease period

Ambient temperature (°C)	400	500	600	700	800
Evaporation rate	0.056	0.19	0.27	0.49	0.61

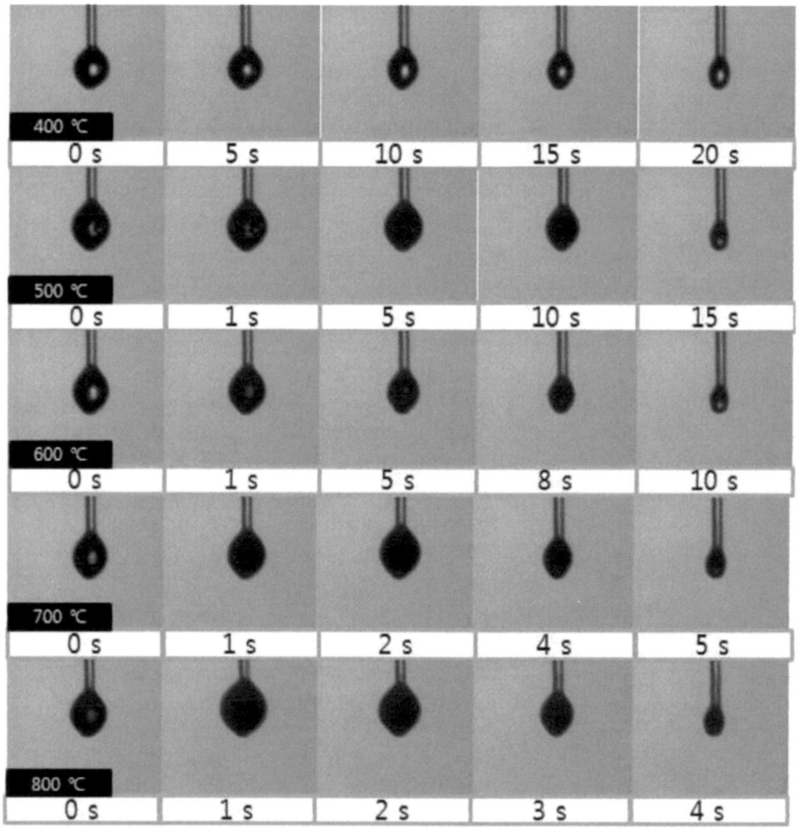

Figure 3. Temporal variation of droplet size at each temperature

Figure 4. Micro explosion phenomenon of a droplet at 800°C

4. Conclusion

This study was focus on the evaporation characteristics of a hydrogen peroxide droplet at elevated temperature. The followings are summarized results.

1) A droplet of hydrogen peroxide follows d2-law after thermal expansion period.
2) The evaporation rate of a hydrogen peroxide droplet increases along increase in ambient temperature.
3) In case of higher ambient temperature, thermal decomposition actively is promoted and water vapor and oxygen gas inside a droplet increases.
4) Micro explosion phenomenon occurs during evaporation process beyond 700 ℃ of ambient temperature.

After this research, other concentrations of hydrogen peroxide will compare with experimental and theoretical results.

5. References

[1] Pasini, L. Torre, L. Romeo, A. Cervone, and L. d'Agostino, "Reduced-Order Model for H2O2 Catalytic Reactor Performance Analysis," Jornal of Propulsion and Power, vol. 26, pp. 446-453, May-June 2010.
[2] D. Krejci, A. Woschnak, C. Schalemann, and K. Ponweiser, "Structural impact of honeycomb catalysts on hydrogen peroxide decomposition for micro propulsion," Chemical Engineering Reaserch and Design, vol. 90, pp. 2302-2315, December 2012.
[3] V. V. Tyurenkova, "Non-equilibrium diffusion combustion of a fuel droplet," Acta Astronautica, vol. 75, pp. 78-84, June-July 2012.
[4] G. Cai, C. Li, and H. Tian, "Numerical and experimental analysis of heat transfer in injector plate of hydrogen peroxide hybrid rocket motor," Acta Astronautica, pp. 286-294, November-December 2016.
[5] H. Ghassemi, S. W. Baek, and Q. S. Khan, "Experimental study on evaporation of kerosene droplets at elevated pressures and temperatures," Combust. Sci. and Tech., vol. 178, pp. 1669-1684, 2006.
[6] I. Javed, S. W. Baek, and K. Waheed, "Evaporation Characteristics of Heptane Droplets with the Addition of Aluminum Nanoparticles at Elevated Temperatures," Combustion and Flame, vol. 160, pp. 170-183, January 2013.
[7] D. M. Kim, S. W. Baek, and J. S. Yoon, "Ignition Characteristics of Kerosene Droplets with the Addition of Aluminum Nanoparticles at Elevated Temperature and Pressure," Combustion and Flame, vol. 173, pp. 106-113, November 2016.
[8] J. R. Yang and S. C. Wong, "An experimental and theoretical study of the effects of heat conduction through the support fiber on the evaporation of a droplet in a weakly convective flow," Int. J. Heat Mass Transfer, vol. 45, pp. 4589-4598, November 2002.
[9] S. R. Turns, An introduction to combustion : concepts and application, 3rd ed., McGraw-Hill, 2012.
[10] M. Huo, S. Lin, H. Liu, and C. F. Lee, "Study on the spary combustion chararteristics of water-emulsified diesel," Fuel, vol. 123, pp. 218-229, May 2014.
[11] H. M. Kim and S. W. Baek, "Combustion of a Single Emulsion Fuel Droplet in a Rapid Compression Machine," Energy, vol. 106, pp. 422-430, July 2016.

IOP Conf. Series: Earth and Environmental Science **67** (2017) 012038 doi:10.1088/1755-1315/67/1/012038

Impact of traffic composition on accessibility as indicator of transport sustainability

Nahdalina, S P Hadiwardoyo and Nahry
Civil Engineering Department, Faculty of Engineering, Universitas Indonesia, Depok 16424, Indonesia

Abstract. Sustainable transport is closely related to quality of life in the community at present and in the future. Some indicators of transport sustainability are accessibility measurement of origin/destination, the operating costs of transport (vehicle operating cost or VOC) and external transportation costs (emission cost). The indicators could be combined into accessibility measurement model. In other case, almost traffic congestion occurred on the condition of mixed traffic. This paper aimed to analyse the indicator of transport sustainability through simulation under condition of various traffic composition. Various composition of truck to total traffic flow are 0%, 10% and 20%. Speed and V/C are calculated from traffic flow to estimate the VOC and emission cost. 5 VOC components and 3 types of emission cost (CO2, CH4 and N2O) are counted to be a travel cost. Accessibility measurement was calculated using travel cost and gravity model approaches. Result of the research shows that the total traffic flow has indirect impact on accessibility measurement if using travel cost approach. Meanwhile, the composition of traffic flow has an affect on accessibility measurement if using gravity model approach.

1. Introduction

Sustainable transport is closely related to quality of life in the community at present and in the future. There are 3 key elements in sustainable transport i.e. economic, social and environment. SUMMA (2004) have classified the output indicator for economic element of sustainable transport. Three of them are accessibility measurement of origin/destination, the operating costs of transport (vehicle operating cost) and external transportation costs (emission cost). The indicators are combined into accessibility measurement model. In other case, traffic congestion occurred on the condition of mixed traffic. This paper is aimed to analyse the indicators under condition of various traffic composition.

2. Literature review

Accessibility is defined as many aspects depend on the objective of the assessment. In the beginning, it is defined as “potential opportunity to interact” [1]. The definition developed until today, such as "the ease of reached or to be reached" in terms of "who", "where" and "how". It has also been reviewed by [2] and [3]. Until now, the objective and approach of the accessibility model have been improved widely. [4] explained briefly the approaches of accessibility i.e. travel cost, potential opportunity (gravity model), constrain based (time-space diagram), utility surplus and combined of two approaches. First, travel cost approach was used by [5]. The advantage of approach is easy to understand. The construction model and its parameters are simple. Second, accessibility model that frequently to use is gravity model or potential opportunity approach. The model created by [1] in the beginning and developed by several researcher i.e. [4], [6], [7], [8] etc. [9], [10] and [11] improve in another way of accessibility model. They develop time-space constrain to calculate accessibility. Later, [12] and [13] analyse the accessibility with place rank approach.

Travel cost is a parameter that frequently used in calculating accessibility. [14] classify travel cost into function of travel time and travel distance, while [15] itemise travel cost into user cost, infrastructure cost, cost generated by destruction of environment, noise cost, accidents and safety cost and time value cost. [16] describes that travel cost of freight vehicle consist of vehicle asset, fuel, lubricant and operation and maintenance cost. In this paper, the calculation of travel cost included vehicle operating cost (VOC) and emission cost.

3. Methodology

3.1. Accessibility measurement model

As described above, there are several models to calculate the accessibility developed currently. The travel cost and gravity model approach are commonly used to analyse the accessibility. The equation of the approach is displayed below.

$$A_i = \Sigma[1/f(C_{id})] \qquad (1)$$

A_i represents the accessibility measured for location i, f(Cid) denotes the impedance function and Cid is for variable of travel cost or travel time or distance. [4] describes that the model is simple and easy to calculate, however the model ignores the quality of location and travel behaviour aspect.

According to [4], gravity approaches is the most useful techniques among the accessibility model. Introduced by [1], he state that accessibility is "potential opportunity to interact" or generalized of "connection between population through distance". It has been widely discussed by [17], [18], [19], [20], and [21]. The equation below shows the form of accessibility.

$$A_i = \Sigma[W_j/f(C_{id},\beta)] \qquad (2)$$

where W_j is mass opportunity, $f(C_{id},\beta)$ is impedance function, C_{id} represent travel cost variable and β denotes as travel cost coefficient. Similar to travel cost approach, this model is simple and easy to calculate, however, it has considered the quality of location.

3.2. Vehicle operating cost (VOC)

In Indonesia, there is a guideline to calculate VOC. The guideline is [22], published by Department of Public Work. According to the standard, VOC calculation divided into 2 parts; Variable cost and Fixed Cost. First part divided into 5 components i.e. fuel, lubricant, spare part, labour of maintenance and tire consumption cost. Main parameters of the components are speed, V/C, terrain and curve of road, and roughness index.

3.3. Formatting author affiliations

The components of emission cost are fuel consumption and emission factor (released by [23]). The default value of emission factor for moving resource displayed in Table 1 [23] and the value of fuel calorific in Indonesia exposed in Table 2 [24].

Refer to Guidelines for National Greenhouse Gas Inventories, general formula to calculate the emission cost is:

$$\text{Emission}_a = \Sigma_a(\text{Fuel}_a * \text{EF}_a) \qquad (3)$$

$$\text{Fuel}_{bcd} = \Sigma_{bcd}(\text{Vehicle}_b * \text{Distance}_c * \text{Fuel consumption}_d). \qquad (4)$$

where fuel is total amount of fuel consumption of all vehicles and distances while EF is emission factor. Fuel consumption derived from a component of VOC and converted into TJ/km before multiplied by emission factor. To estimate the emission cost, total amount of emission multiplied by unit cost of emission included CO2, CH4 and N2O. According to [25], prediction of emission cost at year 2015 are \$37 (CO2), \$970 (CH4) and \$15,000 (N2O) per tons. The costs equal to 481,000 IDR of CO2, 12,600,000 IDR of CH4 and 195,000,000 IDR of N2O.

IOP Conf. Series: Earth and Environmental Science **67** (2017) 012038 doi:10.1088/1755-1315/67/1/012038

Table 1. Default Value of Emission Factor by IPCC 2006 (Kg/TJ) for Moving Resource

Fuel Type	Emission Factor (Kg/TJ)		
	CO_2	CH_4	N_2O
Gasoline	69300	33	3.2
diesel Fuel	74100	(1)3.9	3.9

Table 2. Fuel Calorific Value in Indonesia

Fuel Type	Calorific Value (TJ/liter)	User
Gasoline	33.E-06	Vehicle
diesel Fuel	36.E-06	Vehicle

3.4. Data collection

To calculate the accessibility measurement, the data are required. Type and assumption data can be seen at Table 3.

Table 3. Data Requirement and Default Value of Road Condition

No	Parameter	Unit	Default Value
1	Speed	km/h	Calculated
2	Grade	m/km	2.5, -2.5
3	Curve of road	°/km	15
4	V/C		Calculated
5	Pavement Condition	m/km	3
5	Cumulative trip length	km	100,000
6	Traffic flow	Veh/h	500-4.500

Speed and V/C calculated from actual traffic flow using Indonesian Highway Capacity Manual 1997. Moreover, actual price of VOC component e.g. fuel, lubricant, tire, new car and labour cost of maintenance are needed.

4. Result and discussion

4.1. Traffic condition

Traffic composition divided into passenger car (PC) and trucks, while the truck also split into small truck (truck 2 axle), medium truck (truck 3 axle) and large truck (truck more than 4 axle). Several assumptions were selected to estimate the speed and V/C. First, the truck compositions from total traffic flow are 0%, 10% and 20%, while the compositions of each truck type from the total truck are 50% of small truck, 25% of medium truck and 25% of large truck. Second, road capacity is 5100 pcu/h (three lanes each direction) and free flow speed is 90 km/h. This conditions are typical of urban toll road in Indonesia. The result of V/C and speed calculation for various vehicle compositions can be seen in Table 4.

The changing of vehicle composition has an impact on traffic flow in pcu.as showed in Table 4. So that, the speed and V/C have a different value for different composition of traffic flow. At the composition of truck of 10%, the traffic flow in pcu/h increase approximately of 4%. Meanwhile, the composition of truck of 20%, the traffic flow in pcu/h increase 8%.

4.2. VOC

As described above, VOC has 5 components. Estimation of VOC for each vehicle type has unique characteristic. Figure 1(a) shows the correlation between cost of VOC component and speed for PC, Figure 1(b) shows the comparison value of VOC by type of vehicle.

Table 4. Result of V/C and Speed Calculation Based on Various Traffic Flow and Composition

Traffic Flow (veh/h)	100% PC			90% PC			80% PC		
	Traffic Flow (pcu/h)	*V/C*	*Speed (km/h)*	*Traffic Flow (pcu/h)*	*V/C*	*Speed (km/h)*	*Traffic Flow (pcu/h)*	*V/C*	*Speed (km/h)*
500	500	0.10	88.6	520	0.10	88.5	540	0.11	88.4
1000	1000	0.20	86.6	1040	0.20	86.5	1080	0.21	86.3
1500	1500	0.29	84.3	1560	0.31	84.0	1620	0.32	83.7
2000	2000	0.39	81.5	2080	0.41	81.0	2160	0.42	80.5
2500	2500	0.49	78.2	2600	0.51	77.5	2700	0.53	76.8
3000	3000	0.59	74.5	3120	0.61	73.5	3240	0.64	72.5
3500	3500	0.69	70.3	3640	0.71	69.0	3780	0.74	67.7
4000	4000	0.78	65.6	4160	0.82	64.1	4320	0.85	62.4
4500	4500	0.88	60.5	4680	0.92	58.6	4860	0.95	56.6

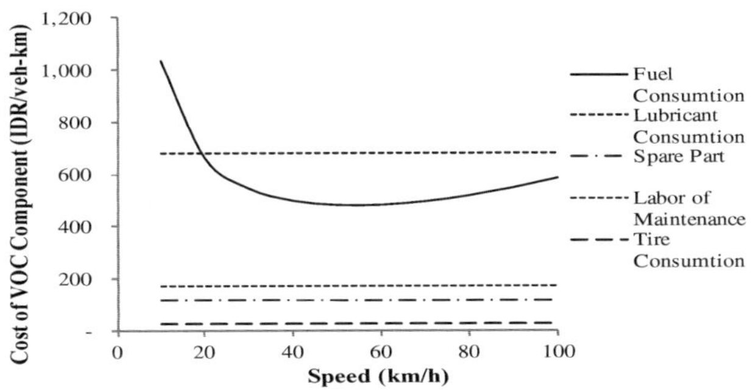

(a)

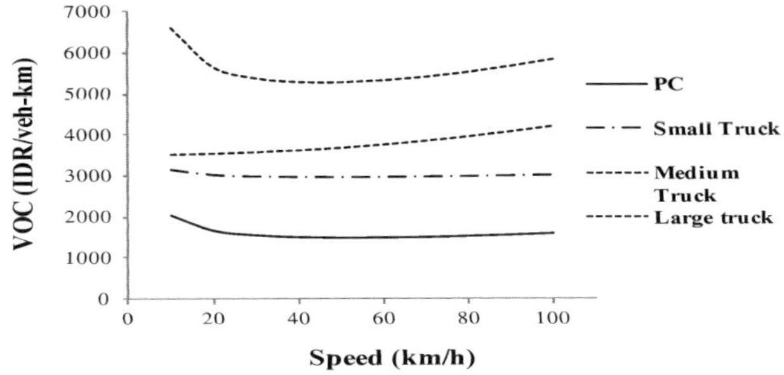

(b)

Figure 1. (a) Typical Relationship between Cost of VOC Component (IDR/veh-km) at Various Speed of PC; VOC of PC (b) VOC at Various Speed by type of vehicle

From Figure 1, it can be seen that fuel consumption and labour of maintenance costs give the highest contribution of VOC. Fuel consumption cost more sensitive than other component, although each type of vehicle has different a pattern. It can be seen that the shape of the curve is not flat. Comparison between type of vehicle, large truck is the costliest than the others.

4.3. Emission cost

In general, emission produced by transportation are CO2, CH4 and N2O. The emission cost of various speed for each type of vehicle displayed in Figure 2.

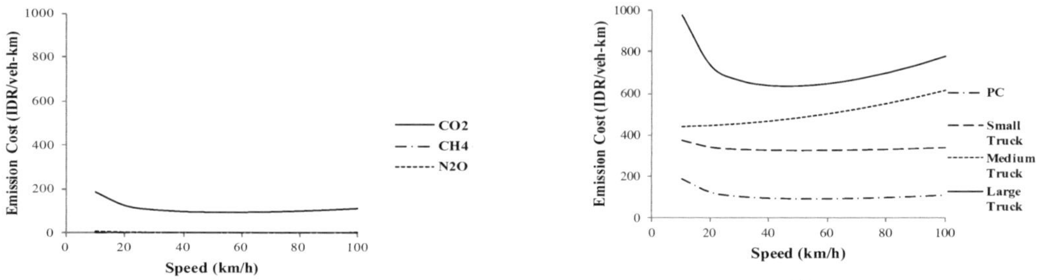

(a) (b)

Figure 2. (a) Comparison of Emission Cost by type of emission for PC, (b) Comparison of Emission Cost by type of Vehicle

Figure 2 chart (a) illustrate the emission cost of PC. A large part of emission composed of CO_2 after that CH_4 and N_2O. Despite, unit emission cost of N_2O higher than CO_2 and CH_4, the total emission cost of CO_2 much higher than 2 others. The curve of emission cost of CO_2 is similar to fuel consumption cost. The minimum cost occurred between 20-40 km/h. Chart (b) shows the comparison of emission cost by type of vehicles. Large truck produces the highest cost than others.

4.4. Travel cost

A lot of travel cost definition developed lately. This paper defined the travel cost as the sum of VOC and emission cost. Figure 4 shows the relationship between travel cost (IDR/km) and traffic flow (veh/h). Figure 3 (a) describe the VOC and emission cost per km of road for composition of 100% PC, meanwhile Figure 3 (b) express the difference of travel cost as truck composition changed.

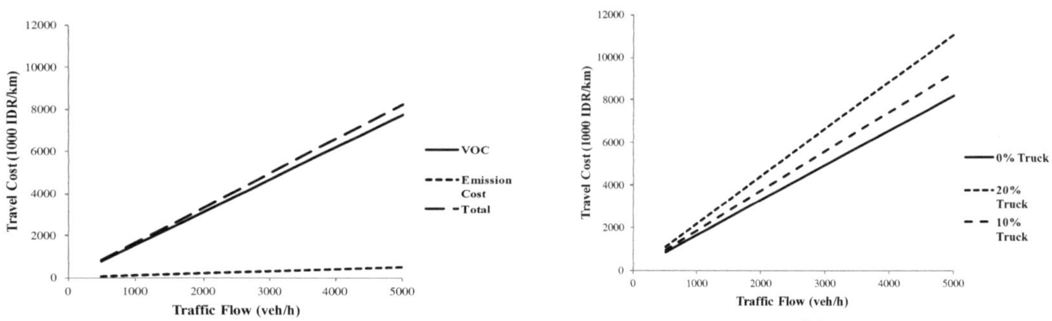

(a) (b)

Figure 3. Comparison of travel cost at various traffic flow (a) travel cost component and (b) various composition of truck

Emission costs only about 5.9% of the travel cost and increased according to the increment of the number of truck.

4.5. Accessibility measurement

Accessibility measurement was conducted by two approaches, i.e. travel cost and gravity model approach. The result of measurement can be seen at Figure 4. Figure 4(a) shows the relationship between accessibility and traffic flow using travel cost approach, while figure 4(b) displays the calculation of accessibility using gravity model approach. Graph (a) explains that the higher traffic flow and truck composition, the accessibility decrease sharply. Graph (b) describes that the additional traffic flow had no significant impact on accessibility, on the other hand, the additional number of trucks can reduce road accessibility.

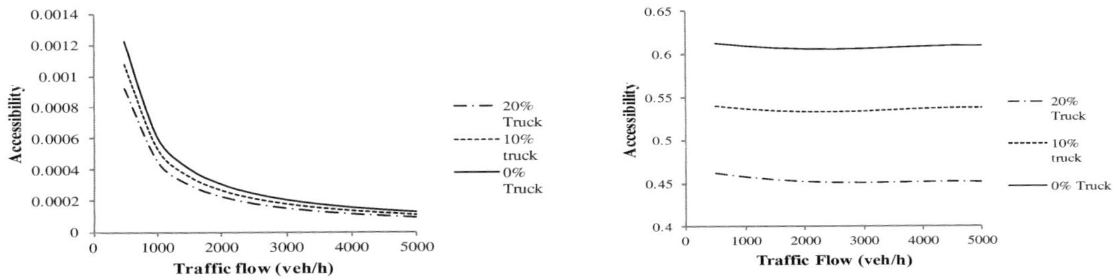

(a) (b)

Figure 4. Accessibility measurement of (a) Travel Cost Approach and (b) Gravity Model Approach.

As described in the equation (1), accessibility measurement has only a parameter of travel cost for travel cost approach, on the other hand, result of Figure 4(a) leads to the accessibility that influenced by traffic flow indirectly. Meanwhile, Figure 4(b) conducts that composition of traffic flow more influenced to the accessibility measurement than total traffic itself.

5. Conclusion

This paper describes the relationship between accessibility measurement and the composition of truck in the point of view of VOC and emission cost that their caused.

In the city that have highly mixed traffic, the changing of vehicle composition has an impact on traffic flow in pcu. In this case, almost 4% every increase 10% of truck composition So that, the speed and V/C have a different value for different composition of traffic flow.

From the research, there are three characteristic of transport sustainability indicators. Firs, that fuel consumption and labour of maintenance costs give the highest contribution of VOC for all value of speed. Fuel consumption cost more sensitive than other component, although each type of vehicle has different pattern. Comparison between type of vehicle, large truck is the costliest than the others.

Second, a large part of emission composed of CO_2. Despite, unit emission cost of N_2O higher than CO_2 and CH_4, the total emission cost of CO_2 much higher than 2 others. The curve of emission cost of CO_2 is similar to fuel consumption cost. The minimum cost occurred between 20-40 km/h. Emission costs only about 5.9% of the travel cost and increased according to the increment the number of truck.

Finally, using travel cost approach, the higher traffic flow and truck composition, the accessibility decrease sharply. In other hand, using gravity model, the additional traffic flow had no significant impact on accessibility. Meanwhile the additional number of trucks can reduce road accessibility.

In a future, the characteristic of indicators can be used as initial assumption of other subject of research, such as impact of freight vehicle restriction on urban road.

6. References

[1] Hansen WG. J Am Inst Plann. Routledge;1959;25:73–6.
[2] Geurs KT, van Wee B. J Transp Geogr. 2004;12(2):127–40.
[3] Duran-Fernandez R, Santos G. Res Transp Econ. 2014;46:55–69.

[4] Baradaran S, Ramjerdi F. J Transp Stat. 2001;4(23):31–48.
[5] Burns LD, Golob TF. Transportation (Amst). 1976;5(2):175–98.
[6] Geurs KT, van Eck JRR. Environ Plan B Plan Des. SAGE Publications; 2003;30(1):69–87.
[7] Yamaguchi K. Transp Res Part E Logist Transp Rev. 2007;43:247–58.
[8] Caschili S, De Montis A, Trogu D. Comput Environ Urban Syst. 2015;49:98–114.
[9] Hägerstraand T. WHAT ABOUT PEOPLE IN REGIONAL SCIENCE? Pap Reg Sci. Blackwell Publishing Ltd; 1970;24(1):7–24.
[10] Miller HJ. Geogr Anal. Blackwell Publishing Ltd; 1999; 31(2):187–212.
[11] Tong L, Zhou X, Miller HJ. Transp Res Part B Methodol. Elsevier Ltd; 2015;
[12] El-Geneidy A, Levinson D. Networks Spat Econ. 2011;11(4):643–59.
[13] Vega A. J Transp Geogr. 2012; 24:411–8.
[14] Zofío JL, Condeço-Melhorado AM, Maroto-Sánchez A, Gutiérrez J.. Transp Res Part A Policy Pract. 2014; 67:141–57.
[15] Levinson DM, Gillen D. Transp Res Part D Transp Environ Researc D. 1998; 3(4):207–23.
[16] Sahin B, Yilmaz H, Ust Y, Guneri AF, Gulsun B. Eur J Oper Res. 2009;193(1):1–11.
[17] Dalvi MQ, Martin KM. Transportation (Amst). 1976;5(1):17–42.
[18] Linneker BJ, Spence NA. Reg Stud. Routledge; 1992;26(1):31–47.
[19] Geertman SCM, Ritsema Van Eck JR. Taylor & Francis; 1995; 9(1):67–80.
[20] Bruinsma F, Rietveld P. Environ Plan A. SAGE Publications; 1998;30(3):499–521.
[21] Levinson DM. J Transp Geogr. 1998;6(1):11–21.
[22] DPU. Pedoman Perhitungan Biaya Opersi kendaraan (Pd T-15-2005- B). Jakarta: Kementerian Pekerjaan; 2005.
[23] Waldron CD, Harnisch J, Lucon O, Mckibbon SR, Saile SB, Wagner F, et al. Volume 2, Chapter 3: Mobile combustion. In: 2006 IPCC Guidelines for National Greenhouse Gas Inventories. WHO; 2006.
[24] KLH (Kementerian Lingkungan Hidup). Buku II - Volume 1 Metodologi Perhitungan Tingkat Emisi Gas Rumah Kaca. In: Pedoman Penyelenggaraan Inventarisasi Gas Rumah Kaca Nasional. Jakarta: Kementerian Lingkungan Hidup; 2012.
[25] Marten AL, Newbold SC. Estimating the Social Cost of Non-CO2 GHG Emissions: Methane and Nitrous Oxide. Washington DC; 2011.

Acknowledgments
The financial support of directorate for research and community services through research grand of PITTA 2016 at Universitas Indonesia is greatly appreciated.

AUTHOR INDEX

Acosta, N J E L....119
Ahmad, Nur Azfahani....162, 169
Ajis, Azizah Md....162
Al-Muttaqii, M....143
Amen, Tareq W.M.....19
Arief, I S....157
Asri, N P....136
Baek, SeungWook....250
Baidowi, A....157
Banlawe, I A P....119
Batdelger, B....190
Bhosale, Rahul....215
Byambatseren, Ch....190
Champreda, V....12
Chavalparit, Orathai....1
Chee, Swee Yong....25
Chenvidhya, D....124
Chuangchote, S....12, 124
Dzulkefli, Muhammad Hilmi....175
Eljamal, Osama....19
Fares, Almomani....215
Fujiani, R....136
Gosh, Ujjal....215
Gupta, R....79
Habib, NoorZainab....25
Hadi, N A....32
Hadiwardoyo, S P....256
Hammed, A M....206
Handayani, L....73
Hara, H....52
Hartono, D M....229
Idris, M A....206
Jami, M S....206
Jamil, Siti Jamiah Tun....169
Jeraputra, Chuttchaval....87
Jia, Chengxia....183
Judd, Simon....215
Kaewkohkiat, Y....242
Kaewsaenee, J....12
Kaida, N....52
Kang, Hongjae....250
Kangsabanik, Subhadip....222
Khalil, Ahmed M.E.....19
Kh'ng, X Y....198
Khreisheh, Majeda....215
Khumkrong, T....124
Kirtikara, K....124
Koh, H L....198
Kumar, Anand....215
Kumaresan, G....111
Kwon, Sejin....250
Laosiripojana, N....12
Lapsongpon, T....66
Leungprasert, S....66
Li, Y....46
Limphitakphong, Nantamol....1
Lin, Xiao-Ming....94
Liu, Jiahao....59
Liu, Pan....183
Liu, Si-Liang....94
Liu, Yao....38
Luo, Qiu....38
Marlinda, L....143
Matsunaga, Nobuhiro....19
Mirzayanti, Y W....151
Moersidik, S S....73
Muhamad, W M N W....32
Murmu, Sneha....222
Nadeem, Humayon....25
Nahdalina....256
Nahry....256
Namara, I....229
Ng, ChoonAun....25
Nordin, N H....52
Nuraini....136
Othman, M K F....32
Othman, Muhammad Anas Bin....162
Pairoj, W....242
Pattanapibul, P....12
Podjojono, B....136
Poh, K.L.....130
Prajitno, D H....143, 151
Promvonge, P....235, 242
Qu, Jiangqi....183
Roesyadi, A....143, 151
Ruangrit, Chaniporn....1
Sam, Suubitaa Spencer....25
Santosh, R....111
Shilpi, M....79
Shurair, Mohamad....215
Sidabutar, N V....229
Sodsri, W....235
Soesilo, T E B....229
Somasundaran, P....79
Song, Shuang....130
Sonomdagva, Ch....190
Sudhakar, P....111
Suherman, I H....157
Tamna, S....235, 242
Tan, W K....198
Teh, S Y....198
Teoh, Wei Ping....25
Thianpong, C....235
Tiptipakorn, Supun....87
Usapein, Parnuwat....1
Vigneswaran, V S....111
Wang, Caiyun....6
Wang, J F....46
Wang, Jian....59
Wardani, A Y....157

AUTHOR INDEX

Won, Jonghan 250
Wu, C H 46
Xiong, Qiming 38
Xu, Chen 103
Yan, Xuesheng...6
Yang, Mu 183
Yi, Ying-Qi...94, 103
Yoshimura, C 66
Yuen, Richard 59
Zhai, Jian-Wei..103
Zhang, Gui...38
Zhang, Qingjing 183
Zhang, Y F 46
Zhang, Yong-Jun 94, 103
Zhou, Bao-Rong 94, 103

Institute of Physics
Dirac House, Temple Back
Bristol BS1 6BE UK

ISSN: 1755-1307
ISBN 978-1-5108-4173-4

9 781510 841734

CURRAN ASSOCIATES INC.
proceedings.com